# Fractional Calculus with Applications in Mechanics

# Fractional Calculus with Applications in Mechanics

*Vibrations and Diffusion Processes*

Teodor M. Atanacković
Stevan Pilipović
Bogoljub Stanković
Dušan Zorica

*Series Editor*
*Noël Challamel*

iSTE

WILEY

First published 2014 in Great Britain and the United States by ISTE Ltd and John Wiley & Sons, Inc.



ISTE Ltd
27-37 St George's Road
London SW19 4EU
UK

www.iste.co.uk

John Wiley & Sons, Inc.
111 River Street
Hoboken, NJ 07030
USA

www.wiley.com



Library of Congress Control Number: 2013955271

British Library Cataloguing-in-Publication Data
A CIP record for this book is available from the British Library
ISBN 978-1-84821-417-0

MIX
Paper from responsible sources
FSC® C013604

Printed and bound in Great Britain by CPI Group (UK) Ltd., Croydon, Surrey CR0 4YY

# Table of Contents

# Preface

The subject of this book is the application of fractional calculus in mechanics. It is written so as to make fractional calculus acceptable to the engineering scientific community as well as to applied mathematicians who intend to use this calculus in their own research. The application of fractional calculus is mostly directed towards various areas of physics, engineering and biology. There are a number of monographs and a huge number of papers that cover various problems in fractional calculus. The list is large and is growing rapidly. The monograph by Oldham and Spanier [OLD 74], published in 1974, had a great influence on the subject. This was the first monograph devoted to fractional operators and their applications in problems of mass and heat transfer. Another important monograph was written by Miller and Ross in 1993 [MIL 93]. The encyclopedic treatise by Samko *et al.* is, up to now, the most prominent book in the field. This book [SAM 87] was first published in Russian in 1987, and the English translation appeared in 1993 [SAM 93]. The monograph by Kilbas *et al.* [KIL 06] contains a detailed introduction to the theory and application of fractional differential equations, mostly given in references. It treats, in a mathematically sound way, the fractional differential equations. Kiryakova in [KIR 94] introduces a generalized fractional calculus. Diethelm in [DIE 10] gives a well-written introduction to fractional calculus before the main exposition on the Caputo-type fractional differential equations. This book has mathematically sound theory and relevant applications. In Russian, besides [SAM 87], we mention monographs by Nahushev [NAH 03], Pshu [PSH 05] and Uchaikin [UCH 08]. Variational calculus with fractional derivatives is analyzed by Klimek in [KLI 09] and by Malinowska and Torres in [MAL 12]. All the above-mentioned monographs have had a great influence in the development of fractional calculus. Also there are other influential books and articles in the field that are worth mentioning. The article by Gorenflo and Mainardi [GOR 97b] had a significant impact in the field of applications of fractional calculus in physics and mechanics. The book by Oustaloup [OUS 95] presents an application of fractional calculus in the control theory. The book edited by Hilfer [HIL 00] and the book by Hermann [HER 11] contain applications of fractional calculus in physics. The article by Butzer and Westphal [BUT 00] contains a complete introduction to fractional calculus. Podlubny's work [POD 99], which has become a standard reference in the field, contains applications of fractional calculus to various problems of mechanics, physics and engineering. Application of fractional calculus in bioengineering is presented by Magin in [MAG 06]. Baleanu *et al.* in [BAL 12a] present the models in which fractional calculus is used, together with the numerical procedures that are used for the solutions. It also contains an extensive review of the relevant literature. Tarasov in [TAR 11] presents, among other topics, application of fractional calculus in statistical and condensed matter physics, as well as in quantum dynamics. Various applications, together with theoretical developments, are presented by Ortigueira in [ORT 11], Baleanu *et al.* in [BAL 12b], Petráš in [PET 11] and Sabatier *et al.* in [SAB 07]. The book by Mainardi [MAI 10] is the standard reference for the application of fractional calculus in viscoelasticity and for the study of wave motion. Finally, the book by Uchaikin [UCH 13] gives detailed motivation for fractional-order differential equations in various branches of physics. It also contains an introduction to the theory of fractional calculus.

Our book is devoted to the application of fractional calculus to (classical) mechanics. We have chosen to concentrate on more sophisticated constitutive equations that complement fundamental physical and geometrical principles. It is assumed that the reader has some basic knowledge of fractional calculus, i.e. calculus of integrals and derivatives of arbitrary real order. The main objective of this book is to complement, in a certain sense, the contents of the other books treating the theory of fractional calculus mentioned above. We will treat non-local elasticity, viscoelasticity, heat conduction (diffusion) problems, elastic and viscoelastic rod theory, waves in viscoelastic rods and the impact of a viscoelastic rod against a rigid wall. The mathematical framework of the problems that we analyze falls into different levels of abstraction. In our papers, on which the most part of the presentation is based, we use an approach to fractional calculus based on the functional analysis. In this way, we are able to use a well-developed method and techniques of the theory of generalized functions, especially of Schwartz space of tempered distributions and the space of exponential distributions, supported by $[0, \infty)$. The use of the generalized function setting only gives us flexibility in proving our results. If one deals with functions in $L^1_{loc}(\mathbb{R})$, equal zero on $(-\infty, 0)$, of polynomial or exponential growth, then their Laplace transform is the classical one. The same holds for the convolution of such functions. Namely, the convolution of such functions is again the locally integrable function, equal zero on $(-\infty, 0)$, that is of polynomial or exponential growth. When we deal with the Fourier transform, our framework is $S'$. The deep connection with the real mechanical models is not lost. Even better explanations and correctness of the proofs can help the reader to understand mathematical models of the discussed problems. Our aim is not to make the book too complicated for the readers with less theoretical background in the quoted mathematical sense. The presentation, for the most part, is intended to avoid unnecessary details and state only major results that a detailed, and often abstract, analysis gives.

The related book Fractional Calculus with Applications in Mechanics: Wave Propagation, Impact and Variational Principles [ATA 14] is complementary to this book, and, indeed, they could have been presented together. However, for practical reasons, it has proved more convenient to present the book separately. There are 13 chapters in the two books combined.

This book, Fractional Calculus with Applications in Mechanics: Vibrations and Diffusion Processes, has a total of seven chapters.

It begins with Part 1, entitled "Mathematical Preliminaries, Definitions and Properties of Fractional Integrals and Derivatives", which presents an introduction to fractional calculus. It comprises two chapters. Chapter 1, entitled "Mathematical Preliminaries", is brief and gives definitions and notions that are used in later parts of the book. Chapter 2, entitled "Basic Definitions and Properties of Fractional Integrals and Derivatives", presents definitions and some of the properties of fractional integrals and derivatives. We give references where the presented results are proved. Some of the results, which are needed in application, of our own research, such as expansion formulas for fractional derivatives and functional dependence of the fractional derivative on the order of derivative, are also presented in this chapter.

Part 2, entitled "Mechanical Systems", is the central part of the book. It comprises five chapters. In Chapter 3, entitled "Restrictions Following from the Thermodynamics for Fractional Derivative Models of a Viscoelastic Body", the analysis of constitutive equations of fractional-order viscoelasticity is presented. The constitutive equations must satisfy two principles: the principle of material frame-indifference (objectivity), which asserts that the response of a material is the same for all observers, and the second law of thermodynamics, which in the case of isothermal processes reduces to dissipation inequality. Sometimes dissipation inequality is called the Volterra theorem for hereditary systems (see [UCH 13]). We analyze, in detail, various constitutive equations and determine the restrictions of the material functions or constants so that the dissipation inequality is satisfied. Material objectivity is discussed for non-local elastic materials in section 3.3, Chapter 3 of the companion book Fractional Calculus with Applications in Mechanics: Wave Propagation, Impact and Variational Principles [ATA 14], for the simple one-dimensional spatial case. In Chapter 4, entitled "Vibrations with Fractional Dissipation", we analyze various vibration problems with single degree of freedom. Problems with single and multiple dissipation terms are analyzed. Also, nonlinear vibrations with symmetrized fractional dissipation is analyzed in detail. Finally, the case of nonlinear vibrations with distributed-order fractional dissipation is analyzed. Existence uniqueness and regularity of solution are examined. Chapter 4 also presents an example from compartmental methods in pharmacokinetics, where the conservation of mass principle is observed. In Chapter 5, entitled "Lateral Vibrations and Stability of Viscoelastic Rods", we present an analysis of lateral vibrations for several choices of constitutive functions for the rod. Thus, the cases of fractional Kelvin–Voigt, Zener and generalized Zener materials are studied. Special attention is given to the stability conditions. The case of Beck's column positioned on a fractional type of viscoelastic foundation is discussed, as well as the case of a compressible rod on a fractional type of viscoelastic foundation. In Chapter 6, entitled "Fractional Diffusion-Wave Equations", we present an analysis of the fractional partial differential equations with the order of time derivatives between 0 and 2. In the case of generalized Burgers/Korteweg–de Vries equation, we consider fractional derivatives with respect to space variable in the range between 2 and 3. By using similarity transformation, we even study some nonlinear cases of generalized heat equations. In Chapter 7, entitled "Fractional Heat Conduction Equations", Cattaneo-type space-time fractional heat conduction equations and fractional Jeffreys-type heat conduction equations are discussed.

The related book Fractional Calculus with Applications in Mechanics: Wave Propagation, Impact and Variational Principles [ATA 14] has a total of six chapters. It begins with Part 1, entitled "Mathematical Preliminaries, Definitions and Properties of Fractional Integrals and Derivatives", which presents an introduction to fractional calculus. It consists of two chapters. Chapter 1, entitled "Mathematical Preliminaries", is brief and gives definitions and notions that are used in later parts of the book. Chapter 2, entitled "Basic Definitions and Properties of Fractional Integrals and Derivatives", presents definitions and some of the properties of fractional integrals and derivatives. We give references where the presented results are proved. Some of the results, of our own research, which are required in applications, such as expansion formulas for fractional derivatives and functional dependence of the fractional derivative on the order of derivative, are also presented in this chapter.

Part 2, entitled "Mechanical Systems", is the central part of this book, containing four chapters. In Chapter 3, entitled "Waves in Viscoelastic Materials of Fractional-Order Type", we present an analysis of waves in fractional viscoelastic materials on infinite domain of space. A wave equation of the fractional Eringen-type is also studied. Stress relaxation, creep and forced oscillations of a viscoelastic of finite length are discussed in detail. In Chapter 4, entitled "Forced Oscillations of a System: Viscoelastic Rod and Body", the problem of oscillation of a rigid body, attached to viscoelastic rod and moving translatory, is analyzed in detail. The case of a light rod (mass of the rod is negligible with respect to the mass of the attached body) and the case of a heavy rod (mass of the rod is comparable with respect to the mass of the attached body) are discussed separately. Also, constitutive equations for solid-like and fluid-like bodies are distinguished. In Chapter 5, entitled "Impact of Viscoelastic Body against the Rigid Wall", we analyze a specific engineering problem of a viscoelastic rod impacting against the rigid wall. The case of a light viscoelastic rod sliding without friction is discussed first. Then, the more complicated case of a light viscoelastic rod attached to a rigid block that slides with dry friction is discussed. Finally, the case of a heavy viscoelastic rod attached to a rigid block that slides without friction is presented. In Chapter 6, entitled "Variational Problems with Fractional Derivatives", we present some results for the optimization of a functional containing fractional derivatives. We formulate the necessary conditions for optimality in standard and generalized problems. We also present dual variational principles for Lagrangians having fractional derivatives. The necessary conditions for optimality are formulated in generalized problems where optimization is performed with respect to the order of the derivative and not only with respect to a given set of admissible functions. Invariance properties of variational principles and Nöther's theorem are discussed in this chapter, as well as the problem of approximation of Euler–Lagrange equations in two different ways. Also, in this chapter, we propose a constrained minimization problem in which the order of the derivative is considered as a constitutive quantity determined by the state and control variables.

A full bibliography of the two related titles is presented together at the end of each book. The bibliography does not pretend to be complete. It only contains references to the papers and books that we used. In every chapter, in the introductory section, we list the references used in that chapter. In this way, the presentation in every chapter is more readable.

We believe that a reader can find enough information for understanding the presented materials and apply the methods used in the book to his/her own investigations. We hope that this book may be useful for graduate students in mechanics and applied mathematics, as well as researchers in those fields.

We are grateful to our colleagues Nenad Grahovac, Alfio Grillo, Diana Dolićanin, Marko Janev, Sanja Konjik, Ljubica Oparnica, Dragan Spasić and Miodrag Žigić who worked with us on some problems presented in this book.

We acknowledge the Serbian Ministry of Science and Education, project numbers 174005 and 174024, for support while writing this book. We thank Professor Nöel Challamel for suggesting that we write this book.

T. M. ATANACKOVIĆ
S. PILIPOVIĆ
B. STANKOVIĆ
D. ZORICA
Novi Sad, December 2013

# PART 1

# Mathematical Preliminaries, Definitions and Properties of Fractional Integrals and Derivatives

# Chapter 1

# Mathematical Preliminaries

## 1.1. Notation and definitions

Sets of natural, integer real and complex numbers are denoted, respectively, by $\mathbb{N}$, $\mathbb{Z}$, $\mathbb{R}$ and $\mathbb{C}$; $\mathbb{N}_0 = \mathbb{N} \cup \{0\}$, $\mathbb{R}_+ = [0, \infty)$.

Let $\Omega$ be an arbitrary subset of $\mathbb{R}$. We denote by $C_b(\Omega)$ the set of continuous functions on $\Omega$ such that

$$\|f\|_{C_b(\Omega)} = \sup_{x \in \Omega} |f(x)| < \infty.$$

It is well known that $C_b(\Omega)$ is a Banach space. If $\Omega$ is open, then we consider compact subsets of $\Omega$, $K \subset\subset \Omega$, continuous functions $f$ on $\Omega$ and the semi-norms

$$\|f\|_K = \sup_{x \in K} |f(x)|.$$

We can take a sequence of compact sets $K_1 \subset K_2 \subset, \ldots,$ so that $\cup_{i=1}^{\infty} K_i = \Omega$. Then, the sequence of semi-norms defines the Fréchet topology on $C(\Omega)$. This topology does not depend on a sequence $\{K_i\}_{i \in \mathbb{N}}$, with the given property. If $K$ is compact, then $C(K)$ always denotes the set of continuous functions on $K$ with the sup-norm over $K$.

Let $\Omega$ be open in $\mathbb{R}^n$. Then, we consider $C^k(\Omega) \subset C(\Omega)$: the space of functions having all the derivatives, up to order $k \in \mathbb{N}_0$, continuous. The Fréchet topology is defined by the semi-norms

$$\|f\|_{k,K} = \sup_{\substack{x \in K, \\ |i| \leq k}} |f^{(i)}(x)|, \quad K \subset\subset \Omega, \ |i| = i_1 + \ldots + i_n. \qquad [1.1]$$

The same topology is obtained if we again take, for compact sets in [1.1], a sequence of compact sets $K_1 \subset K_2 \subset \ldots$, so that $\cup_{i=1}^{\infty} K_i = \Omega$.

If $K$ is compact, we use the notation $C^k(K)$ for the Banach space of functions with all derivatives continuous up to order $k$, with the norm [1.1]. In the case $k = 0$, we use the notation $C^0(\Omega) = C(\Omega)$ and $C^0(\mathbb{R}) = C(\mathbb{R})$. If $k = \infty$, then we call $C^\infty(\Omega)$ the space of smooth functions. It is the Fréchet space with the sequence of semi-norms $\|f\|_{p,K_p}$, $K_p \subset K_{p+1}$, $K_p \subset\subset \Omega$, $p \in \mathbb{N}$, $\cup_{p=1}^{\infty} K_p = \Omega$. Its subspace $C_0^\infty(\mathbb{R})$ consists of compactly supported smooth functions, that is of the smooth functions equal to zero outside the compact sets.

Analytic functions on $(a,b) \subset \mathbb{R}$ are smooth functions on $(a,b)$, so that their Taylor series converges in any point $a_0$ of $(a,b)$ on a suitable interval around $a_0$. A space $\mathcal{A}((a,b))$ of such functions is a Fréchet space under the convergence structure from $C^\infty((a,b))$.

$BV_{loc}(\mathbb{R}_+)$ denotes the space of functions $f$ of locally bounded variations on $\mathbb{R}_+$. This means: for every interval $[a,b] \subset \mathbb{R}_+$, there exists a constant $M$ such that $\sum_{i=0}^{n} |f(t_{i+1}) - f(t_i)| < M$ for every finite choice of points $t_0 = a, \ldots, t_n = b$.

$L^p((a,b)) = L^p([a,b])$, $p \geq 1$, is the space of measurable functions for which $\left(\int_a^b |f(x)|^p \mathrm{d}x\right)^{\frac{1}{p}} < \infty$. We shorten the notation and use the symbol $L^p(a,b)$. In $L^p(a,b)$, $p \geq 1$, the norm is defined as

$$\|f\|_p = \left(\int_{(a,b)} |f(x)|^p \mathrm{d}x\right)^{\frac{1}{p}}.$$

In $L^\infty(a,b)$, we have $\|f\|_\infty = \operatorname{vrai\,sup}_{x \in (a,b)} |f(x)|$.

More precisely, above we consider spaces of Lebesgue measurable functions on a Lebesgue measurable set $A \subset \mathbb{R}^n$ (above $A = (a,b)$) and identify them through the equivalence relation: $f \sim g$ over $A$ if $f(x) = g(x)$, $x \in A \setminus N$ where $N$ is of zero Lebesgue measure. This relation determines the classes $[f]$, ..., and in the following, we put $f$ for $[f]$. In this sense, when $f \sim g$, we say that these functions are equal almost everywhere on $A$ ($f = g$ almost everywhere (a.e.) on $A$) and we just identify $f$ and $g$. So, the notation above $\text{vrai}\sup_{x\in(a,b)} f(x)$, for a measurable function bounded almost everywhere on $(a, b)$, means: supremum up to a set of points in $(a, b)$ with the zero measure. In the sequel, we will consider equality almost everywhere, as well as the integration in the sense of Lebesgue. $L^1_{loc}(a,b) = L^1_{loc}((a,b))$, $a, b \in \mathbb{R}^n$, $a < b$, is the space of measurable functions $f$ on $(a,b) \in \mathbb{R}^n$ such that for every compact set $K \subset (a,b)$, there holds $\int_K |f(x)|\mathrm{d}x < \infty$. It is clear that $L^1_{loc}([a,b)) \neq L^1_{loc}(a,b)$.

If $p$ and $q$ are real numbers such that $p, q \geq 1$ and $\frac{1}{p} + \frac{1}{q} = 1$ (for $p = 1$, $q = \infty$), and if $f \in L^p(a,b)$, $g \in L^q(a,b)$, then $fg \in L^1(a,b)$ and

$$\int_{(a,b)} |f(x)g(x)|\mathrm{d}x \leq \|f\|_p \|g\|_p. \qquad \text{[Hölder inequality]}$$

A real-valued function $f$ defined on $[a, b] \subset \mathbb{R}$ is said to be absolutely continuous on $[a, b]$, if for given $\varepsilon > 0$, there is a $\delta > 0$ such that

$$\sum_{i=1}^{n} |f(x_i') - f(x_i)| < \varepsilon$$

for every finite collection $\{(x_i', x_i)\}_{i\in\mathbb{N}}$ of non-overlapping intervals with

$$\sum_{i=1}^{n} |x_i' - x_i| < \delta.$$

The space of absolutely continuous functions is denoted by $AC([a,b]) = AC^1([a,b])$. There holds $C([a,b]) \subset AC([a,b])$. Moreover, $f \in AC([a,b])$, if and only if there exists an integrable function $g$ over $[a,b]$ such that

$$f(x) = c + \int_a^x g(t)\mathrm{d}t, \quad g = f' \text{ a.e. on } [a,b].$$

$AC^n([a,b])$, $n \in \mathbb{N}$, $n \geq 2$, is the space of functions $f$, which have continuous derivatives up to the order $n$ 1 on $[a,b]$ and $f^{(n-1)} \in AC([a,b])$. Notation $AC^n_{loc}([0,\infty))$ means that the function $f \in AC^n([0,b])$, for every $b > 0$.

A function $f$ on $[a,b]$ is Hölder continuous at $x_0 \in [a,b]$ if there exist $A > 0$ and $\lambda > 0$, such that

$$|f(x) - f(x_0)| \leq A|x - x_0|^{\lambda}$$

in a neighborhood of $x_0$. Hölder-type spaces on an interval $[a,b]$ are defined as subspaces of integrable functions on this interval with the following properties:

– $\mathcal{H}^{\lambda} \equiv \mathcal{H}^{\lambda}([a,b]) = \{f \mid |f(x_1) - f(x_2)| \leq A|x_1 - x_2|^{\lambda},\ x_1, x_2 \in [a,b]\},\ \lambda \in (0,1]$;

– $\mathcal{H} \equiv \mathcal{H}([a,b]) = \cup_{0<\lambda\leq 1}\mathcal{H}^{\lambda}([a,b])$;

– $\mathcal{H}^* \equiv \mathcal{H}^*([a,b]) = \Big\{f \mid f(x) = \frac{f^*(x)}{(x-a)^{1-\epsilon_1}(b-x)^{1-\epsilon_2}},\ x \in (a,b),\ \epsilon_1, \epsilon_2 > 0,\ f^* \in \mathcal{H}^{\lambda}([a,b]),\ \lambda \in (0,1]\Big\}$;

– $\mathcal{H}_0^{\lambda}(\epsilon_1, \epsilon_2) = \{f \in \mathcal{H}^* \mid f^*(0) = f^*(b) = 0\}$;

– $\mathcal{H}_{\alpha}^* = \cup_{\alpha<\lambda\leq 1,\ \epsilon_1,\epsilon_2>0}\mathcal{H}_0^{\lambda}(\epsilon_1, \epsilon_2)$;

– $h^{\lambda} \equiv h^{\lambda}([a,b]) = \Big\{f \mid \frac{f(x_1 - x_2)}{|x_1 - x_2|^{\lambda}} \to 0,\ x_2 \to x_1\Big\};\ h^{\lambda} \subset \mathcal{H}^{\lambda}$.

## 1.2. Laplace transform of a function

Let $f \in L^1_{loc}(\mathbb{R})$ and $f(t) = 0,\ t \in (-\infty, 0)$. The Laplace transform of $f$ is defined by

$$\mathcal{L}[f(t)](s) = \tilde{f}(s) = \lim_{A\to\infty}\int_0^A f(t)\mathrm{e}^{-st}\mathrm{d}t, \qquad [1.2]$$

for those complex numbers $s$ for which this limit exists. It is well known that the existence of the limit in [1.2] at $s = s_0$ implies the existence of this limit for any $s \in \mathbb{C}$ with the property $\operatorname{Re} s > \operatorname{Re} s_0$. We can consider the integral $\int_0^{\infty}|f(t)|\mathrm{e}^{-t\operatorname{Re} s}\mathrm{d}t$. If it is finite (we say an integral exists, or converges) for $s = s_1$, then $f$ is called an absolutely convergent Laplace transformable function. In this case

$$\mathcal{L}[f(t)](s) = \int_0^{\infty} f(t)\mathrm{e}^{-st}\mathrm{d}t$$

is absolutely convergent for any $s \in \mathbb{C}$ such that $\operatorname{Re} s > \operatorname{Re} s_1$.

The number $a_e = \inf\{\mathrm{Re}\, s_0 \in \mathbb{R}\}$ representing the infimum of those $s_0 \in \mathbb{C}$ for which the Laplace transform is defined is called the abscissa of existence. The abscissa of absolute convergence $a_a$ is defined in the same way. We have $a_a \geq a_e$.

It is clear that $\tilde{f}(s)$ exists (absolutely exists) for every $s \in \mathbb{C}$, $\mathrm{Re}\, s > a_e$ ($s \in \mathbb{C}$, $\mathrm{Re}\, s > a_a$). It is an analytic function in the half-plane $\mathrm{Re}\, s > a_e$, since, by partial integration, it can be represented as an absolutely convergent Laplace transform.

In the following, we consider the following class of Laplace transformable functions. Function $f \in L^1_{loc}([0,\infty))$ is called exponentially bounded if there exist constant $C = C_f > 0$, $r = r_f \in \mathbb{R}$ and $c = c_f \geq 0$ such that

$$|f(t)| \leq C\mathrm{e}^{rt}, \quad t > c. \qquad [1.3]$$

We denote by $L^{exp}([0,\infty))$ the space of such functions. The growth order $r$ is greater or equal than the abscissa of absolute convergence, $r \geq a_a$.

The Laplace transform is a linear operation on the space of exponentially bounded functions. If a function and its derivatives on $[0,\infty)$ up to order $k$ are of exponential growth, then

$$\mathcal{L}\left[f^{(k)}(t)\right](s) = s^k \tilde{f}(s) - s^{k-1} f(0) - \ldots - f^{k-1}(0), \quad \mathrm{Re}\, s > c,$$

for suitable $c > 0$. Let us mention several useful properties of the Laplace transform, based on appropriate assumptions and on corresponding domains

$$\begin{aligned}
\mathcal{L}\left[f(t)\mathrm{e}^{at}\right](s) &= \tilde{f}(s-a), \quad \mathcal{L}\left[f(at)\right](s) = \frac{1}{a}\tilde{f}\left(\frac{s}{a}\right), \\
\mathcal{L}\left[t^n f(t)\right](s) &= (-1)^n \tilde{f}^{(n)}(s), \quad \mathcal{L}\left[f(t) * g(t)\right](s) = \tilde{f}(s)\tilde{g}(s),
\end{aligned}$$

where the convolution of two locally integrable functions on $[0,\infty)$ is defined by

$$f(t) * g(t) = \int_0^t f(t-\tau)g(\tau)\mathrm{d}\tau, \quad t \geq 0.$$

The inverse Laplace transform is defined by

$$f(t) = \mathcal{L}^{-1}\left[\tilde{f}(s)\right](t) = \frac{1}{2\pi\mathrm{i}} \lim_{q\to\infty} \int_{p-\mathrm{i}q}^{p+\mathrm{i}q} \tilde{f}(s)\mathrm{e}^{st}\mathrm{d}s, \quad t \geq 0,$$

where $p > r$ (see [1.3]).

## 1.3. Spaces of distributions

The reader of this book has to have a knowledge of the theory of the generalized functions; here we call them distributions, as they are commonly known. This theory is a powerful tool used in mathematical theory and applications. Apart from books that discuss the basic theory, for example [SCH 51, VLA 73], there are a number of application-oriented textbooks such as [DUI 10].

We refer to [SCH 51, VLA 73] for the material of this section. By $\mathcal{D}(\mathbb{R}^n) = C_0^\infty(\mathbb{R}^n)$ and $\mathcal{S}(\mathbb{R}^n)$, the well-known Schwartz spaces are denoted. Norms in the space $\mathcal{D}_K(\mathbb{R}^n)$ of smooth functions supported by $K$ are

$$p_{K,m}(\varphi) = \sup_{x\in K,|\alpha|\leq m} |\varphi^{(\alpha)}(x)|, \quad m \in \mathbb{N}_0,$$

while in $\mathcal{S}(\mathbb{R}^n)$ are

$$q_m(\varphi) = \sup_{x\in\mathbb{R},|\alpha|\leq m} (1+|x|)^m|\varphi^{(\alpha)}(x)|, \quad m \in \mathbb{N}_0.$$

Then, $\mathcal{D}(\mathbb{R})$ is the inductive limit

$$\mathcal{D}(\mathbb{R}) = \operatorname*{ind\,lim}_{K_n\subset\subset\mathbb{R}} \mathcal{D}_{K_n},$$

where $K_n$, $n \in \mathbb{N}$, is an increasing sequence of compact sets so that $\cup_n K_n = \mathbb{R}$.

The corresponding duals, spaces of continuous linear functionals, $\mathcal{D}'(\mathbb{R}^n)$ and its subspace $\mathcal{S}'(\mathbb{R}^n)$, with the strong topologies, are the space of distributions and the space of tempered distributions. The space of compactly supported distributions is denoted by $\mathcal{E}'(\mathbb{R}^n)$. It is the dual space for the Fréchet space see section 1.1.

Operations of multiplication and differentiation in $\mathcal{D}'(\mathbb{R})$ are defined in a usual way

$$\langle af,\varphi\rangle = \langle f,a\varphi\rangle,\ \ \langle f^{(k)},\varphi\rangle = (-1)^k\langle f,\varphi^{(k)}\rangle,\ \ a\in C^\infty(\mathbb{R}),\ \varphi\in\mathcal{D}(\mathbb{R}),\ k\in\mathbb{N}.$$

Note $a(x)\,\delta(x-x_0) = a(x_0)\,\delta(x-x_0)$, where $\delta$ is the Dirac distribution,

$$\langle\delta(x-x_0),\varphi(x)\rangle = \varphi(x_0),\quad \varphi\in\mathcal{D}(\mathbb{R}).$$

We note that $\mathcal{D}' = \mathcal{D}'(\mathbb{R})$ contains regular elements defined by $f \in L_{loc}(\mathbb{R})$; they are denoted by $f_{reg}$ and defined by

$$f_{reg} : \varphi \mapsto \langle f_{reg}, \varphi \rangle = \int_{\mathbb{R}} f(x)\varphi(x)\mathrm{d}x, \quad \varphi \in \mathcal{D}(\mathbb{R}).$$

We can see that $\varphi_n \to 0$ in $\mathcal{D}$ implies $\langle f_{reg}, \varphi_n \rangle \to 0$, $n \to \infty$.

Polynomially bounded and locally integrable functions on $\mathbb{R}$ define, in the same way, regular tempered distributions. We will usually denote, by the same symbol $f$, a function and a corresponding distribution $f_{reg}$. Only if we want to explain in detail the relation between them do we use the symbol $f_{reg}$.

The Fourier transform of a function $\varphi \in \mathcal{S}$ ($\mathcal{S} = \mathcal{S}(\mathbb{R})$) is defined by

$$\mathcal{F}[\varphi(x)](\xi) = \hat{\varphi}(\xi) = \int_{\mathbb{R}} \varphi(x)\mathrm{e}^{-\mathrm{i}\xi x}\mathrm{d}x, \quad \xi \in \mathbb{R}.$$

The Fourier transform is an isomorphism on $\mathcal{S}$. If $f \in \mathcal{S}'$ ($\mathcal{S}' = \mathcal{S}'(\mathbb{R})$), then

$$\langle \mathcal{F}[f], \varphi \rangle = \langle f, \mathcal{F}[\varphi] \rangle, \quad \varphi \in \mathcal{S},$$

defines the Fourier transform of a tempered distribution. The Fourier transform is an isomorphism on $\mathcal{S}'$. The inverse Fourier transform of $\varphi \in \mathcal{S}$ is defined by

$$\varphi(x) = \mathcal{F}^{-1}[\hat{\varphi}(\xi)](x) = \frac{1}{2\pi} \int_{\mathbb{R}} \varphi(\xi)\mathrm{e}^{\mathrm{i}\xi x}\mathrm{d}\xi, \quad x \in \mathbb{R}.$$

If $\varphi, \psi \in \mathcal{S}$, their convolution is defined by

$$\varphi(x) * \psi(x) = \int_{\mathbb{R}} \varphi(x - \zeta)\,\psi(\zeta)\,\mathrm{d}\zeta, \quad x \in \mathbb{R}.$$

If $\operatorname{supp}\varphi, \operatorname{supp}\psi \in [0, \infty)$, which means that $\varphi = \psi = 0$ on $(-\infty, 0)$, then

$$\varphi(t) * \psi(t) = \int_0^t \varphi(t - \tau)\,\psi(\tau)\,\mathrm{d}\tau, \quad t \geq 0 \text{ and } \varphi(t) * \psi(t) = 0, \quad t < 0.$$

We know

$$\mathcal{F}\left[\varphi\left(x\right)*\psi\left(x\right)\right]\left(\xi\right)=\hat{\varphi}\left(\xi\right)*\hat{\psi}\left(\xi\right),\quad \varphi,\psi\in\mathcal{S}$$

and, as a consequence,

$$\mathcal{F}\left[f\left(x\right)*g\left(x\right)\right]\left(\xi\right)=\hat{f}\left(\xi\right)*\hat{g}\left(\xi\right),\quad f,g\in\mathcal{S}'.$$

Sobolev space $W^{k,p}(\mathbb{R})$, $p\in[1,\infty]$, $k\in\mathbb{N}_0$, is defined as the space of $L^p$-functions $f$ with the property that all the distributional derivatives of $f$ up to order $k$ are elements of $L^p(\mathbb{R})$. It is a Banach space with the norm

$$\|f\|_{W^{k,p}}=\sum_{j=0}^{p}\left\|f^{(j)}\right\|_{L^p}.$$

Clearly, $W^{k,p}(\mathbb{R})\subset\mathcal{S}'(\mathbb{R})$.

$\mathcal{D}_{L^p}$, $1\leq p<\infty$ is a space of smooth functions with all derivatives belonging to $L^p$. Note $\mathcal{D}_{L^p}\subset\mathcal{D}_{L^q}$ if $p\leq q$. $\dot{\mathcal{B}}$ is a subspace of $\mathcal{D}_{L^\infty}=\mathcal{B}$, defined as follows: $\varphi\in\dot{\mathcal{B}}$ if and only if $|\varphi^{(\alpha)}(x)|\to 0$ as $|x|\to\infty$ for every $\alpha\in\mathbb{N}_0$.

$\mathcal{D}'_{L^p}$, $1<p\leq\infty$ is the dual space of $\mathcal{D}_{L^q}$, $1\leq q<\infty$, $\frac{1}{p}+\frac{1}{q}=1$. $\mathcal{D}'_{L^1}$ is the dual of $\dot{\mathcal{B}}$ and $\mathcal{D}'_{L^\infty}$ is denoted by $\mathcal{B}'$ (see [SCH 51]).

$\mathcal{S}'_+$ denotes a subspace of tempered distributions consisting of distributions with supports in $[0,\infty)$. Note that $\mathcal{S}'_+$ is a convolution algebra.

The following structural theorem holds: $f\in\mathcal{S}'_+$ if and only if there exists a continuous function $F$ on $\mathbb{R}$ such that $F(x)=0$, $x<0$, $|F(x)|\leq C(1+|x|)^k$ for some $C>0$, $k>0$ and there exists $p\in\mathbb{N}_0$ such that

$$f(x)=F^{(p)}(x),\quad x\in\mathbb{R}, \qquad [1.4]$$

where the derivative is taken in the sense of distributions.

Let $f\in\mathcal{S}'_+$. Its Laplace transform is defined by

$$\mathcal{L}\left[f\left(t\right)\right]\left(s\right)=\tilde{f}(s)=\langle f(t),\mathrm{e}^{-st}\rangle=s^p\mathcal{L}\left[F\left(t\right)\right]\left(s\right),\quad \operatorname{Re}s>0, \qquad [1.5]$$

where we assume that $f$ is of the form [1.4]. Clearly, $\tilde{f}(s)$ is a holomorphic function for $\mathrm{Re}\, s > 0$. We will often consider equations, with solutions $u$ determining the tempered distributions, by the use of the Laplace transform. If we assume that $u$ is of exponential growth, then we have $\tilde{u}(s)$, $\mathrm{Re}\, s > s_0$, for some $s_0 > 0$.

We consider the family $\{f_\alpha\}_{\alpha\in\mathbb{R}} \in \mathcal{S}'_+$ (see [VLA 84])

$$f_\alpha(t) = \begin{cases} \frac{t^{\alpha-1}}{\Gamma(\alpha)} H(t), & t \in \mathbb{R},\ \alpha > 0, \\ \frac{\mathrm{d}^m}{\mathrm{d}t^m} f_{\alpha+m}(t), & \alpha \leq 0,\ \alpha + m > 0,\ m \in \mathbb{N}, \end{cases} \tag{1.6}$$

where the $m$th derivative is understood in the distributional sense. Family $\{\check{f}_\alpha\}_{\alpha\in\mathbb{R}} \in \mathcal{S}'_-$ is defined by $\check{f}_\alpha(t) = f_\alpha(-t)$. The Heaviside function is defined as

$$H(t) = \begin{cases} 0, & t < 0, \\ 1, & t > 0. \end{cases}$$

Operators $f_\alpha *$ and $\check{f}_\alpha *$ are convolution operators

$$f_\alpha *, \check{f}_\alpha * : \mathcal{S}'_+ \to \mathcal{S}'_+ \text{ and } \check{f}_\alpha * : \mathcal{S}'_- \to \mathcal{S}'_-.$$

The semi-group property holds for $f_\alpha$

$$f_\alpha * f_\beta = f_{\alpha+\beta},\quad \alpha, \beta \in \mathbb{R}.$$

The Laplace transform of $f_\alpha$ is

$$\mathcal{L}[f_\alpha(t)] = \frac{1}{s^\alpha} \tilde{f}(s),\quad \mathrm{Re}\, s > 0.$$

EXAMPLE 1.1.– Let $f$ be an absolutely continuous function on $[0, \infty)$ so that $f(0^+) = p \neq 0$. Assume that $f$ and $f'$ have the (classical) Laplace transform (denoted by $\mathcal{L}_c$) in the domain $\mathrm{Re}\, \lambda > \lambda_0 > 0$. Put $f_{reg}$, and $(f')_{reg}$, for the corresponding distributions. Note that $f_{reg} = fH$, with symbol $\mathcal{L}_d$ for [1.5]. We have

$$\begin{aligned} \mathcal{L}_d\left[(f')_{reg}(t)\right](s) &= \mathcal{L}_d[f'(t)H(t)](s) = \mathcal{L}_d[(f(t)H(t))'](s) \\ &\quad -\mathcal{L}_d[f(0)\delta(t)](s) \\ &= s\mathcal{L}_d[f_{reg}(t)](s) - f(0), \end{aligned}$$

where

$$\mathcal{L}_d\left[(f')_{reg}(t)\right](s) = \mathcal{L}_d\left[(f(t)\,H(t))'\right](s) = s\mathcal{L}_d\left[f_{reg}(t)\right](s), \quad \operatorname{Re} s > \lambda_0,$$

while in the classical case,

$$\mathcal{L}_c\left[f'(t)\right](s) = \mathcal{L}_c\left[f(t)\right](s) - f(0), \quad \operatorname{Re} s > \lambda_0.$$

EXAMPLE 1.2.– Let $u(x,t)$, $x \in \mathbb{R}^n$, $t > 0$, be a classical solution of the wave equation

$$\frac{\partial^2}{\partial t^2}u(x,t) - \frac{\partial^2}{\partial x^2}u(x,t) = f(x,t),$$
$$u(x,0) = u_0(x), \quad \frac{\partial}{\partial t}u(x,0) = v_0(x),$$

that is the second derivative above are locally integrable functions on $[0,\infty) \times \mathbb{R}^n$, equal to zero for $t < 0$, $u_0, v_0$ are locally integrable functions on $\mathbb{R}^n$ and $f \in L^1_{loc}([0,\infty) \times \mathbb{R}^n)$ so that it has the classical Laplace transform with respect to $t$ in the domain $\operatorname{Re} s > 0$.

Writing

$$u_{reg}(x,t) = u(x,t)\,H(t), \quad \text{and} \quad f_{reg}(x,t) = f(x,t)\,H(t)$$

for the corresponding distributions, we rewrite the wave equation in the space of distributions as

$$H(t)\frac{\partial^2}{\partial t^2}u(x,t) - H(t)\frac{\partial^2}{\partial x^2}u(x,t) = H(t)f(x,t),$$
$$\frac{\partial^2}{\partial t^2}\left(H(t)u(x,t)\right) - \delta(t)\frac{\partial}{\partial t}u(x,t) - \delta'(t)u(x,t) - \frac{\partial^2}{\partial x^2}\left(H(t)u(x,t)\right)$$
$$= H(t)f(x,t),$$
$$\frac{\partial^2}{\partial t^2}u_{reg}(x,t) - \delta(t)v_0(x) - \delta'(t)u_0(x) - \frac{\partial^2}{\partial x^2}u_{reg}(x,t) = f_{reg}(x,t),$$
$$t > 0, \ x \in \mathbb{R},$$

where the last equation is written in the space of distributions. So with the application of the distributional Laplace transform with respect to $t$, for $\operatorname{Re} s > 0$, we have

$$s^2\tilde{u}_{reg}(x,s) - v_0(x) - su_0(x) - \frac{\partial^2}{\partial x^2}\tilde{u}_{reg}(x,s) = \tilde{f}_{reg}(x,s).$$

The space $\mathcal{K}(\mathbb{R})$ is the space of smooth functions $\varphi$ with the property

$$\sup_{x\in\mathbb{R},\alpha\leq m}\left|\varphi^{(\alpha)}(x)\right| \mathrm{e}^{m|x|} < \infty, \quad m \in \mathbb{N}_0. \qquad [1.7]$$

The space $\mathcal{K}'(\mathbb{R})$ is the dual of $\mathcal{K}(\mathbb{R})$ and elements of $\mathcal{K}'(\mathbb{R})$ are of the form $f = \sum_{\alpha=0}^{r}\Phi_\alpha^{(\alpha)}$, where $\Phi_\alpha$ are continuous functions with the property $|\Phi_\alpha(t)| \leq Ce^{k_0|t|}$, $\alpha \leq r,\ t \in \mathbb{R}$, for some $C > 0$, $r \in \mathbb{N}_0$ and some $k_0 \in \mathbb{N}_0$. $\mathcal{K}'_+(\mathbb{R}) = \mathcal{K}'_+$ is a subspace of $\mathcal{K}'(\mathbb{R})$ consisting of elements supported by $[0,\infty)$ (see [ABD 99, HAS 61]). Its elements are of the form

$$f(x) = (\Phi(x)\mathrm{e}^{kx})^{(p)}, \quad x \in \mathbb{R}, \qquad [1.8]$$

where $\Phi$ is a continuous bounded function such that $\Phi(t) = 0,\ t \leq 0$. Note that $\mathcal{S}$ and $\mathcal{S}'_+$ are subspaces of $\mathcal{K}'(\mathbb{R})$ and $\mathcal{K}'_+$, respectively. The construction implies that elements of $\mathcal{K}'_+$ have the Laplace transform, that is if $f$ is of the form [1.8], then its Laplace transform $\tilde{f}$ is an analytic function in the domain $\operatorname{Re} s > k$.

The Lizorkin space of test functions $\Phi$ is introduced so that Riesz integro-differentiation (and therefore symmetrized fractional derivative) is well defined (see [SAM 93]). Let

$$\Psi = \{\psi \mid \psi \in \mathcal{S}(\mathbb{R}),\ \psi^{(j)}(0) = 0,\ j = 0,1,2...\},$$

and consider the space $\Phi$ consisting of the Fourier transforms of functions in $\Psi$, i.e. $\Phi = \mathcal{F}[\Psi]$. Then, $\Phi$ consists of those functions $\varphi \in \mathcal{S}(\mathbb{R})$ that are orthogonal to polynomials

$$\int_{\mathbb{R}} x^k\varphi(x)\mathrm{d}x = 0, \quad k \in \mathbb{N}_0.$$

The space $\Psi'$ and the space of Lizorkin generalized functions $\Phi'$ are dual spaces of $\Psi$ and $\Phi$, respectively. Recall, for $f \in \Phi'$, we have

$$\langle \mathcal{F}[f], \psi\rangle = \langle f, \mathcal{F}[\psi]\rangle, \quad \psi \in \Psi.$$

Let $f \in C^{\infty}(\mathbb{R} \setminus \{0\})$ be such that it has all the derivatives bounded by the polynomials in $\mathbb{R} \setminus \{0\}$. Then, product $f \cdot u$ is defined by

$$\langle f \cdot u, \psi \rangle = \langle u, f \cdot \psi \rangle, \quad \psi \in \Psi.$$

## 1.4. Fundamental solution

Let $P$ be a linear partial integro-differential operator with constant coefficients. A fundamental solution of $P$, denoted by $E$, is a distributional solution to the equation $Pu = \delta$. Once the fundamental solution is determined, we find a solution to $Pu = f$ as $u = E * f$, if this convolution exits.

The Cauchy problem for the second-order linear partial integro-differential operator with constant coefficients $P$ is given by

$$Pu(x,t) = f(x,t), \quad x \in \mathbb{R}, \ t > 0, \qquad [1.9]$$

$$u(x,0) = u_0(x), \quad \frac{\partial}{\partial t} u(x,0) = v_0(x), \qquad [1.10]$$

where $f$ is continuous for $t \geq 0$, $u_0 \in C^1(\mathbb{R})$ and $v_0 \in C(\mathbb{R})$. A classical solution $u(x,t)$ to the Cauchy problem [1.9], [1.10] is of class $C^2$ for $t > 0$ and of class $C^1$ for $t \geq 0$, satisfies equation [1.9] for $t > 0$, and initial conditions [1.10] when $t \to 0$. If functions $u$ and $f$ are continued by zero for $t < 0$, then the following equation is satisfied in $\mathcal{D}'(\mathbb{R}^2)$:

$$Pu(x,t) = f(x,t) + u_0(x)\delta'(t) + v_0(x)\delta(t). \qquad [1.11]$$

The explanation is given in example 1.2 in the case of the wave equation. The problem of finding generalized solutions (in $\mathcal{D}'(\mathbb{R}^2)$) of equation [1.11] that vanish for $t < 0$ will be called the generalized Cauchy problem for the operator $P$. If there is a fundamental solution $E$ of the operator $P$ and if $f \in \mathcal{D}'(\mathbb{R}^2)$ vanishes for $t < 0$, then there exists a unique solution to the corresponding generalized Cauchy problem and is given by

$$u(x,t) = E(x,t) * (f(x,t) + u_0(x)\delta'(t) + v_0(x)\delta(t)),$$

if the convolution $E * f$ exits. We refer to [DAU 00, TRE 75, VLA 84] for more details.

## 1.5. Some special functions

The Euler gamma function is defined by

$$\Gamma(z) = \int_0^\infty t^{z-1} e^{-t} dt, \quad \mathrm{Re}\, z > 0.$$

The gamma function can also be represented by

$$\Gamma(z) = \lim_{n\to\infty} \frac{n! n^z}{\prod_{k=0}^n (z+k)}, \quad \mathrm{Re}\, z > 0.$$

It satisfies $\Gamma(z+1) = z\Gamma(z)$, $\mathrm{Re}\, z > 0$. By the analytic continuation, we have that $\Gamma(z)$, $z \neq -n$, $n \in \mathbb{N}_0$, is an analytic function. Gamma function has simple poles at $z = n$, $n \in \mathbb{N}_0$. Having $\Gamma(1) = 1$, we obtain $\Gamma(n+1) = n!, \quad n \in \mathbb{N}$. We refer to [POD 99] for the properties of the gamma function.

We refer to [GOR 97b, MAI 00] for the theory of Mittag-Leffler functions presented in this section. The one-parameter Mittag-Leffler function is defined by

$$E_\alpha(z) = \sum_{k=0}^\infty \frac{z^k}{\Gamma(\alpha k+1)}, \quad z \in \mathbb{C}, \ \alpha > 0. \tag{1.12}$$

The one-parameter Mittag-Leffler function is an entire function of order $\rho = \frac{1}{\alpha}$ and type 1. In some special cases of $\alpha$, the one-parameter Mittag-Leffler function becomes

$$E_2(z^2) = \cosh z, \quad E_2(-z^2) = \cos z, \quad z \in \mathbb{C},$$
$$E_{\frac{1}{2}}(\pm\sqrt{z}) = e^z \left(1 + \mathrm{erf}(\pm\sqrt{z})\right), \quad z \in \mathbb{C},$$

with $\mathrm{erf}\, z = \frac{2}{\sqrt{\pi}} \int_0^z e^{-u^2} du$ being the error function.

The asymptotics of [1.12] are as follows:

$$E_\alpha(z) \approx \frac{1}{\alpha} e^{\sqrt{z}} - \sum_{k=1}^\infty \frac{z^{-k}}{\Gamma(1-\alpha k)}, \quad |z| \to \infty, \ |\arg z| < \frac{\alpha\pi}{2}, \ \alpha \in (0,2),$$

$$E_\alpha(z) \approx -\sum_{k=1}^\infty \frac{z^{-k}}{\Gamma(1-\alpha k)}, \quad |z| \to \infty, \ \arg z \in \left(\frac{\alpha\pi}{2}, 2\pi - \frac{\alpha\pi}{2}\right), \ \alpha \in (0,2),$$

$$E_\alpha(z) \approx \frac{1}{\alpha} \sum_m e^{\sqrt[\alpha]{z} e^{\frac{2\pi i m}{\alpha}}} - \sum_{k=1}^\infty \frac{z^{-k}}{\Gamma(1-\alpha k)}, \quad |z| \to \infty, \ \arg z \in (-\pi,\pi), \ \alpha \geq 2, \ m \in \mathbb{N}, \ \arg(z + 2\pi m) \in \left(-\tfrac{\alpha\pi}{2}, \tfrac{\alpha\pi}{2}\right).$$

The two-parameter Mittag-Leffler function is defined by

$$E_{\alpha,\beta}(z) = \sum_{k=0}^{\infty} \frac{z^k}{\Gamma(\alpha k + \beta)}, \quad z \in \mathbb{C}, \ \alpha > 0, \ \beta \in \mathbb{C}.$$

It is an entire function of order $\rho = \frac{1}{\alpha}$ and type 1. In some special cases of $\alpha$ and $\beta$, it becomes

$$E_{1,2}(z) = \frac{\mathrm{e}^z - 1}{z}, \quad E_{2,2}(z) = \frac{\sinh\sqrt{z}}{\sqrt{z}}, \quad z \in \mathbb{C}.$$

We define one- and two-parameter Mittag-Leffler-type functions, respectively, by

$$e_\alpha(t,\lambda) = E_\alpha(-\lambda t^\alpha) \ \text{ and } \ e_{\alpha,\beta}(t,\lambda) = t^{\beta-1} E_{\alpha,\beta}(-\lambda t^\alpha), \quad t \geq 0, \ \lambda \in \mathbb{C}.$$

In applications, we will often omit the parameter $\lambda$. According to [MAI 00], if $\alpha \in (0,1)$ and $\lambda > 0$, we have $e_\alpha \in C^\infty((0,\infty)) \cap C([0,\infty))$ and $\frac{\mathrm{d}}{\mathrm{d}t} e_\alpha \in C^\infty((0,\infty)) \cap L^1_{loc}([0,\infty))$. Also, $e_\alpha$ is a completely monotonic function, i.e. $(-1)^k \frac{\mathrm{d}^k}{\mathrm{d}t^k} e_\alpha(t) > 0$.

The Laplace transforms of $e_\alpha$ and $e_{\alpha,\beta}$ are

$$\mathcal{L}\left[e_\alpha(t,\lambda)\right](s) = \frac{s^{\alpha-1}}{s^\alpha + \lambda}, \quad \mathcal{L}\left[e_{\alpha,\beta}(t,\lambda)\right](s) = \frac{s^{\alpha-\beta}}{s^\alpha + \lambda}, \quad \operatorname{Re} s > \sqrt[\alpha]{|\lambda|},$$

respectively.

Functions $e_\alpha$ and $e_{\alpha,\beta}$ admit integral representations given by

$$e_\alpha(t,\lambda) = \frac{1}{\pi} \int_0^\infty \frac{\lambda q^{\alpha-1} \sin(\alpha\pi)}{q^{2\alpha} + 2\lambda q^\alpha \cos(\alpha\pi) + \lambda^2} \mathrm{e}^{-qt} \mathrm{d}q, \quad t \geq 0, \ \alpha \in (0,1), \quad \lambda > 0,$$

$$e_{\alpha,\beta}(t,\lambda) = \frac{1}{\pi} \int_0^\infty \frac{\lambda \sin((\beta-\alpha)\pi) + q^\alpha \sin(\beta\pi)}{q^{2\alpha} + 2\lambda q^\alpha \cos(\alpha\pi) + \lambda^2} q^{\alpha-\beta} \mathrm{e}^{-qt} \mathrm{d}q, \quad t \geq 0, \ 0 < \alpha \leq \beta < 1, \ \lambda > 0.$$

Chapter 2

# Basic Definitions and Properties of Fractional Integrals and Derivatives

## 2.1. Definitions of fractional integrals and derivatives

In this section, we review some basic properties of fractional integrals and derivatives, which we will need later in the analysis of concrete problems. This section contains results from various books and papers [ALM 12, ATA 14a, ATA 13a, ATA 07a, ATA 09b, ATA 09d, ATA 08b, BUT 00, CAN 87, CAP 67, CAP 71b, DIE 10, HER 11, KIL 04, KIL 06, KIR 94, NAH 03, ODI 07, POO 12a, POO 12b, POO 13, ROS 93, SAM 95, SAM 93, TAR 06, TRU 99, UCH 08, WES 03].

### 2.1.1. *Riemann–Liouville fractional integrals and derivatives*

There are many possible generalizations of the notion of a derivative of a function that would lead to the answer of the question: what is $\frac{\mathrm{d}^n}{\mathrm{d}x^n}y(x)$ when $n$ is any real number? We start from the Cauchy formula for an $n$-fold primitive of a function $f$ given as

$$ {}_a\mathrm{I}_t^n f(t) = \frac{1}{(n-1)!}\int_a^t (t-\tau)^{n-1} f(\tau)\,\mathrm{d}\tau, \quad t \in [a,b], \; n \in \mathbb{N}, \qquad [2.1]$$

where it is assumed that $f(t) = 0$, for $t < a$. Note that $(n-1)! = \Gamma(n)$, where $\Gamma$ is the Euler gamma function (see section 1.5).

DEFINITION 2.1.– *The left Riemann–Liouville fractional integral of order $\alpha \in \mathbb{C}$ is formally given by*

$$
{}_a\mathrm{I}_t^\alpha f(t) = \frac{1}{\Gamma(\alpha)} \int_a^t (t-\tau)^{\alpha-1} f(\tau)\,\mathrm{d}\tau, \quad t \in [a,b], \ \mathrm{Re}\,\alpha > 0. \qquad [2.2]
$$

In the special case of positive real $\alpha$ ($\alpha \in \mathbb{R}_+$) and $f \in L^1(a,b)$, the integral ${}_a\mathrm{I}_t^\alpha f$ exists for almost all $t \in [a,b]$. Also, ${}_a\mathrm{I}_t^\alpha f \in L^1(a,b)$ (see [DIE 10, p. 13]). For $\alpha = 0$, we define ${}_a\mathrm{I}_t^0 f = f$. This definition is motivated by the following reasoning. Suppose that $f \in C^1([a,b])$. Then, after integration by parts, from [2.2], we have

$$
{}_a\mathrm{I}_t^\alpha f(t) = \frac{(t-a)^\alpha}{\Gamma(\alpha+1)} f(a) + \frac{1}{\Gamma(\alpha+1)} \int_a^t (t-\tau)^\alpha f^{(1)}(\tau)\,\mathrm{d}\tau,
$$

so that

$$
\lim_{\alpha\to 0} {}_a\mathrm{I}_t^\alpha f(t) = f(a) + \int_a^t f^{(1)}(\tau)\,\mathrm{d}\tau = f(t).
$$

DEFINITION 2.2.– *The right Riemann–Liouville fractional integral of order $\alpha \in \mathbb{C}$ is formally given by*

$$
{}_t\mathrm{I}_b^\alpha f(t) = \frac{1}{\Gamma(\alpha)} \int_t^b (\tau-t)^{\alpha-1} f(\tau)\,\mathrm{d}\tau, \quad t \in [a,b], \ \mathrm{Re}\,\alpha > 0. \qquad [2.3]
$$

The existence is the same as in the case of the left Riemann–Liouville fractional integral given above.

In the special case when $f(t) = (t-a)^{\beta-1}$ and $g(t) = (b-t)^{\beta-1}$, $t \in [a,b]$, $\alpha, \beta \in \mathbb{C}$, we have

$$
\begin{aligned}
{}_a\mathrm{I}_t^\alpha (t-a)^{\beta-1} &= \frac{\Gamma(\beta)}{\Gamma(\beta+\alpha)} (t-a)^{\beta+\alpha-1}, \quad \mathrm{Re}\,\alpha > 0, \ \mathrm{Re}\,\beta > 0, \\
{}_t\mathrm{I}_b^\alpha (b-t)^{\beta-1} &= \frac{\Gamma(\beta)}{\Gamma(\beta+\alpha)} (b-t)^{\beta+\alpha-1}, \quad \mathrm{Re}\,\alpha > 0, \ \mathrm{Re}\,\beta > 0.
\end{aligned}
$$

Operators ${}_a\mathrm{I}_t^\alpha$ and ${}_t\mathrm{I}_b^\alpha$ with $\mathrm{Re}\,\alpha > 0$ are bounded operators from $L^p(a,b)$ into $L^p(a,b)$, $p \geq 1$. The following estimates hold:

$$
\|\mathrm{I}_t^\alpha f\|_{L^p(a,b)} \leq \frac{(b-a)^{\mathrm{Re}\,\alpha}}{|\Gamma(\alpha)|\,\mathrm{Re}\,\alpha} \|f\|_{L^p(a,b)}, \quad \|{}_t\mathrm{I}_b^\alpha f\|_{L^p(0,b)} \leq \frac{(b-a)^{\mathrm{Re}\,\alpha}}{|\Gamma(\alpha)|\,\mathrm{Re}\,\alpha} \|f\|_{L^p(a,b)}, \qquad [2.4]
$$

see [SAM 93, p. 48]. If $\alpha \in (0,1)$ and $1 < p < \frac{1}{\alpha}$, then the operators ${}_0\mathrm{I}_t^{\alpha}$ and ${}_t\mathrm{I}_b^{\alpha}$ are bounded from $L^p(a,b)$ into $L^q(a,b)$ for $q = \frac{p}{1-\alpha p}$ (see [SAM 93, p. 66]).

Introducing the function

$$f_\alpha(t) = \begin{cases} \frac{1}{\Gamma(\alpha)}(t-a)^{\alpha-1}, & t > a, \\ 0, & t < a, \end{cases} \qquad \mathrm{Re}\,\alpha > 0, \qquad [2.5]$$

we conclude that the integral [2.2] may be written in the form of convolution as

$${}_a\mathrm{I}_t^{\alpha} y(t) = f_\alpha(t) * y(t) = \int_a^t f_\alpha(t-\tau)\, y(\tau)\, \mathrm{d}\tau. \qquad [2.6]$$

REMARK 2.1.– Expression [2.6] may be used to define the generalized fractional integral with the different choice of $f_\alpha$. For example, in [KIL 04], various generalizations of the fractional integral were presented, including the generalization that uses the two-parameter Mittag-Leffler function

$$K^{(\alpha)} f(t) = \frac{1}{\Gamma(\alpha)} t^{\alpha-1} E_{\rho,\alpha}(\omega t^{\rho}) * f(t), \quad \omega \in \mathbb{R}.$$

The fractional integral of purely imaginary order is defined as

$${}_a\mathrm{I}_t^{\mathrm{i}\theta} y(t) = \frac{\mathrm{d}}{\mathrm{d}t}\left({}_a\mathrm{I}_t^{1+\mathrm{i}\theta} y(t)\right) = \frac{1}{\Gamma(1+\mathrm{i}\theta)} \frac{\mathrm{d}}{\mathrm{d}t} \int_a^t (t-\tau)^{\mathrm{i}\theta}\, y(\tau)\, \mathrm{d}\tau, \qquad [2.7]$$

with $\theta \neq 0$.

The asymptotic behavior of the left Riemann–Liouville fractional integral may be characterized as follows.

PROPOSITION 2.1.– [UCH 08, p. 165] Suppose that $f \in L^1_{loc}([0,\infty))$ is an analytic function in $(0,\infty)$. Then

$${}_a\mathrm{I}_t^{\alpha} f(t) \sim {}_0\mathrm{I}_t^{\alpha} f(t) + \frac{a}{\pi}\Gamma(\alpha+1)\sin(\alpha\pi)\, f(0)\, t^{-\alpha-1} \sim {}_0\mathrm{I}_t^{\alpha} f(t), \quad \text{as } t \to \infty. \qquad [2.8]$$

If ${}_a\mathrm{I}_t^{\alpha} f$ is used to model a hereditary process, then a physical meaning of [2.8] is that for large times, the importance of the initial state of the system is small.

DEFINITION 2.3.– *The left and right Riemann–Liouville fractional derivatives* ${}_{a}\mathrm{D}_{t}^{\alpha}f$ *and* ${}_{t}\mathrm{D}_{b}^{\alpha}f$ *of the order* $\alpha \in \mathbb{C}$, $\operatorname{Re}\alpha \geq 0$, $n-1 \leq \operatorname{Re}\alpha < n$, $n \in \mathbb{N}$, *with the appropriate assumptions on* $f$ *(see below), are defined as*

$$
\begin{aligned}
{}_{a}\mathrm{D}_{t}^{\alpha}f(t) &= \frac{\mathrm{d}^{n}}{\mathrm{d}t^{n}}\left({}_{a}\mathrm{I}_{t}^{n-\alpha}f(t)\right) = \frac{1}{\Gamma(n-\alpha)}\frac{\mathrm{d}^{n}}{\mathrm{d}t^{n}}\int_{a}^{t}\frac{f(\tau)}{(t-\tau)^{\alpha-n+1}}\mathrm{d}\tau, \quad t\in(a,b),\\
{}_{t}\mathrm{D}_{b}^{\alpha}f(t) &= (-1)^{n}\frac{\mathrm{d}^{n}}{\mathrm{d}t^{n}}\left({}_{t}\mathrm{I}_{b}^{n-\alpha}f(t)\right) = (-1)^{n}\frac{1}{\Gamma(n-\alpha)}\frac{\mathrm{d}^{n}}{\mathrm{d}t^{n}}\int_{t}^{b}\frac{f(\tau)}{(\tau-t)^{\alpha-n+1}}\mathrm{d}\tau,\\
&\qquad t\in(a,b).
\end{aligned}
\qquad [2.9]
$$

If $f \in AC^{n}([a,b])$ and $n-1 \leq \operatorname{Re}\alpha < n$, $n \in \mathbb{N}$, then ${}_{a}\mathrm{D}_{t}^{\alpha}f$ and ${}_{t}\mathrm{D}_{b}^{\alpha}f$ exist almost everywhere on $[a,b]$ and

$$
{}_{a}\mathrm{D}_{t}^{\alpha}f(t) = \sum_{k=0}^{n-1}\frac{f^{(k)}(a)}{\Gamma(1+k-\alpha)}(t-a)^{k-\alpha} + \frac{1}{\Gamma(n-\alpha)}\int_{a}^{t}\frac{f^{(n)}(\tau)}{(t-\tau)^{\alpha-n+1}}\mathrm{d}\tau, \qquad [2.10]
$$

$$
{}_{t}\mathrm{D}_{b}^{\alpha}f(t) = \sum_{k=0}^{n-1}(-1)^{k}\frac{f^{(k)}(b)}{\Gamma(1+k-\alpha)}(b-t)^{k-\alpha} + \frac{(-1)^{n}}{\Gamma(n-\alpha)}\int_{t}^{b}\frac{f^{(n)}(\tau)}{(\tau-t)^{\alpha-n+1}}\mathrm{d}\tau, \qquad [2.11]
$$

see [KIL 06, p. 73]. From the definitions, it follows that in the special case when $f(t) = (t-a)^{\beta-1}$, $t > a$, and $f(t) = (b-t)^{\beta-1}$, $t < b$, $\beta \in \mathbb{C}$, we have

$$
\begin{aligned}
{}_{a}\mathrm{D}_{t}^{\alpha}(t-a)^{\beta-1} &= \frac{\Gamma(\beta)}{\Gamma(\beta-\alpha)}(t-a)^{\beta-\alpha-1} \quad \text{and}\\
{}_{t}\mathrm{D}_{b}^{\alpha}(b-t)^{\beta-1} &= \frac{\Gamma(\beta)}{\Gamma(\beta-\alpha)}(b-t)^{\beta-\alpha-1}.
\end{aligned}
\qquad [2.12]
$$

Again, from [2.12], for constant function $f = C$, we have

$$
{}_{a}\mathrm{D}_{t}^{\alpha}C = \frac{C}{\Gamma(1-\alpha)}(t-a)^{-\alpha} \quad \text{and} \quad {}_{t}\mathrm{D}_{b}^{\alpha}C = \frac{C}{\Gamma(1-\alpha)}(b-t)^{-\alpha}.
$$

Also, ${}_a\mathrm{D}_t^\alpha f(t) = 0$ and ${}_t\mathrm{D}_b^\alpha g(t) = 0$, $n-1 \le \mathrm{Re}\,\alpha < n$, if and only if, respectively,

$$f(t) = \sum_{k=1}^{n} c_k (t-a)^{\alpha-k} \text{ and } g(t) = \sum_{k=1}^{n} d_k (b-t)^{\alpha-k}, \qquad [2.13]$$

where $c_k$ and $d_k$, $k = 1, \ldots, n$, are arbitrary constants. Thus, functions $f$ and $g$ in [2.13] play the role of constants for the left and right Riemann–Liouville fractional derivatives, respectively.

Let $\alpha = k + \gamma$, $k \in \mathbb{N}_0$, $\gamma \in [0,1)$. Then, ${}_0\mathrm{D}_t^\alpha$ and ${}_t\mathrm{D}_b^\alpha$ may be written as

$$\begin{aligned} {}_0\mathrm{D}_t^\alpha f(t) &= \frac{1}{\Gamma(1-\gamma)} \frac{\mathrm{d}^{k+1}}{\mathrm{d}t^{k+1}} \int_0^t \frac{f(\tau)}{(t-\tau)^\gamma} \mathrm{d}\tau, \quad t > 0, \\ {}_t\mathrm{D}_b^\alpha f(t) &= (-1)^{k+1} \frac{1}{\Gamma(1-\gamma)} \frac{\mathrm{d}^{k+1}}{\mathrm{d}t^{k+1}} \int_t^b \frac{f(\tau)}{(\tau-t)^\gamma} \mathrm{d}\tau, \quad t < b. \end{aligned}$$

Sometimes, in short, it is written ${}_a\mathrm{D}_t^\alpha f = f^{(\alpha)}$.

Let $\alpha \in [0,1)$. Then, for $t > a$ and $t < b$, we have

$$\begin{aligned} {}_a\mathrm{D}_t^\alpha f(t) &= \frac{1}{\Gamma(1-\alpha)} \frac{\mathrm{d}}{\mathrm{d}t} \int_a^t \frac{f(\tau)}{(t-\tau)^\alpha} \mathrm{d}\tau \text{ and} \\ {}_t\mathrm{D}_b^\alpha f(t) &= -\frac{1}{\Gamma(1-\alpha)} \frac{\mathrm{d}}{\mathrm{d}t} \int_t^b \frac{f(\tau)}{(\tau-t)^\alpha} \mathrm{d}\tau. \end{aligned} \qquad [2.14]$$

In the case when $\alpha$ is purely imaginary, i.e. $\alpha = \mathrm{i}\theta$, the left Riemann–Liouville fractional derivative is defined as

$${}_a\mathrm{D}_t^{\mathrm{i}\theta} f(t) = \frac{1}{\Gamma(1-\mathrm{i}\theta)} \frac{\mathrm{d}}{\mathrm{d}t} \int_a^t \frac{f(\tau)}{(t-\tau)^{\mathrm{i}\theta}} \mathrm{d}\tau, \quad t \ge a.$$

Consider the problem of determining $\lim_{\alpha\to 1^-} {}_a\mathrm{D}_t^\alpha f$. Then, we have the following proposition.

PROPOSITION 2.2.– [NAH 03, p. 174] Suppose that $f \in C^1([0,T])$. Then, $\lim_{\alpha\to 1^-} {}_a\mathrm{D}_t^\alpha f = f^{(1)}$.

We put $\frac{\mathrm{d}^n}{\mathrm{d}t^n}(\cdot) = \mathrm{D}^n(\cdot)$. The index rule holds for the integer-order integrals and derivatives

$$\begin{aligned} \left({}_a\mathrm{I}_t^n\, {}_a\mathrm{I}_t^m\right) f(t) &= \left({}_a\mathrm{I}_t^m\, {}_a\mathrm{I}_t^n\right) f(t) = {}_a\mathrm{I}_t^{n+m} f(t), \quad n, m \in \mathbb{N}_0, \\ \left(\mathrm{D}^n \mathrm{D}^m\right) f(t) &= \left(\mathrm{D}^m \mathrm{D}^n\right) f(t) = \mathrm{D}^{m+n} f(t), \quad n, m \in \mathbb{N}_0. \end{aligned} \qquad [2.15]$$

The semi-group property $[2.15]_1$ holds for fractional integrals only.

PROPOSITION 2.3.– [DIE 10, p. 14] The fractional integral ${}_a\mathrm{I}_t^\alpha$ as a mapping from $L^1(a,b) \to L^1(a,b)$ forms a commutative semi-group with respect to orders of integrals. The identity operator ${}_a\mathrm{I}_t^0$ is the neutral element. Thus, if $\operatorname{Re}\alpha, \operatorname{Re}\beta > 0$

$$\begin{aligned} \left({}_a\mathrm{I}_t^\alpha\, {}_a\mathrm{I}_t^\beta\right) f(t) &= \left({}_a\mathrm{I}_t^\beta\, {}_a\mathrm{I}_t^\alpha\right) f(t) = {}_a\mathrm{I}_t^{\alpha+\beta} f(t), \\ \left({}_t\mathrm{I}_b^\alpha\, {}_t\mathrm{I}_b^\beta\right) f(t) &= \left({}_t\mathrm{I}_b^\beta\, {}_t\mathrm{I}_b^\alpha\right) f(t) = {}_t\mathrm{I}_b^{\alpha+\beta} f(t), \end{aligned}$$

holds for almost all $t \in [a,b]$ (almost everywhere (a.e.) in $[a,b]$) if $f \in L^p(a,b)$, $1 \leq p \leq \infty$.

Also, it can be shown that for $\operatorname{Re}\alpha > 0$, $f \in L^p(a,b)$, $1 \leq p \leq \infty$, the composition of fractional derivatives and fractional integrals holds, for almost all $t \in (a,b)$ (see [SAM 93, p. 44]),

$$\left({}_a\mathrm{D}_t^\alpha\, {}_a\mathrm{I}_t^\alpha\right) f(t) = f(t), \quad \text{and} \quad \left({}_t\mathrm{D}_b^\alpha\, {}_t\mathrm{I}_b^\alpha\right) f(t) = f(t),$$

showing that ${}_a\mathrm{D}_t^\alpha$, ${}_t\mathrm{D}_b^\alpha$ are the left inverses of ${}_a\mathrm{I}_t^\alpha$, ${}_t\mathrm{I}_b^\alpha$, respectively. However by applying ${}_a\mathrm{D}_t^\alpha$ and ${}_t\mathrm{D}_b^\alpha$ to the right of ${}_a\mathrm{I}_t^\alpha$ and ${}_t\mathrm{I}_b^\alpha$, we have different situation. To examine the resulting relations, we define the following spaces:

$$\begin{aligned} {}_a\mathrm{I}_t^\alpha\left(L^p\right) &= \left\{f \mid f = {}_a\mathrm{I}_t^\alpha \varphi,\ \varphi \in L^p(a,b)\right\} \quad \text{and} \\ {}_t\mathrm{I}_b^\alpha\left(L^p\right) &= \left\{g \mid g = {}_t\mathrm{I}_b^\alpha \phi,\ \phi \in L^p(a,b)\right\}. \end{aligned} \qquad [2.16]$$

PROPOSITION 2.4.– [KIL 06, p. 74] Let $\operatorname{Re}\alpha > 0$, $n-1 < \operatorname{Re}\alpha < n$. Then, the following holds:

i) If $f \in {}_a\mathrm{I}_t^\alpha\left(L^p\right)$, $1 \leq p \leq \infty$, then

$$\left({}_a\mathrm{I}_t^\alpha\, {}_a\mathrm{D}_t^\alpha\right) f(t) = f(t), \quad \text{a.e., in } [a,b]. \qquad [2.17]$$

ii) If $f \in L^1(a,b)$, ${}_a\mathrm{I}_t^{n-\alpha} f \in AC^n([a,b])$, then

$$\left({}_a\mathrm{I}_t^{\alpha}\, {}_a\mathrm{D}_t^{\alpha}\right) f(t) = f(t) - \sum_{j=1}^{n} \frac{(t-a)^{\alpha-j}}{\Gamma(\alpha-j+1)} \left[\frac{\mathrm{d}^{n-j}}{\mathrm{d}t^{n-j}} \left({}_a\mathrm{I}_t^{n-\alpha} f\right)\right]_{t=a} \quad [2.18]$$

holds for almost all $t \in [a,b]$.

We state the results about the index rule for the fractional derivatives.

PROPOSITION 2.5.– [KIL 06, p. 75] Let $\alpha, \beta > 0$, $n-1 \leq \alpha < n$, $m-1 \leq \beta < m$ and $\alpha + \beta < n$. Let $f \in L^1(a,b)$ and ${}_a\mathrm{I}_t^{m-\alpha} f \in AC^m([a,b])$. Then, the following index rule holds:

$$\left({}_a\mathrm{D}_t^{\alpha}\, {}_a\mathrm{D}_t^{\beta}\right) f(t) = {}_a\mathrm{D}_t^{\alpha+\beta} f(t) - \sum_{j=1}^{m} \frac{(t-a)^{-j-\alpha}}{\Gamma(1-j-\alpha)} \left[{}_a\mathrm{D}_t^{\beta-j} f(t)\right]_{t=a}, \quad t \in [a,b]. \quad [2.19]$$

There are special cases when the index rule holds (see [KIL 06, p. 74]).

The composition rule for the left Riemann–Liouville derivative and the right Riemann–Liouville integral takes a rather complicated form (see [NAH 03, p. 22]). Suppose that $f$ is Hölder continuous in $[a,b]$ and $f \in L^1(a,b)$. Then, for $\alpha \in (0,1)$, we have

$$\left({}_a\mathrm{D}_t^{\alpha}\, {}_t\mathrm{I}_b^{\alpha}\right) f(t) = f(t)\cos(\alpha\pi) + \mathrm{S}_{ab}^{\alpha} f(t)\cos(\alpha\pi), \quad \text{a.e. in } [a,b],$$

where

$$\mathrm{S}_{ab}^{\alpha} f(t) = \frac{1}{\pi} \int_a^b \left|\frac{u-a}{t-a}\right|^{\alpha} \frac{f(u)}{u-t} \mathrm{d}u. \quad [2.20]$$

The integral in [2.20] should be taken as a Cauchy principal value.

In the variational problems, important result is integration by parts formula. We state it as follows.

PROPOSITION 2.6.– [SKM, pp. 46 and 67]

i) Suppose $0 < \alpha < 1$, $f \in L^p(a,b)$, $g \in L^q(a,b)$. Then

$$\int_a^b f(t) \left({}_a\mathrm{I}_t^{\alpha} g(t)\right) \mathrm{d}t = \int_a^b \left({}_t\mathrm{I}_b^{\alpha} f(t)\right) g(t)\, \mathrm{d}t, \quad [2.21]$$

for $p \geq 1$, $q \geq 1$ and $\frac{1}{p} + \frac{1}{q} \leq 1 + \alpha$.

ii) Suppose $0 < \operatorname{Re}\alpha < 1$, $f \in {}_t\mathrm{I}_b^\alpha(L^p)$ and $g \in {}_a\mathrm{I}_t^\alpha(L^q)$. Then

$$\int_a^b f(t)\left({}_a\mathrm{D}_t^\alpha g(t)\right)\mathrm{d}t = \int_a^b \left({}_t\mathrm{D}_b^\alpha f(t)\right) g(t)\,\mathrm{d}t, \qquad [2.22]$$

for $p \geq 1$, $q \geq 1$ and $\frac{1}{p} + \frac{1}{q} \leq 1 + \alpha$.

For a generalization of the integration by parts formula [2.22], see [11.81].

Fractional derivatives could be expressed in terms of integer-order derivatives through expansion formula.

PROPOSITION 2.7.– [SAM 93, p. 278] Suppose $\alpha \in \mathbb{R}_+$ and that $f$ is an analytic function on $(a, b)$. Then

$${}_a\mathrm{D}_t^\alpha f(t) = \sum_{n=0}^{\infty} \binom{\alpha}{n} \frac{(t-a)^{n-\alpha}}{\Gamma(n+1-\alpha)} f^{(n)}(t), \quad t \in (a, b), \qquad [2.23]$$

where

$$\binom{\alpha}{n} = (-1)^{n-1} \frac{\alpha\Gamma(n-\alpha)}{\Gamma(1-\alpha)\Gamma(n+1)},$$

denotes the binomial coefficients.

The fractional derivatives can be expressed in terms of a function and its moments. The following expansion formula may be proved.

PROPOSITION 2.8.– [ATA 14a] Let $f \in C^1([0, T])$ and $0 < \alpha < 1$. Then

$${}_0\mathrm{D}_t^\alpha f(t) = \frac{f(t)}{t^\alpha} A(N) - \sum_{p=1}^{N} C_{p-1} \frac{V_{p-1}(f)(t)}{t^{p+\alpha}} + Q_{N+1}(f)(t), \quad t \in (0, T], \qquad [2.24]$$

where

$$A(N) = \frac{1}{\Gamma(1-\alpha)} + \frac{1}{\Gamma(\alpha)\Gamma(1-\alpha)} \sum_{p=1}^{N} \frac{\Gamma(p+\alpha)}{p!} = \frac{\sin(\alpha\pi)}{\alpha\pi} \frac{\Gamma(N+1+\alpha)}{\Gamma(N+1)}, \qquad [2.25]$$

$$C_{p-1} = \frac{\Gamma(p+\alpha)}{\Gamma(\alpha)\Gamma(1-\alpha)\Gamma(p)}, \qquad [2.26]$$

$$V_{p-1}(f)(t) = \int_0^t \tau^{p-1} f(\tau)\mathrm{d}\tau, \quad t \in [0, T], \; p \in \mathbb{N}, \qquad [2.27]$$

and the reminder term $Q_{N+1}(f)$ satisfies the estimate

$$|Q_{N+1}(f)(t)| \leq \frac{C \cdot M_t}{\Gamma(\alpha)\Gamma(1-\alpha)} \cdot \frac{t^{1-\alpha}}{N^{\alpha_1}}, \quad t \in [0,T], \ 0 < \alpha_1 < 1-\alpha, \quad [2.28]$$

with $M_t = \max_{0 \leq \tau \leq t} |y^{(1)}(\tau)|$ and certain constant $C > 0$.

Thus

$$\lim_{N \to \infty} Q_{N+1}(f)(t) = 0 \text{ uniformly on } [0,T]$$

and the approximation formula for the left Riemann–Liouville fractional derivative becomes

$${}_0\mathrm{D}_t^{\alpha} f(t) \approx \frac{f(t)}{t^{\alpha}} \mathcal{A}(N) - \sum_{p=1}^{N} \mathcal{C}_{p-1} \frac{V_{p-1}(f)(t)}{t^{p+\alpha}}, \quad t \in (0,T]. \quad [2.29]$$

From the expansion formula [2.24], approximation to the right fractional derivative could be derived.

PROPOSITION 2.9.– [ATA 14a] Let $g \in C^1([0,T])$ and $0 < \alpha < 1$. Then, the right Riemann–Liouville fractional derivative can be approximated by

$${}_t\mathrm{D}_T^{\alpha} g(t) \approx \frac{g(t)}{t^{\alpha}} \mathcal{A}(N) + \sum_{p=1}^{N} \mathcal{C}_{p-1} t^{p-1} W_{p-1}(g)(t), \quad t \in (0,T], \quad [2.30]$$

where $A(N, \alpha)$ and $C_{p-1}(\alpha)$ are defined by [2.25] and [2.26], respectively, and $W_{p-2}(g)$ is

$$W_{p-1}(g)(t) = \int_t^T \frac{g(\tau)}{\tau^{p+\alpha}} \mathrm{d}\tau.$$

REMARK 2.2.– Expansion formula [2.24] may be expressed in a different form (see [ATA 08b]) in which the first derivative of a function appears. Thus, for $t \in (0,T]$,

$${}_0\mathrm{D}_t^{\alpha} f(t) = \mathcal{A}(N) \frac{f(t)}{t^{\alpha}} + \mathcal{B}(N) t^{1-\alpha} f^{(1)}(t) - \sum_{p=1}^{N} \mathcal{C}_{p-1} \frac{V_{p-1}(f)(t)}{t^{p+\alpha}} + R_{N+1}(t), [2.31]$$

where

$$\mathcal{A}(N)=\frac{1}{\Gamma(1-\alpha)}-\frac{1}{\Gamma(\alpha-1)\Gamma(2-\alpha)}\sum_{p=2}^{N}\frac{\Gamma(p-1+\alpha)}{(p-1)!}=\frac{\sin(\alpha\pi)}{\alpha\pi}\frac{\Gamma(N+\alpha)}{\Gamma(N)},$$

$$\mathcal{B}(N)=\frac{1}{\Gamma(2-\alpha)}\left(1+\frac{1}{\Gamma(\alpha-1)}\sum_{p=1}^{N}\frac{\Gamma(p-1+\alpha)}{p!}\right)=\frac{\Gamma(N+\alpha)}{\Gamma(\alpha)\Gamma(2-\alpha)\Gamma(N+1)},$$

$$\mathcal{C}_{p-1}=\frac{\Gamma(p+\alpha)}{\Gamma(\alpha)\Gamma(1-\alpha)\Gamma(p)},$$

$$R_{N+1}(t)=\frac{t^{-\alpha}}{\Gamma(\alpha)\Gamma(1-\alpha)}\int_0^t y^{(2)}(\tau)\left(\sum_{p=N+1}^{\infty}\frac{\Gamma(p+\alpha)}{p!}\left(\frac{\tau}{t}\right)^p \mathrm{d}\tau\right),\quad t\in(0,T] \qquad [2.32]$$

and

$$V_p^{(1)}(f)(t)=t^p f(t),\quad V_p(f)(0)=0,\ \ t\in[0,T],\ p\in\mathbb{N}.$$

There are several definitions of the fractional derivatives of variable order. In [ROS 93, SAM 95], the following definition was proposed. The left Riemann–Liouville fractional derivative of variable $\alpha$ for $0\le\alpha(t)<1$ is

$${}_0\mathrm{D}_t^{\alpha(t)}f(t)=\frac{1}{\Gamma(1-\alpha(t))}\frac{\mathrm{d}}{\mathrm{d}t}\int_0^t\frac{f(\tau)}{(t-\tau)^{\alpha(t)}}\mathrm{d}\tau,\quad t\in[0,T]. \qquad [2.33]$$

In [ATA 13a], the following expansion formula for [2.33] is proved.

PROPOSITION 2.10.– [ATA 13a] Suppose that $f\in C^2([0,T])$, $\alpha\in C^1([0,T])$. Then, the fractional derivative of the order $\alpha$, defined by [2.33], may be written as

$$\begin{aligned}{}_0\mathrm{D}_t^{\alpha(t)}f(t) &= A_1\left(f(t),f^{(1)}(t),\alpha(t),t\right)\\ &\quad -\alpha^{(1)}(t)A_2\left(f(t),f^{(1)}(t),\alpha(t),t\right)+R_1^N(t)\\ &\quad +R_2^{N,M}(t),\quad t\in(0,T],\end{aligned}$$

where

$$A_1\left(f(t),f^{(1)}(t),\alpha(t),t\right)$$
$$=\frac{f(t)}{t^{\alpha(t)}}\left(\frac{1}{\Gamma(1-\alpha(t))}-\frac{1}{\Gamma(\alpha(t)-1)\Gamma(2-\alpha(t))}\sum_{p=2}^{N}\frac{\Gamma(p-1+\alpha(t))}{(p-1)!}\right)$$
$$+\frac{f^{(1)}(t)t^{1-\alpha(t)}}{\Gamma(2-\alpha(t))}\left(1+\frac{1}{\Gamma(\alpha(t)-1)}\sum_{p=1}^{N}\frac{\Gamma(p-1+\alpha(t))}{p!}\right)$$
$$-\frac{1}{\Gamma(\alpha(t)-1)\Gamma(2-\alpha(t))}\sum_{p=0}^{N}\frac{\Gamma(p+1+\alpha(t))}{p!}\frac{V_p(f)(t)}{t^{p+1+\alpha(t)}},\quad t\in(0,T],$$

and ($t\in[0,T]$)

$$A_2\left(f(t),f^{(1)}(t),\alpha(t),t\right)$$
$$=\frac{1}{\Gamma(1-\alpha(t))}\left(\frac{f(t)\,t^{1-\alpha(t)}}{1-\alpha(t)}\left(\ln t-\frac{1}{1-\alpha(t)}\right)\right.$$
$$-\frac{f^{(1)}(t)t^{2-\alpha(t)}}{2-\alpha(t)}\left(\ln t-\frac{1}{2-\alpha(t)}\right)$$
$$+\frac{t^{1-\alpha(t)}\ln t}{\Gamma(\alpha(t))}\sum_{k=0}^{N}\frac{\Gamma(k+\alpha(t))}{k!}\left(\frac{tf^{(1)}(t)}{(k+1)(k+2)}-\frac{f(t)}{k+1}+\frac{V_k(f)(t)}{t^{k+1}}\right)$$
$$+\frac{t^{1-\alpha(t)}}{\Gamma(\alpha(t))}\sum_{p=1}^{M}\frac{1}{p}\sum_{k=0}^{N}\frac{\Gamma(k+\alpha(t))}{k!}\left(\frac{tf^{(1)}(t)}{(k+p+1)(k+p+2)}-\frac{f(t)}{k+p+1}\right.$$
$$\left.\left.+\frac{V_{k+p}(f)(t)}{t^{k+p+1}}\right)\right),$$

with $V_p(f)(t)=\int_0^t\tau^pf(\tau)\,\mathrm{d}\tau$ being moments of the function $f$ and satisfying

$$V_p^{(1)}(f)(t)=t^pf(t),\quad V_p(0)=0,\ \ t\in[0,T],\ p=0,1,\ldots\ . \qquad [2.34]$$

Also, there exists $N_\varepsilon\in\mathbb{N}$ such that for any $\varepsilon>0$, and for $N,M>N_\varepsilon$, it holds that

$$\left|R_1^N(t)+R_2^{N,M}(t)\right|<\varepsilon.$$

Thus, the approximation formula for ${}_0\mathrm{D}_t^{\alpha(t)} f(t)$, $t \in (0,T]$, becomes

$$
\begin{aligned}
{}_0\mathrm{D}_t^{\alpha(t)} f(t) &\approx \widehat{{}_0\mathrm{D}_t^{\alpha(t)} f(t)} \\
&= \frac{f(t)}{t^{\alpha(t)}}\left(\frac{1}{\Gamma(1-\alpha(t))} - \frac{1}{\Gamma(\alpha(t)-1)\Gamma(2-\alpha(t))}\sum_{p=2}^{N}\frac{\Gamma(p-1+\alpha(t))}{(p-1)!}\right) \\
&+ \frac{f^{(1)}(t)t^{1-\alpha(t)}}{\Gamma(2-\alpha(t))}\left(1+\frac{1}{\Gamma(\alpha(t)-1)}\sum_{p=1}^{N}\frac{\Gamma(p-1+\alpha(t))}{p!}\right) \\
&- \frac{1}{\Gamma(\alpha(t)-1)\Gamma(2-\alpha(t))}\sum_{p=2}^{N}\frac{\Gamma(p-1+\alpha(t))}{(p-2)!}\frac{V_{p-2}(f)(t)}{t^{p-1+\alpha(t)}} \\
&- \frac{\alpha^{(1)}(t)}{\Gamma(1-\alpha(t))}\left(f(t)\left(\frac{1}{1-\alpha(t)}t^{1-\alpha(t)}\ln t - \frac{t^{1-\alpha(t)}}{(1-\alpha(t))^2}\right)\right. \\
&- \frac{f^{(1)}(t)t^{2-\alpha(t)}}{2-\alpha(t)}\left(\ln t - \frac{1}{2-\alpha(t)}\right) \\
&+ \frac{t^{1-\alpha(t)}\ln t}{\Gamma(\alpha(t))}\sum_{k=0}^{N}\frac{\Gamma(k+\alpha(t))}{k!}\left(\frac{tf^{(1)}(t)}{(k+1)(k+2)} - \frac{f(t)}{k+1} + \frac{V_k(f)(t)}{t^{k+1}}\right) \\
&+ \frac{t^{1-\alpha(t)}}{\Gamma(\alpha(t))}\sum_{p=1}^{M}\frac{1}{p}\sum_{k=0}^{N}\frac{\Gamma(k+\alpha(t))}{k!}\left(\frac{tf^{(1)}(t)}{(k+p+1)(k+p+2)} - \frac{f(t)}{k+p+1}\right. \\
&\left.\left.+ \frac{V_{k+p}(f)(t)}{t^{k+p+1}}\right)\right). \qquad [2.35]
\end{aligned}
$$

Note that for the case $\alpha = const.$, expressions [2.29] and [2.35] coincide.

REMARK 2.3.– The procedure of expressing fractional derivatives in terms of function, its first derivative and moments of function is extended in different directions in a series of papers [POO 12a, POO 12b, POO 13].

2.1.1.1. *Laplace transform of Riemann–Liouville fractional integrals and derivatives*

Suppose that $f$ is exponentially bounded (see section 1.2), that is $f \in L^1(0,\infty)$, $|f(t)| \le Ae^{s_0 t}$, $t > 0$, where $A > 0$, $s_0 > 0$. Then

$$\mathcal{L}\left[{}_0\mathrm{I}_t^{\alpha} f(t)\right](s) = \frac{1}{s^{\alpha}}\tilde{f}(s), \quad \operatorname{Re} s > s_0, \qquad [2.36]$$

see [KIL 06, p. 84]. Expression [2.36] follows from the well-known property of the Laplace transform of convolution and $\mathcal{L}\left[\frac{t^{\alpha-1}}{\Gamma(\alpha)}\right](s) = \frac{1}{s^{\alpha}}$ (see section 1.3).

For the fractional derivatives, we have the following result.

PROPOSITION 2.11.– [KIL 06, p. 84] Let $n-1 < \operatorname{Re}\alpha < n$, $f \in AC^n_{loc}([0,\infty))$ and $f$ be of exponential growth. Suppose that there exist finite limits

$$\lim_{t\to 0}\left(\mathrm{D}^k\,{}_0\mathrm{I}_t^{n-\alpha} f(t)\right) \text{ and } \lim_{t\to\infty}\left(\mathrm{D}^k\,{}_0\mathrm{I}_t^{n-\alpha} f(t)\right) = 0, \quad k = 0, 1, \ldots, n-1.$$

Then

$$\mathcal{L}\left[{}_0\mathrm{D}_t^{\alpha} f(t)\right](s) = s^{\alpha}\tilde{f}(s) - \sum_{k=0}^{n-1} s^{n-k-1}\left[\mathrm{D}^k\,{}_0\mathrm{I}_t^{n-\alpha} f(t)\right]_{t=0}, \quad \operatorname{Re} s > s_0. \qquad [2.37]$$

For $0 < \alpha < 1$

$$\mathcal{L}\left[{}_0\mathrm{D}_t^{\alpha} f(t)\right](s) = s^{\alpha}\tilde{f}(s) - \left[{}_0\mathrm{I}_t^{1-\alpha} f(t)\right]_{t=0} = s^{\alpha}\tilde{f}(s), \quad \operatorname{Re} s > s_0. \qquad [2.38]$$

Relation [2.38] could be used for the (heuristic) definition of the fractional derivative.

The Leibnitz rule for fractional derivatives does not hold in its usual form. It could be shown that for analytic functions, we have the following.

PROPOSITION 2.12.– [SAM 93, p. 280] Suppose that $f$ and $g$ are analytic for $t > 0$ and $\alpha > 0$. Then

$${}_a\mathrm{D}_t^{\alpha}\left(f(t)\,g(t)\right) = \sum_{k=0}^{\infty}\binom{\alpha}{k}\left(\mathrm{D}^k g(t)\right)\left({}_a\mathrm{D}_t^{\alpha-k} f(t)\right), \quad t > a. \qquad [2.39]$$

Note that in [2.39], on the right-hand side we have integer-order derivatives of $g$ and fractional-order derivatives of $f$. There is an apparent lack of symmetry in the derivatives of the two functions. The left-hand side of [2.39] does not depend on the order of the functions $f$ and $g$, while on the right-hand side there are only integer derivatives of $g$ and non-integer derivatives (integrals) of $f$. It could be shown that the two functions $f$ and $g$ can be interchanged without changing the value of the fractional derivative of their product.

### 2.1.2. *Riemann–Liouville fractional integrals and derivatives on the real half-axis*

The Riemann–Liouville fractional integrals and derivatives defined on a finite interval $[a,b]$ can be naturally extended to a half-line $\mathbb{R}_+$ as

$$\begin{aligned} \mathrm{I}_+^{\alpha} f(t) &= \frac{1}{\Gamma(\alpha)}\int_0^t (t-\tau)^{\alpha-1} f(\tau)\,\mathrm{d}\tau, \quad t > 0,\ \operatorname{Re}\alpha > 0, \\ \mathrm{I}_-^{\alpha} f(t) &= \frac{1}{\Gamma(\alpha)}\int_t^{\infty} (\tau-t)^{\alpha-1} f(\tau)\,\mathrm{d}\tau, \quad t > 0,\ \operatorname{Re}\alpha > 0, \qquad [2.40] \end{aligned}$$

and

$$\begin{aligned}
\mathrm{D}_{+}^{\alpha} f(t) &= \frac{1}{\Gamma(n-\alpha)} \frac{\mathrm{d}^n}{\mathrm{d}t^n} \int_0^t \frac{f(\tau)}{(t-\tau)^{\alpha-n+1}} \mathrm{d}\tau, \ t>0, \ n-1 \le \operatorname{Re}\alpha < n, \\
\mathrm{D}_{-}^{\alpha} f(t) &= (-1)^n \frac{1}{\Gamma(n-\alpha)} \frac{\mathrm{d}^n}{\mathrm{d}t^n} \int_t^\infty \frac{f(\tau)}{(t-\tau)^{\alpha-n+1}} \mathrm{d}\tau, \ t>0, \\
&\qquad n-1 \le \operatorname{Re}\alpha < n.
\end{aligned}$$

Operator $[2.40]_2$ is sometimes called the Weyl integral.

Fourier transform of the Riemann–Liouville fractional integrals $\mathrm{I}_{+}^{\alpha} f$ and $\mathrm{I}_{-}^{\alpha} f$ are given as

$$\mathcal{F}\left[\mathrm{I}_{+}^{\alpha} f(t)\right](\omega) = \frac{1}{(\mathrm{i}\omega)^\alpha} \hat{f}(\omega) \text{ and } \mathcal{F}\left[\mathrm{I}_{-}^{\alpha} f(t)\right](\omega) = \frac{1}{(-\mathrm{i}\omega)^\alpha} \hat{f}(\omega), \quad \omega \in \mathbb{R}, \quad [2.41]$$

for $0 < \operatorname{Re}\alpha < 1$ and for $f \in L^1(\mathbb{R})$. Equation [2.41] cannot be extended directly to the case $\operatorname{Re}\alpha \ge 1$. For the Riemann–Liouville derivatives, we have

$$\mathcal{F}\left[\mathrm{D}_{+}^{\alpha} f(t)\right](\omega) = (\mathrm{i}\omega)^\alpha \hat{f}(\omega) \text{ and } \mathcal{F}\left[\mathrm{D}_{-}^{\alpha} f(t)\right](\omega) = (-\mathrm{i}\omega)^\alpha \hat{f}(\omega), \quad \omega \in \mathbb{R}. \quad [2.42]$$

In [2.41] and [2.42], we have $(\pm\mathrm{i}\omega)^\alpha = |\omega|^\alpha e^{\mp \frac{\alpha\pi}{2} \operatorname{sgn}\omega}$. In the case when $f \in \mathcal{S}'$, we have that [2.41] and [2.42] remain the same (see [VLA 73, p. 110]).

### 2.1.3. *Caputo fractional derivatives*

We present the definition of fractional derivative from Caputo [CAP 67] and Caputo and Mainardi [CAP 71b]. The left Caputo fractional derivative of a function of order $\alpha$, denoted by ${}_a^C\mathrm{D}_t^\alpha f$, is

$${}_a^C\mathrm{D}_t^\alpha f(t) = \begin{cases} \frac{1}{\Gamma(n-\alpha)} \int_a^t \frac{f^{(n)}(\tau)}{(t-\tau)^{\alpha+1-n}} \mathrm{d}\tau, & n-1 \le \alpha < n, \\ \frac{\mathrm{d}^n}{\mathrm{d}t^n} f(t), & \alpha = n, \end{cases} \quad t \in [a,b]. \quad [2.43]$$

Similarly, the right Caputo derivative is defined as

$${}_t^C\mathrm{D}_b^\alpha f(t) = \begin{cases} (-1)^n \frac{1}{\Gamma(n-\alpha)} \int_t^b \frac{f^{(n)}(\tau)}{(\tau-t)^{\alpha+1-n}} \mathrm{d}\tau, & n-1 \le \alpha < n, \\ (-1)^n \frac{\mathrm{d}^n}{\mathrm{d}t^n} f(t), & \alpha = n, \end{cases} \quad t \in [a,b]. \quad [2.44]$$

It is easy to see that

$$ {}_{a}^{C}\mathrm{D}_{t}^{\alpha}f\left(t\right)={}_{a}\mathrm{I}_{t}^{n-\alpha}\left(\frac{\mathrm{d}^{n}}{\mathrm{d}t^{n}}f(t)\right) \text{ and } {}_{t}^{C}\mathrm{D}_{b}^{\alpha}f\left(t\right)=\left(-1\right)^{n}{}_{t}\mathrm{I}_{b}^{n-\alpha}\left(\frac{\mathrm{d}^{n}}{\mathrm{d}t^{n}}f(t)\right), $$

where ${}_{a}\mathrm{I}_{t}^{n-\alpha}$ and ${}_{t}\mathrm{I}_{b}^{n-\alpha}$ are the Riemann–Liouville fractional integrals [2.2] and [2.3], respectively.

Observe that [2.43] for $a=0$ can be written as

$$ {}_{0}^{C}\mathrm{D}_{t}^{\alpha}f\left(t\right)=\frac{t^{n-1-\alpha}}{\Gamma\left(n-\alpha\right)}*\frac{\mathrm{d}^{n}}{\mathrm{d}t^{n}}f(t),\;\; t>0,\; n-1\leq\mathrm{Re}\,\alpha<n. \qquad [2.45] $$

Note that the Caputo derivative of a constant function is zero

$$ {}_{a}^{C}\mathrm{D}_{t}^{\alpha}C=0 \text{ and } {}_{t}^{C}\mathrm{D}_{b}^{\alpha}C=0. \qquad [2.46] $$

For $n-1\leq\alpha<n$, the Caputo derivatives ${}_{a}^{C}\mathrm{D}_{t}^{\alpha}$ and ${}_{t}^{C}\mathrm{D}_{b}^{\alpha}$ are operators mapping $C^{n}([a,b])$ into

$$ \begin{aligned} C_{a}\left([a,b]\right) &= \left\{f\mid f\in C\left([a,b]\right),\; f\left(a\right)=0\right\},\;\; \|f\|_{C_{a}}=\|f\|_{C}, \\ C_{b}\left([a,b]\right) &= \left\{f\mid f\in C\left([a,b]\right),\; f\left(b\right)=0\right\},\;\; \|f\|_{C_{b}}=\|f\|_{C}, \end{aligned} $$

respectively.

PROPOSITION 2.13.– [KIL 06, p. 94] Let $n-1\leq\mathrm{Re}\,\alpha<n$, $\alpha\neq\mathbb{N}$. Then, the Caputo derivatives ${}_{a}^{C}\mathrm{D}_{t}^{\alpha}$ and ${}_{t}^{C}\mathrm{D}_{b}^{\alpha}$ are bounded operators from $C^{n}([a,b])$ into $C_{a}\left([a,b]\right)$ and $C_{b}\left([a,b]\right)$, respectively, and the following estimates hold

$$ \begin{aligned} \|{}_{a}^{C}\mathrm{D}_{t}^{\alpha}f\|_{C_{a}} &\leq \frac{(b-a)^{n-\mathrm{Re}\,\alpha}}{|\Gamma\left(n-\alpha\right)|\left(n-\mathrm{Re}\,\alpha+1\right)}\|f\|_{C^{n}}, \\ \|{}_{t}^{C}\mathrm{D}_{b}^{\alpha}f\|_{C_{b}} &\leq \frac{(b-a)^{n-\mathrm{Re}\,\alpha}}{|\Gamma\left(n-\alpha\right)|\left(n-\mathrm{Re}\,\alpha+1\right)}\|f\|_{C^{n}}. \end{aligned} $$

In general, the Caputo and the Riemann–Liouville fractional derivatives do not coincide. The connections between them are given as

$$\begin{aligned} {}_a^C\mathrm{D}_t^\alpha f(t) &= {}_a\mathrm{D}_t^\alpha \left( f(t) - \sum_{k=0}^{n-1} \frac{(t-a)^k}{k!} f^{(k)}(a) \right), \quad t \in [a,b], \\ {}_t^C\mathrm{D}_b^\alpha f(t) &= {}_t\mathrm{D}_b^\alpha \left( f(t) - \sum_{k=0}^{n-1} \frac{(b-t)^k}{k!} f^{(k)}(b) \right), \quad t \in [a,b]. \end{aligned}$$

In particular, if $0 < \operatorname{Re}\alpha < 1$, for $t \in [a,b]$, we have

$${}_a^C\mathrm{D}_t^\alpha f(t) = {}_a\mathrm{D}_t^\alpha (f(t) - f(a)) \quad \text{and} \quad {}_t^C\mathrm{D}_b^\alpha f(t) = {}_t\mathrm{D}_b^\alpha (f(t) - f(b)),$$

or ($t \in [a,b]$)

$$\begin{aligned} {}_a^C\mathrm{D}_t^\alpha f(t) &= {}_a\mathrm{D}_t^\alpha f(t) - \frac{f(a)}{\Gamma(1-\alpha)(t-a)^\alpha} \quad \text{and} \\ {}_t^C\mathrm{D}_b^\alpha f(t) &= {}_t\mathrm{D}_b^\alpha f(t) - \frac{f(b)}{\Gamma(1-\alpha)(b-t)^\alpha}. \end{aligned} \qquad [2.47]$$

Thus, the Caputo fractional derivatives are regularized Riemann–Liouville fractional derivatives.

The expansion formula [2.24] for the Caputo derivative ${}_0^C\mathrm{D}_t^\alpha f$, $0 < \alpha < 1$, with [2.47], becomes ($t \in (0,T]$)

$$\begin{aligned} {}_0\mathrm{D}_t^\alpha f(t) &= \frac{f(t)}{t^\alpha}\mathcal{A}(N,\alpha) - \frac{f(0)}{t^\alpha \Gamma(1-\alpha)} - \sum_{p=1}^{N} C_{p-1}(\alpha) \frac{V_{p-1}(f)(t)}{t^{p+\alpha}} \\ &\quad + Q_{N+1}(f)(t), \end{aligned} \qquad [2.48]$$

where $\mathcal{A}$, $C_{p-1}$ and $V_p$ are given by [2.25], [2.26] and [2.27], respectively.

The case when $\alpha$ is close to 1 is called the case of low-level fractionality. For this case, we refer to [HER 11, TAR 06]. We have, for $t \in [a,b]$,

$${}_a^C\mathrm{D}_t^{1-\varepsilon} f(t) = f^{(1)}(t) + \varepsilon \left( f^{(1)}(0) \ln t + \int_0^t f^{(2)}(\xi) \ln(t-\xi)\, \mathrm{d}\xi \right) + o(\varepsilon^2), \quad \varepsilon \to 0^+, \qquad [2.49]$$

where $o$ is the Landau symbol "small $o$". Recall that $a(x) = o(|x|^{\alpha})$ means $\frac{a(x)}{|x|^{\alpha}} \to 0$ as $|x| \to 0$, or $|x| \to \infty$.

We have the following integration by parts formula for the Caputo derivatives.

PROPOSITION 2.14.– [ALM 12, p. 112] Let $n-1 \leq \alpha < n$ and let $f, g \in C^n([a,b])$. Then

$$\begin{aligned} \int_a^b g(t)\left({}_t^C\mathrm{D}_b^{\alpha} f(t)\right) \mathrm{d}t &= \int_a^b \left({}_a^C\mathrm{D}_t^{\alpha} g(t)\right) f(t)\, \mathrm{d}t \\ &+ \sum_{j=0}^{n-1} (-1)^{n+j} \left[\left({}_a^C\mathrm{D}_t^{\alpha+j-n} g(t)\right)\left({}_a^C\mathrm{D}_t^{n-1-j} f(t)\right)\right]_{t=a}^{t=b}. \end{aligned} \quad [2.50]$$

The Laplace transform of the left Caputo derivative is given as follows.

PROPOSITION 2.15.– [KIL 06, p. 98] Suppose that $n-1 < \alpha \leq n$ and let $f$ be such that $f \in C^n(\mathbb{R}_+)$, $|f(t)|, \left|f^{(1)}(t)\right|, \ldots, \left|f^{(n)}(t)\right| \leq Be^{s_0 t}$, $B, s_0 > 0$, $t > 0$. Suppose that $\lim_{t\to\infty} f^{(k)}(t) = 0$, for $k = 0, 1, \ldots, n-1$. Then

$$\mathcal{L}\left[{}_0^C\mathrm{D}_t^{\alpha} f(t)\right](s) = s^{\alpha}\tilde{f}(s) - \sum_{k=0}^{n-1} s^{\alpha-k-1} f^{(k)}(0), \quad \mathrm{Re}\, s > s_0. \quad [2.51]$$

For $0 < \alpha < 1$, expression [2.51] becomes

$$\mathcal{L}\left[{}_0^C\mathrm{D}_t^{\alpha} f(t)\right](s) = s^{\alpha}\tilde{f}(s) - s^{\alpha-1} f(0), \quad \mathrm{Re}\, s > s_0. \quad [2.52]$$

### 2.1.4. *Riesz potentials and Riesz derivatives*

Let $0 < \alpha < 1$. Consider the following integrals:

$$\mathrm{R}^{\alpha} f(x) = \frac{1}{2\Gamma(\alpha)\cos\frac{\alpha\pi}{2}} \int_{-\infty}^{\infty} \frac{1}{|x-\zeta|^{1-\alpha}} f(\zeta)\, \mathrm{d}\zeta, \quad x \in \mathbb{R}, \quad [2.53]$$

$$\mathrm{H}^{\alpha} f(x) = \frac{1}{2\Gamma(\alpha)\sin\frac{\alpha\pi}{2}} \int_{-\infty}^{\infty} \frac{\mathrm{sgn}(x-\zeta)}{|x-\zeta|^{1-\alpha}} f(\zeta)\, \mathrm{d}\zeta, \quad x \in \mathbb{R}. \quad [2.54]$$

Then, $\mathrm{R}^{\alpha} f$ is called the Riesz potential of $f$ of the order $\alpha$ on $\mathbb{R}$, while $\mathrm{H}^{\alpha} f$ is the conjugate Riesz potential of the order $\alpha$ on $\mathbb{R}$. The following proposition holds for [2.53] and [2.54]. Note that equality is almost everywhere (see section 1.1).

PROPOSITION 2.16.– [UCH 08, p. 200] The Riemann–Liouville fractional integral and the Riesz potential, for $x \in \mathbb{R}$, are connected as follows:

$$
\begin{aligned}
{}_{-\infty}\mathrm{I}_x^{\alpha} f(x) &= \cos\frac{\alpha\pi}{2}\,\mathrm{R}^{\alpha} f(x) + \sin\frac{\alpha\pi}{2}\,\mathrm{H}^{\alpha} f(x), \\
{}_{x}\mathrm{I}_{\infty}^{\alpha} f(x) &= \cos\frac{\alpha\pi}{2}\,\mathrm{R}^{\alpha} f(x) - \sin\frac{\alpha\pi}{2}\,\mathrm{H}^{\alpha} f(x), \\
\mathrm{R}^{\alpha} f(x) &= \frac{1}{2\cos\frac{\alpha\pi}{2}}\left({}_{-\infty}\mathrm{I}_x^{\alpha} f(x) + {}_{x}\mathrm{I}_{\infty}^{\alpha} f(x)\right), \\
\mathrm{H}^{\alpha} f(x) &= \frac{1}{2\sin\frac{\alpha\pi}{2}}\left({}_{-\infty}\mathrm{I}_x^{\alpha} f(x) - {}_{x}\mathrm{I}_{\infty}^{\alpha} f(x)\right).
\end{aligned}
\tag{2.55}
$$

Also, if $\alpha, \beta > 0$, $\alpha + \beta < 1$, then

$$
\mathrm{R}^{\alpha}\,\mathrm{R}^{\beta} f(x) = \mathrm{R}^{\alpha+\beta} f(x) \quad \text{and} \quad \mathrm{H}^{\alpha}\,\mathrm{H}^{\beta} f(x) = -\mathrm{R}^{\alpha+\beta} f(x), \quad x \in \mathbb{R}. \tag{2.56}
$$

Integrals [2.53] and [2.54] exist with the appropriate assumptions on $f$. For example, if $f \in L^1(\mathbb{R}) \cap L^2(\mathbb{R})$, then $\mathrm{R}^{\alpha} f$ and $\mathrm{H}^{\alpha} f$ exist almost everywhere and belong to $L^1_{loc}(\mathbb{R})$, as stated in [BUT 00, p. 46]. More generally, the following result holds.

PROPOSITION 2.17.– [KIL 06, p. 129] Let $1 \leq p \leq \infty$, $1 \leq q \leq \infty$. Then, $\mathrm{R}^{\alpha}$ is a bounded operator from $L^p(\mathbb{R})$ into $L^q(\mathbb{R})$ if and only if

$$
0 < \alpha < 1, \quad 1 < p < \frac{1}{\alpha}, \quad \frac{1}{q} = \frac{1}{p} - \frac{1}{\alpha}.
$$

The Fourier transforms of $\mathrm{R}^{\alpha}$ and $\mathrm{H}^{\alpha}$ are

$$
\begin{aligned}
\mathcal{F}\left[\mathrm{R}^{\alpha} f(x)\right](\omega) &= \frac{1}{|\omega|^{\alpha}}\hat{f}(\omega) \quad \text{and} \\
\mathcal{F}\left[\mathrm{H}^{\alpha} f(x)\right](\omega) &= -\mathrm{i}\frac{\operatorname{sgn}(\omega)}{|\omega|^{\alpha}}\hat{f}(\omega), \quad \omega \in \mathbb{R}\setminus\{0\},
\end{aligned}
\tag{2.57}
$$

see [BUT 00].

There is another important property of $\mathrm{R}^{\alpha}$. Namely, it maps the Lizorkin space of test functions $\Phi$ into itself, i.e. $\mathrm{R}^{\alpha}(\Phi) = \Phi$ (see [SAM 93, p. 493]).

DEFINITION 2.4.– *[BUT 00, UCH 08] The Riesz fractional derivative of the order $\alpha$ is defined as*

$$^{R}\mathrm{D}^{\alpha} f(x) = \frac{\mathrm{d}}{\mathrm{d}x}\mathrm{H}^{1-\alpha} f(x) = \frac{1}{2\Gamma(1-\alpha)\cos\frac{\alpha\pi}{2}}\frac{\mathrm{d}}{\mathrm{d}x}\int_{-\infty}^{\infty}\frac{\mathrm{sgn}\,(x-\zeta)}{|x-\zeta|^{\alpha}} f(\zeta)\,\mathrm{d}\zeta, \quad x\in\mathbb{R}, \qquad [2.58]$$

*while the conjugate Riesz derivative of the order $\alpha$ is defined as*

$$^{RC}\mathrm{D}^{\alpha} f(x) = \frac{\mathrm{d}}{\mathrm{d}x}\mathrm{R}^{1-\alpha} f(x) = \frac{1}{2\Gamma(1-\alpha)\sin\frac{\alpha\pi}{2}}\frac{\mathrm{d}}{\mathrm{d}x}\int_{-\infty}^{\infty}\frac{1}{|x-\zeta|^{\alpha}} f(\zeta)\,\mathrm{d}\zeta, \quad x\in\mathbb{R}. \qquad [2.59]$$

It is easy to see that the Fourier transforms of $^{R}\mathrm{D}^{\alpha} f$ and $^{RC}\mathrm{D}^{\alpha} f$, for $0<\alpha<1$, are

$$\begin{aligned}\mathcal{F}\left[{}^{R}\mathrm{D}^{\alpha} f(x)\right](\omega) &= |\omega|^{\alpha}\hat{f}(\omega) \;\text{ and}\\ \mathcal{F}\left[{}^{RC}\mathrm{D}^{\alpha} f(x)\right](\omega) &= \mathrm{i}\,|\omega|^{\alpha}\,\mathrm{sgn}\,(\omega)\,\hat{f}(\omega), \quad \omega\in\mathbb{R},\end{aligned} \qquad [2.60]$$

see [BUT 00]. Therefore, $^{R}\mathrm{D}^{\alpha}$ inverts $\mathrm{R}^{\alpha}$, while $^{RC}\mathrm{D}^{\alpha}$ inverts $\mathrm{H}^{\alpha}$, if the functions involved satisfy certain additional conditions. Namely, $^{R}\mathrm{D}^{\alpha}\,\mathrm{R}^{\alpha} f = f$ holds for functions $f$ belonging to the Lizorkin space $\Phi$ (see [SAM 93, p. 150]).

In various applications, there is a need to study the following potential:

$$^{R}_{a}\mathrm{I}^{\alpha}_{b} f(x) = \int_{a}^{b}\frac{1}{|x-\zeta|^{1-\alpha}} f(\zeta)\,\mathrm{d}\zeta, \quad x\in(a,b), \qquad [2.61]$$

where $0<\alpha<1$. The Carleman equation involves the potential of the type [2.61]. The inverse of $^{R}_{a}\mathrm{I}^{\alpha}_{b}$, i.e. the solution to the equation

$$^{R}_{a}\mathrm{I}^{\alpha}_{b} f(x) = \int_{a}^{b}\frac{1}{|x-\zeta|^{1-\alpha}} f(\zeta)\,\mathrm{d}\zeta = g(x), \quad x\in(a,b), \qquad [2.62]$$

for a specified class of functions $f$ and $g$ is given in [SAM 93, p. 627]. Let $g(x) = (x-a)^n$, $n = 0, 1, 2, \ldots$. Then, the solution to [2.62], for $x \in (a,b)$, becomes

$$\begin{aligned} f(x) &= \frac{n!}{\pi}(b-a)^n \sin\frac{\alpha\pi}{2} \frac{\Gamma(1-\alpha)}{\Gamma(n+1-\alpha)} \frac{1}{((x-a)(b-x))^{\frac{\alpha}{2}}} \\ &\quad \times \sum_{k=0}^{n} \binom{n-\frac{\alpha}{2}}{n-k} \left(\frac{b-x}{x-a}\right)^k . \end{aligned} \qquad [2.63]$$

If $g = 0$, then the solution to [2.62] is $f = 0$ (see [SAM 93, eq. (30.84)], or [NAH 03, p. 47]).

### 2.1.5. *Symmetrized Caputo derivative*

Let $0 \leq \beta < 1$, $-\infty \leq a < b \leq \infty$. The symmetrized Caputo fractional derivative of an absolutely continuous function $f$ is defined as

$${}_a^C\mathcal{E}_b^\beta f(x) = \frac{1}{2}\left({}_a^C\mathrm{D}_x^\beta - {}_x^C\mathrm{D}_b^\beta\right) f(x) = \frac{1}{2}\frac{1}{\Gamma(1-\beta)} \int_a^b \frac{f^{(1)}(\theta)}{|x-\theta|^\beta} \mathrm{d}\theta, \quad x \in [a,b]. \qquad [2.64]$$

For $a = -\infty$ and $b = \infty$, we write $\mathcal{E}_x^\beta$ instead of ${}_a^C\mathcal{E}_b^\beta$ and then

$$\mathcal{E}_x^\beta f(x) = \frac{1}{2}\frac{1}{\Gamma(1-\beta)} |x|^{-\beta} * f^{(1)}(x), \quad x \in \mathbb{R}. \qquad [2.65]$$

Note that $\mathcal{E}_x^0 f(x) = 0$ and $\mathcal{E}_x^\beta f(x) \to f^{(1)}(x)$, as $\beta \to 1$. This yields that the symmetrized fractional derivative generalizes the first derivative of a function, but it does not generalize the derivative of zero order: the zeroth-order symmetrized fractional derivative of a function is not a function itself, but zero.

For $f = const.$, we have that ${}_a^C\mathcal{E}_b^\beta f = 0$, and conversely, the fact that $f = const.$ is the unique solution to equation ${}_a^C\mathcal{E}_b^\beta f = 0$ is shown in [ATA 09f].

In studying [2.64], we need some properties of the function $|x|^{-\beta}$. It is an element of the Lizorkin space $\Psi'$ for all $\beta \in \mathbb{R}$ and for $\beta \in [0,1)$ it holds that

$$\begin{aligned} \mathcal{F}\left[|x|^{-\beta}\right](\xi) &= 2\Gamma(1-\beta)\sin\frac{\beta\pi}{2}\frac{1}{|\xi|^{1-\beta}}, \quad \xi \in \mathbb{R}, \\ \mathcal{F}\left[\mathcal{E}_x^\beta f(x)\right](\xi) &= \mathrm{i}\frac{\xi}{|\xi|^{1-\beta}} \sin\frac{\beta\pi}{2}\hat{f}(\xi), \quad \xi \in \mathbb{R}. \end{aligned}$$

### 2.1.6. *Other types of fractional derivatives*

In this section, we present other types of fractional integrals and derivatives. For extensive review of definitions, see [KIR 94].

#### 2.1.6.1. *Canavati fractional derivative*

There is another definition of fractional derivatives that is useful in deriving inequalities. This is the Canavati fractional derivative. It is "between" the Riemann–Liouville derivative and the Caputo derivative. Let $n-1<\alpha<n$. Then, the Canavati derivative of order $\alpha$ is defined as

$$ {}_{a}^{Can}\mathrm{D}_{t}^{\alpha} f(t)=\frac{1}{\Gamma(n-\alpha)}\frac{\mathrm{d}}{\mathrm{d}t}\int_{a}^{t}\frac{f^{(n-1)}(\tau)}{(t-\tau)^{\alpha-n+1}}\mathrm{d}\tau, \qquad [2.66]$$

see [CAN 87]. The Canavati derivative [2.66] is used in [ANA 09] for $f\in C^{\alpha}([a,b])$ where

$$C^{\alpha}([a,b])=\left\{f\in C^{n-1}([a,b])\mid {}_{a}\mathrm{I}_{t}^{n-1}f^{(n-1)}\in C^{1}([a,b])\right\}.$$

#### 2.1.6.2. *Marchaud fractional derivatives*

The left Marchaud fractional derivative of the order $0<\alpha<1$ for $f\in\mathcal{H}^{\lambda}([a,b])$, $\lambda>\alpha$ (see section 1.1) is defined as

$$ {}_{a}^{M}\mathrm{D}_{t}^{\alpha} f(t)=\frac{f(t)}{\Gamma(1-\alpha)(t-a)^{\alpha}}+\frac{\alpha}{\Gamma(1-\alpha)}\int_{a}^{t}\frac{f(t)-f(\tau)}{(t-\tau)^{1+\alpha}}\mathrm{d}\tau,\ t\in[a,b]. \qquad [2.67]$$

The right Marchaud fractional derivative is defined as

$$ {}_{t}^{M}\mathrm{D}_{b}^{\alpha} f(t)=\frac{f(t)}{\Gamma(1-\alpha)(b-t)^{\alpha}}+\frac{\alpha}{\Gamma(1-\alpha)}\int_{t}^{b}\frac{f(t)-f(\tau)}{(\tau-t)^{1+\alpha}}\mathrm{d}\tau,\ t\in[a,b]. \qquad [2.68]$$

The integrals in [2.67] and [2.68] are assumed to be convergent. To make it precise, let

$$\psi_{\varepsilon}(t)=\int_{a}^{t-\varepsilon}\frac{f(t)-f(\tau)}{(t-\tau)^{1+\alpha}}\mathrm{d}\tau,\ \ \varepsilon>0. \qquad [2.69]$$

Then

$$ {}_{t}^{M}\mathrm{D}_{t}^{\alpha} f(t)=\frac{f(t)}{\Gamma(1-\alpha)(t-a)^{\alpha}}+\lim_{\varepsilon\to 0}\psi_{\varepsilon}(t). \qquad [2.70]$$

If $f \in L^p(a,b)$, then the limit in [2.70] is considered in the norm of $L^p(a,b)$.

For functions belonging to $C^1([a,b])$, the Marchaud derivatives coincide with the corresponding Riemann–Liouville derivatives.

2.1.6.3. *Grünwald–Letnikov fractional derivatives*

The left Grünwald–Letnikov fractional derivative of the order $\alpha$ is, according to [KIL 06, p. 122], formally defined as

$$ {}_{a}^{G-L}\mathrm{D}_{t}^{\alpha} f(t) = \lim_{h\to 0}\left(\frac{1}{h^{\alpha}}\sum_{j=0}^{\left[\frac{t-a}{h}\right]}(-1)^{j}\binom{\alpha}{j}f(t-jh)\right), \quad t>a,\ \alpha>0. \qquad [2.71]$$

Similarly, the right Grünwald–Letnikov fractional derivative of the order $\alpha$ is defined as

$$ {}_{t}^{G-L}\mathrm{D}_{b}^{\alpha} f(t) = \lim_{h\to 0}\left(\frac{1}{h^{\alpha}}\sum_{j=0}^{\left[\frac{t-a}{h}\right]}(-1)^{j}\binom{\alpha}{j}f(t+jh)\right), \quad t>a,\ \alpha>0. \qquad [2.72]$$

There is a connection between the Marchaud and the Grünwald–Letnikov fractional derivatives.

PROPOSITION 2.18.– [SAM 93, p. 386] Let $f \in L^p(a,b)$, $1 \le p < \infty$. Then, limit [2.71] exists in the sense of $L^p(a,b)$ convergence, if and only if there exists the Marchaud fractional derivative in sense [2.70]. Both limits, if they exist, are equal almost everywhere.

## 2.2. Some additional properties of fractional derivatives

### 2.2.1. *Fermat theorem for fractional derivative*

Let $0<\alpha<1$. As a motivation, following [SAM 93, p. 111], we start from

$$\begin{aligned} {}_{0}\mathrm{D}_{t}^{\alpha} y(t) &= \frac{1}{\Gamma(1-\alpha)}\frac{\mathrm{d}}{\mathrm{d}t}\int_0^t \frac{y(t-\tau)}{\tau^{\alpha}}\mathrm{d}\tau \\ &= \frac{1}{\Gamma(1-\alpha)}\frac{y(0)}{t^{\alpha}} + \frac{1}{\Gamma(1-\alpha)}\int_0^t y^{(1)}(t-\tau)\left(\alpha\int_{\tau}^{t}\xi^{-1-\alpha}\mathrm{d}\xi + \frac{1}{t^{\alpha}}\right)\mathrm{d}\tau \\ &= \frac{y(t)}{\Gamma(1-\alpha)\,t^{\alpha}} + \frac{\alpha}{\Gamma(1-\alpha)}\int_0^t \frac{y(t)-y(t-\tau)}{\tau^{1+\alpha}}\mathrm{d}\tau, \quad t>0. \qquad [2.73] \end{aligned}$$

Let $0 < \alpha < 1$. Similarly, for the Caputo derivative of an integrable function, we have

$$
\begin{aligned}
{}_0^C\mathrm{D}_t^\alpha y(t) &= {}_0\mathrm{D}_t^\alpha y(t) - \frac{y(0)}{\Gamma(1-\alpha)t^\alpha} \\
&= \frac{y(t)-y(0)}{\Gamma(1-\alpha)t^\alpha} + \frac{\alpha}{\Gamma(1-\alpha)}\int_0^t \frac{y(t)-y(t-\tau)}{\tau^{1+\alpha}}\mathrm{d}\tau, \quad t>0. \qquad [2.74]
\end{aligned}
$$

Suppose that $y$ is an increasing positive function with the maximum at $t^* \in (0,t)$. Then

$$
y(t^*) - y(t) \geq 0, \quad t \in [0, t^*].
$$

From [2.73] and [2.74], we conclude that

$$
\left[{}_0\mathrm{D}_t^\alpha y(t)\right]_{t=t^*} \geq \frac{y(t^*)}{\Gamma(1-\alpha)(t^*)^\alpha} > 0 \text{ and } \left[{}_0^C\mathrm{D}_t^\alpha y(t)\right]_{t=t^*} \geq \frac{y(t^*)-y(0)}{\Gamma(1-\alpha)(t^*)^\alpha} > 0. \qquad [2.75]
$$

The above results may be used to prove the following proposition.

PROPOSITION 2.19.– [NAH 03, p. 104] Suppose that $y$ is an integrable function on $[A,B]$. Suppose further that there exists $\delta > 0$ such that $y \in \mathcal{H}^\lambda([t^*-\delta, t^*])$, $\lambda > \alpha$, and that $y$ attains a maximum at a point $t^* \in [A,B]$. Then, for any $\alpha \in [0,1]$ and $a \in [t^*-\delta, t^*]$, $a \neq t^*$, we have

$$
\left[{}_a\mathrm{D}_t^\alpha y(t)\right]_{t=t^*} \leq \frac{y(t^*)}{\Gamma(1-\alpha)(t^*-a)^\alpha} \text{ and } \left[{}_a^C\mathrm{D}_t^\alpha y(t)\right]_{t=t^*} \leq \frac{y(x^*)-y(a)}{\Gamma(1-\alpha)(t^*-a)^\alpha}. \qquad [2.76]
$$

Thus, at the point of maximum, fractional derivatives either satisfy [2.76] or do not exist.

It could be easily shown that for a minimum of a function, the inequalities in [2.76] become.

### 2.2.2. *Taylor theorem for fractional derivatives*

The mean value theorem for the Riemann–Liouville fractional derivative reads as follows.

PROPOSITION 2.20.– [TRU 99] Let $\alpha \in [0,1]$ and suppose that $f \in C([a,b])$, such that ${}_a\mathrm{D}_t^\alpha f \in C([a,b])$. Then

$$f(t) = (t-a)^{\alpha-1}\left[(t-a)^{1-\alpha} f(t)\right]_{t=a^+} + [{}_a\mathrm{D}_t^\alpha f(t)]_{t=\xi} \frac{(t-a)^\alpha}{\Gamma(\alpha+1)}, t \in (a,b], \qquad [2.77]$$

with $a \leq \xi \leq b$.

The generalization of the Taylor formula for the Riemann–Liouville fractional derivative has several different forms. To state the formula, we need the following definition.

DEFINITION 2.5.– *A function $f : [a,b] \to \mathbb{R}$ is said to be $\alpha$-continuous, $0 \leq \alpha \leq 1$, at $t_0$ if there exists $\lambda \in [0, 1-\alpha)$ such that $g(t) = |t-t_0|^\lambda f(t)$ is continuous at $t_0$.*

Function $f$ is $\alpha$-continuous in $[a,b]$ if it is $\alpha$-continuous for every $t \in [a,b]$. Let $C_\alpha = \{f \mid [a,b] \to \mathbb{R},\ f$ is $\alpha$-continuous$\}$. Note that $C_1([a,b]) = C([a,b])$. Let ${}_a\mathrm{I}_b^\alpha(a,b) = \{f \mid [a,b] \to \mathbb{R},\ {}_a\mathrm{I}_t^n f(t)$ exists and it is finite for all $t \in [a,b]\}$. A function $f$ is singular of order $\alpha$ at $t = t^*$ if $\lim_{t\to t^*} \frac{f(t)}{(t-t^*)^{\alpha-1}} = k < \infty$ and $k \neq 0$. Finally, we use ${}_a\mathrm{D}_t^{j\alpha}$ to denote the application of ${}_a\mathrm{D}_t^\alpha$ $j$-times, i.e. ${}_a\mathrm{D}_t^{j\alpha} = \underbrace{{}_a\mathrm{D}_t^\alpha \cdots {}_a\mathrm{D}_t^\alpha}_{j\text{-times}}$.

PROPOSITION 2.21.– [TRU 99] Let $\alpha \in [0,1]$, $n \in \mathbb{N}$. Let $f$ be a continuous function in $(a,b]$ satisfying the following conditions:

i) ${}_a\mathrm{D}_t^{j\alpha} f \in C([a,b])$ and ${}_a\mathrm{D}_t^{j\alpha} f \in {}_a\mathrm{I}_b^\alpha(a,b)$, for all $j = 1,\ldots,n$;

ii) ${}_a\mathrm{D}_t^{(n+1)\alpha} f$ is continuous in $[a,b]$;

iii) if $\alpha < \frac{1}{2}$, then for each $j \in \mathbb{N}$, $1 \leq j \leq n$, such that $(j+1)\alpha < 1$, ${}_a\mathrm{D}_t^{(j+1)\alpha} f$ is $\gamma$-continuous at $t = a$ for some $\gamma$, $1-(j+1)\alpha \leq \gamma \leq 1$, or it is singular of order $\alpha$ at $t = a$.

Then, for $t \in (a,b]$,

$$f(t) = \sum_{j=0}^{n} \frac{c_j}{\Gamma((j+1)\alpha)} (t-a)^{(j+1)\alpha-1} + \frac{\left[{}_a\mathrm{D}_t^{(n+1)\alpha} f(t)\right]_{t=\xi}}{\Gamma((n+1)\alpha+1)} (t-a)^{(n+1)\alpha},$$

$$a \leq \xi \leq b, \qquad [2.78]$$

where

$$c_j = \Gamma(\alpha)\left[(t-a)_a^{1-\alpha}\, \mathrm{D}_t^{j\alpha} f(t)\right]_{t=a^+}, \quad j = 0,1,\ldots,n. \qquad [2.79]$$

The Taylor formula for the Caputo derivative is given in the following proposition.

PROPOSITION 2.22.– [ODI 07, p. 289] Suppose that ${}_{a}^{C}\mathrm{D}_{t}^{j\alpha}f \in C((a,b])$ for $j = 0,1,\ldots,n+1$, where $0<\alpha\leq 1$. Then

$$f(t) = \sum_{j=0}^{n}\frac{(t-a)^{j\alpha}}{\Gamma(j\alpha+1)}\left[{}_{a}^{C}\mathrm{D}_{t}^{j\alpha}f(t)\right]_{t=a^{+}} + \frac{\left[{}_{a}^{C}\mathrm{D}_{t}^{(n+1)\alpha}f(t)\right]_{t=\xi}}{\Gamma((n+1)\alpha+1)}(t-a)^{(n+1)\alpha}, \quad t\in(a,b], \qquad [2.80]$$

with $a\leq\xi\leq b$.

REMARK 2.4.– [ODI 07, p. 288] In the special case for the Caputo derivative, the corresponding result is stated as follows. Suppose that $f\in C([a,b])$ and ${}_{a}^{C}\mathrm{D}_{t}^{\alpha}f \in C((a,b])$, for $0<\alpha\leq 1$. Then

$$f(t) = f(a) + \frac{\left[{}_{a}^{C}\mathrm{D}_{t}^{\alpha}f(t)\right]_{t=\xi}}{\Gamma(\alpha+1)}(t-a)^{\alpha}, \quad t\in(a,b], \qquad [2.81]$$

where $a\leq\xi\leq b$.

## 2.3. Fractional derivatives in distributional setting

Throughout this section, we will assume that functions that appear determine the tempered distributions.

### 2.3.1. *Definition of the fractional integral and derivative*

We introduce the following definition.

DEFINITION 2.6.– *The convolution operator $f_\alpha *$ in $\mathcal{S}'_+$ ($\check{f}_\alpha *$ in $\mathcal{S}'_-$) is the operator of fractional integration for $\alpha>0$ and the operator of the left (right) fractional differentiation for $\alpha<0$*

$$\begin{aligned} {}_{D}\mathrm{I}_{t}^{\alpha}u &= f_\alpha * u, \quad \alpha>0, \\ {}_{D}\mathrm{D}_{t}^{\alpha}u &= f_{-\alpha}*u = \frac{\mathrm{d}^m}{\mathrm{d}t^m}f_{m-\alpha}*u = f_{m-\alpha}*\frac{\mathrm{d}^m}{\mathrm{d}t^m}u = \frac{\mathrm{d}^m}{\mathrm{d}t^m}\left[f_{m-\alpha}*u\right], \quad \alpha>0, \\ &\qquad m\in\mathbb{N}, \\ {}_{D}\check{\mathrm{D}}_{t}^{\alpha}u &= \check{f}_{-\alpha}*u = (-1)^m\frac{\mathrm{d}^m}{\mathrm{d}t^m}f_{m-\alpha}*u = (-1)^m f_{m-\alpha}*\frac{\mathrm{d}^m}{\mathrm{d}t^m}u, \quad \alpha>0,\ m\in\mathbb{N}, \end{aligned} \qquad [2.82]$$

*where $f_\alpha$ is given by [1.6] and $\check{f}_\alpha(t) = f_\alpha(-t)$.*

Operator ${}_D\mathrm{I}_t^\alpha$ coincides with the operator of fractional derivation ${}_D\mathrm{D}_t^\alpha$ for $-\alpha \in \mathbb{N}$ and it is the operator of fractional integration for $\alpha \in \mathbb{N}$.

We have that the Laplace transforms of the distributional fractional integral and derivative are

$$\mathcal{L}\left[{}_D\mathrm{I}_t^\alpha u\left(t\right)\right]\left(s\right) = \frac{1}{s^\alpha}\tilde{u}(s) \text{ and } \mathcal{L}\left[{}_D\mathrm{D}_t^\alpha u\left(t\right)\right]\left(s\right) = s^\alpha \tilde{u}\left(s\right), \quad \mathrm{Re}\, s > 0.$$

We derive, in the following proposition, a connection between the Caputo fractional derivative of a function $u$ belonging to $AC_{loc}^m\left([0,\infty)\right)$ and the distributional fractional derivative of a distribution $u_{reg}$ belonging to $\mathcal{S}'_+$. Also, we derive the connection between the corresponding Laplace transforms. Recall that notation $u_{reg}$ means that we consider $u$ as a distribution, i.e. regular distribution $u_{reg}$ is determined by $u$.

PROPOSITION 2.23.– [ATA 09d] Let $\alpha \in (m-1, m]$, $m \in \mathbb{N}$, $u \in AC_{loc}^m\left([0,\infty)\right)$ and put

$$u_{reg}\left(t\right) = u\left(t\right) H\left(t\right), \quad t \in \mathbb{R}. \qquad [2.83]$$

i) Then

$${}_D\mathrm{D}_t^\alpha u_{reg}\left(t\right) = {}_0^C\mathrm{D}_t^\alpha u\left(t\right) + \sum_{j=0}^{m-1} \frac{\mathrm{d}^{m-1-j}}{\mathrm{d}t^{m-1-j}} f_{m-\alpha}\left(t\right) \frac{\mathrm{d}^j}{\mathrm{d}t^j} u\left(0\right), \quad t>0, \qquad [2.84]$$

where ${}_0^C\mathrm{D}_t^\alpha$ is defined by [2.45].

ii) Also

$$\mathcal{L}\left[{}_0^C\mathrm{D}_t^\alpha u\left(t\right)\right]\left(s\right) = \mathcal{L}\left[{}_D\mathrm{D}_t^\alpha u_{reg}\left(t\right)\right]\left(s\right) - \sum_{j=0}^{m-1} s^{\alpha-1-j} \frac{\mathrm{d}^j}{\mathrm{d}t^j} u\left(0\right), \quad s \in \mathbb{C}_+. \qquad [2.85]$$

Using the notation

$$\mathcal{L}\left[{}_D\mathrm{D}_t^\alpha u_{reg}\left(t\right)\right]\left(s\right) = s^\alpha \mathcal{L}\left[u_{reg}\left(t\right)\right]\left(s\right) = s^\alpha \tilde{u}\left(s\right), \quad s \in \mathbb{C}_+,$$

[2.85] can be written as

$$\mathcal{L}\left[{}_0^C\mathrm{D}_t^\alpha u\left(t\right)\right]\left(s\right) = s^\alpha \tilde{u}\left(s\right) - \sum_{j=0}^{m-1} s^{\alpha-1-j} \frac{\mathrm{d}^j}{\mathrm{d}t^j} u\left(0\right).$$

### 2.3.2. *Dependence of fractional derivative on order*

Recall that ${}_0\mathrm{D}_t^\alpha u = {}_D\mathrm{D}_t^\alpha u$ if $u \in AC^m([a,b])$, $m-1 \leq \alpha < m$, $m \in \mathbb{N}_0$.

We examine the mapping

$$\alpha \mapsto {}_D\mathrm{D}_t^\alpha u, \quad \alpha \in (-\infty, \infty),$$

for given $u \in L^1_{loc}(\mathbb{R})$, such that $u(t) = 0$ for $t < 0$, i.e. $u \in L^1_{loc}([0,\infty))$.

PROPOSITION 2.24.– [ATA 07a] Let $u \in L^1_{loc}(\mathbb{R})$, $u(t) = 0$, $t < 0$, so that it determines a tempered distribution. Then, $\alpha \mapsto {}_D\mathrm{D}_t^\alpha u$ is a smooth mapping from $(-\infty,\infty)$ to $\mathcal{S}'_+$. Also, for every $\alpha \in \mathbb{R}$ with $k > \alpha$,

$$\frac{\partial}{\partial\alpha} {}_D\mathrm{D}_t^\alpha u(t) = \frac{\mathrm{d}^k}{\mathrm{d}t^k}(f_k * u)(\alpha, t), \quad t \in \mathbb{R}, \qquad [2.86]$$

where the derivatives are understood in the sense of the tempered distributions,

$$f_k(\alpha, t) = \frac{t^{k-\alpha-1}}{\Gamma(k-\alpha)}[\psi(k-\alpha) - \ln(t)], \quad \alpha \in (-\infty, k), \ t > 0, \qquad [2.87]$$

$f_k(\alpha,t) = 0$, $t \leq 0$, and $\psi(x) = \frac{\mathrm{d}}{\mathrm{d}x}\ln\Gamma(x)$, $x > 0$, is the Euler function.

Note that a locally integrable function $f$ determines a tempered distribution in $\mathcal{S}'_+$ if it is polynomially bounded on $\mathbb{R}$ as $t \to \infty$. Thus, in sections 2.3.2 and 2.3.3 we will assume that $f$ and its derivatives are polynomially bounded on $\mathbb{R}$ as $t \to \infty$.

PROPOSITION 2.25.– [ATA 07a] Let $u \in L^1_{loc}(\mathbb{R})$, $u(t) = 0$, $t < 0$, so that it determines a tempered distribution. Then

$$\left[\frac{\partial}{\partial\alpha} {}_D\mathrm{D}_t^\alpha u(t)\right]_{\alpha=0} = -(c + \ln t)\,u(t) + \int_0^t \frac{u(t) - u(t-\tau)}{\tau}\mathrm{d}\tau, \quad t > 0,$$

where $c = 0.5772$ is Euler's constant.

Note that for an analytic function, it holds that

$$\left[\frac{\partial}{\partial\alpha} {}_D\mathrm{D}_t^\alpha u(t)\right]_{\alpha=0} = -(c + \ln t)\,u(t) + \sum_{n=0}^{\infty} \frac{(-1)^k u^{(k)}(t)}{k\Gamma(k+1)}, \quad t > 0,$$

given in [WES 03, p. 112].

REMARK 2.5.– [ATA 07a] Proposition 2.25 allows the following representation of fractional derivative ${}_D\mathrm{D}_t^\alpha u$ for $\alpha$ small enough:

$$\begin{aligned} {}_D\mathrm{D}_t^\alpha u(t) &= u(t) + \alpha \left[ \frac{\partial}{\partial \alpha} {}_D\mathrm{D}_t^\alpha u(t) \right]_{\alpha=0} + o(\alpha) \\ &= u(t) + \alpha \left[ -(c + \ln t) u(t) + \int_0^t \frac{u(t) - u(t-\tau)}{\tau} \mathrm{d}\tau \right] + o(\alpha), t > 0. \end{aligned} \qquad [2.88]$$

A relation similar to [2.88] was used in [DIE 02a] for the study of fractional differential equations through the change of the order of integration.

PROPOSITION 2.26.– [ATA 07a] Let $\alpha \in \mathbb{R}$, $u \in L^1_{loc}(\mathbb{R})$, $u(t) = 0$, $t < 0$, so that it determines a tempered distribution. Then

$$\mathcal{L}\left[ \frac{\partial}{\partial \alpha} {}_D\mathrm{D}_t^\alpha u(t) \right](s) = s^\alpha \tilde{u}(s) \ln s, \quad \mathrm{Re}\, s > 0. \qquad [2.89]$$

### 2.3.3. *Distributed-order fractional derivative*

Let $u$ be an element of $\mathcal{S}'_+$. Then, it is proved in [ATA 09b] that the mappings

$$\alpha \mapsto {}_D\mathrm{D}_t^\alpha u : \mathbb{R} \mapsto \mathcal{S}'_+ \quad \text{and} \quad \alpha \mapsto \langle {}_D\mathrm{D}_t^\alpha u(t), \varphi(t) \rangle : \mathbb{R} \mapsto \mathbb{R} \qquad [2.90]$$

are smooth (see proposition 2.24). We define the distributed-order fractional derivative by the use of the distributional fractional derivative.

DEFINITION 2.7.– *[ATA 09b] Let* $\phi \in \mathcal{E}'$, $\mathrm{supp}\, \phi \subset [0, 2]$ *and* $u \in \mathcal{S}'_+$. *Then, the distributed-order fractional derivative of* $u$

$${}_D\mathrm{D}_\phi u(\cdot) = \int_{\mathrm{supp}\, \phi} \phi(\alpha)\, {}_D\mathrm{D}_t^\alpha u(\cdot)\, \mathrm{d}\alpha \qquad [2.91]$$

*is defined as element of* $\mathcal{S}'_+$ *by*

$$\left\langle \int_{\mathrm{supp}\, \phi} \phi(\alpha)\, {}_D\mathrm{D}_t^\alpha u(t)\, \mathrm{d}\alpha, \varphi(t) \right\rangle = \langle \phi(\alpha), \langle {}_D\mathrm{D}_t^\alpha u(t), \varphi(t) \rangle \rangle, \quad \varphi \in \mathcal{S}. \qquad [2.92]$$

Let $u \in AC^m_{loc}([0, \infty))$, $\alpha \in [0, m]$. Recall that the Caputo fractional derivative ${}_0^C\mathrm{D}_t^\alpha$ is defined on intervals $\alpha \in (j-1, j]$, $j \in \{1, \ldots, m\}$ and

$$\lim_{\alpha \to (j-1)+0} {}_0^C\mathrm{D}_t^\alpha u(t) = \frac{\mathrm{d}^{j-1}}{\mathrm{d}t^{j-1}} u(t) - \frac{\mathrm{d}^{j-1}}{\mathrm{d}t^{j-1}} u(0), \quad t > 0.$$

Moreover, $\lim_{\alpha\to j-0}\,{}_0^C\mathrm{D}_t^{\alpha}u(t) = \frac{\mathrm{d}^j}{\mathrm{d}t^j}u(t)$, $t > 0$. Thus, $[0,m] \ni \alpha \mapsto {}_0^C\mathrm{D}_t^{\alpha}u$ is continuous in intervals $\alpha \in (j-1,j)$, $j \in \{1,\dots,m\}$, left continuous at $j$, $j \in \{1,\dots,m\}$ and it has jumps that appear in the limit from the right at points $j-1$, $j \in \{1,\dots,m\}$. For fixed $\alpha \in [0,m]$, the function $[0,\infty) \ni t \mapsto {}_0^C\mathrm{D}_t^{\alpha}u(t)$ is locally integrable on $[0,\infty)$.

For the sake of the next proposition, we introduce the following definition. Recall that we have assumed that function $u$, equal to zero on $(-\infty,0)$, has classical derivatives that are polynomially bounded as $t\to\infty$.

DEFINITION 2.8.– *[ATA 09d] Let $u \in AC^2_{loc}([0,\infty))$.*

*i) Let $\alpha \mapsto \phi(\alpha)$ be continuous in $[0,2]$. Then, we define the distributed-order fractional derivative as*

$$\mathrm{D}_{\phi}u(t) = \int_0^2 \phi(\alpha)\,{}_0^C\mathrm{D}_t^{\alpha}u(t)\,\mathrm{d}\alpha,\quad t>0.$$

*ii) Let $\alpha = \{\alpha_j\}_{j\in\{0,\dots,k\}}$, $\alpha_j \in [0,2]$, $j \in \{0,\dots,k\}$. Then, we define the distributed-order fractional derivative as*

$$\mathrm{D}_{\phi}u(t) = \sum_{j=0}^{k} a_j\,{}_0^C\mathrm{D}_t^{\alpha_j}u(t),\quad t>0.$$

*iii) If $\phi$ is a continuous function in $[\mu,\eta] \subset [0,2]$ and $\phi(\alpha) = 0$, $\alpha \in [0,2]\setminus[\mu,\eta]$, then we define the distributed-order fractional derivative as*

$$\mathrm{D}_{\phi}u(\cdot,t) = \int_{\mu}^{\eta} \phi(\alpha)\,{}_0^C\mathrm{D}_t^{\alpha}u(\cdot,t)\,\mathrm{d}\alpha. \qquad [2.93]$$

Let us derive connections between the distributed-order fractional derivative of $u \in AC^2_{loc}([0,\infty))$ and the corresponding distribution $u_{reg} \in \mathcal{S}'_+$ (in the sense of [2.83]) in cases that are analyzed above.

PROPOSITION 2.27.– [ATA 09d]

i) If $\phi$ belongs to $C([0,2])$ and $u \in AC^2_{loc}([0,\infty))$, then

$$\begin{aligned} {}_D\mathrm{D}_{\phi}u_{reg}(t) &= \mathrm{D}_{\phi}u(t) + u(0)\int_0^2 \phi(\alpha)\,f_{1-\alpha}(t)\,\mathrm{d}\alpha \\ &\quad + \frac{\mathrm{d}}{\mathrm{d}t}u(0)\int_1^2 \phi(\alpha)\,f_{2-\alpha}(t)\,\mathrm{d}\alpha. \end{aligned} \qquad [2.94]$$

ii) Let $\phi = \sum_{j=0}^{k} a_j \delta(\cdot - \alpha_j)$, $\operatorname{supp}\phi \subset [0,2]$, $a_j \in \mathbb{R}_+$, $0 \leq \alpha_k \leq \alpha_j \leq \alpha_0 \leq 2$, $\phi \in \mathcal{E}'$. Let $l \leq k$ be chosen so that $\alpha_l > 1$ and $\alpha_{l+1} \leq 1$. Then

$$\begin{aligned} {}_D\mathrm{D}_\phi u_{reg}(t) &= \mathrm{D}_\phi u(t) + u(0)\left(\sum_{j=l+1}^{k} a_j f_{1-\alpha_j}(t)\right. \\ &\quad \left. + \sum_{j=0}^{l} a_j \frac{\mathrm{d}}{\mathrm{d}t} f_{2-\alpha_j}(t)\right) + \frac{\mathrm{d}}{\mathrm{d}t} u(0) \sum_{j=0}^{l} a_j f_{2-\alpha_j}(t). \end{aligned} \qquad [2.95]$$

iii) Both cases will be summarized by the use of [2.92] as

$$\begin{aligned} {}_D\mathrm{D}_\phi u_{reg}(t) &= \mathrm{D}_\phi u(t) + u(0)\left(\int_{\alpha\in[0,1]} \phi(\alpha) f_{1-\alpha}(t)\,\mathrm{d}\alpha\right. \\ &\quad \left. + \int_{\alpha\in(1,2]} \phi(\alpha) \frac{\mathrm{d}}{\mathrm{d}t} f_{2-\alpha}(t)\,\mathrm{d}\alpha\right) \\ &\quad + \frac{\mathrm{d}}{\mathrm{d}t} u(0) \int_{\alpha\in(1,2]} \phi(\alpha) f_{2-\alpha}(t)\,\mathrm{d}\alpha. \end{aligned} \qquad [2.96]$$

REMARK 2.6.– We use the order of points $0 \leq \alpha_k \leq \alpha_j \leq \alpha_0 \leq 2$ because we will consider two cases separately. The first case is when $\alpha_j \leq 1$, $j \in \{0, \ldots, k\}$. The second case is when some of the $\alpha_j$ are in $(1,2]$. So, this notation is helpful from this point of view.

REMARK 2.7.– If $\operatorname{supp}\phi \subset [0,1]$ and $u$ belongs to $AC^1_{loc}([0,\infty))$, then [2.96] reduces to

$${}_D\mathrm{D}_\phi u_{reg}(t) = \mathrm{D}_\phi u(t) + u(0) \int_{\alpha\in[0,1]} \phi(\alpha) f_{1-\alpha}(t)\,\mathrm{d}\alpha. \qquad [2.97]$$

In the next proposition, we apply the Laplace transform to ${}_D\mathrm{D}_\phi u$, $u \in \mathcal{S}'_+$.

PROPOSITION 2.28.– [ATA 09b] Let $\phi \in \mathcal{E}'$, $\operatorname{supp}\phi \subset [0,2]$ and $u \in \mathcal{S}'_+$. Then:

i) $u \mapsto {}_D\mathrm{D}_\phi u$ is linear and continuous mapping from $\mathcal{S}'_+$ to $\mathcal{S}'_+$.

ii)

$$\mathcal{L}\left[{}_D\mathrm{D}_\phi u\right](s) = \left\langle \phi(\alpha), s^\alpha \tilde{u}(s) \right\rangle, \quad s \in \mathbb{C}_+. \qquad [2.98]$$

iii) Let $\phi \in C([\mu,\eta])$, $[\mu,\eta] \subset [0,2]$ and $\phi(\alpha) = 0$, $\alpha \in [0,2] \setminus [\mu,\eta]$. Then

$$\mathcal{L}[{}_{D}\mathrm{D}_{\phi}u](s) = \tilde{u}(s) \int_{\mu}^{\eta} \phi(\alpha) s^{\alpha} \mathrm{d}\alpha, \quad s \in \mathbb{C}_{+}. \qquad [2.99]$$

Further, we use proposition 2.28 in order to derive the Laplace transform of a function $u \in AC_{loc}^{2}([0,\infty))$, using the connection between its distributed-order fractional derivative and distributed-order fractional derivative of the corresponding distribution $u_{reg} \in \mathcal{S}'_{+}$ (in the sense of [2.83]). Again, we have two cases.

PROPOSITION 2.29.– [ATA 09d]

i) Let $\phi \in C([0,2])$ and $u \in AC_{loc}^{2}([0,\infty))$, then $(s \in \mathbb{C}_{+})$

$$\begin{aligned} \mathcal{L}[\mathrm{D}_{\phi}u(t)](s) &= \tilde{u}(s) \int_{0}^{2} \phi(\alpha) s^{\alpha} \mathrm{d}\alpha - u(0) \frac{1}{s} \int_{0}^{2} \phi(\alpha) s^{\alpha} \mathrm{d}\alpha - \frac{\mathrm{d}}{\mathrm{d}t} u(0) \frac{1}{s^{2}} \\ &\quad \times \int_{1}^{2} \phi(\alpha) s^{\alpha} \mathrm{d}\alpha. \end{aligned} \qquad [2.100]$$

ii) Let $\phi = \sum_{j=0}^{k} a_{j} \delta(\cdot - \alpha_{j})$, $\operatorname{supp}\phi \subset [0,2]$, $a_{j} \in \mathbb{R}_{+}$, $0 \leq \alpha_{k} \leq \alpha_{j} \leq \alpha_{0} \leq 2$, $\alpha_{l} > 1$ and $\alpha_{l+1} \leq 1$. Let $u \in AC_{loc}^{2}([0,\infty))$, then $(s \in \mathbb{C}_{+})$

$$\mathcal{L}[\mathrm{D}_{\phi}u(t)](s) = \tilde{u}(s) \sum_{j=0}^{k} a_{j} s^{\alpha_{j}} - u(0) \frac{1}{s} \sum_{j=0}^{k} a_{j} s^{\alpha_{j}} - \frac{\mathrm{d}}{\mathrm{d}t} u(0) \frac{1}{s^{2}} \sum_{j=0}^{l} a_{j} s^{\alpha_{j}}. \qquad [2.101]$$

iii) Both [2.100] and [2.101] are summarized by

$$\begin{aligned} \mathcal{L}[\mathrm{D}_{\phi}u(t)](s) &= \tilde{u}(s) \int_{\alpha \in [0,2]} \phi(\alpha) s^{\alpha} \mathrm{d}\alpha - u(0) \frac{1}{s} \int_{\alpha \in [0,2]} \phi(\alpha) s^{\alpha} \mathrm{d}\alpha \\ &\quad - \frac{\mathrm{d}}{\mathrm{d}t} u(0) \frac{1}{s^{2}} \int_{\alpha \in (1,2]} \phi(\alpha) s^{\alpha} \mathrm{d}\alpha, \quad s \in \mathbb{C}_{+}. \end{aligned} \qquad [2.102]$$

REMARK 2.8.– If $\operatorname{supp}\phi \subset [0,1]$ and $u$ belongs to $AC_{loc}^{1}([0,\infty))$, then equation [2.102] reduces to

$$\mathcal{L}[\mathrm{D}_{\phi}u(t)](s) = \tilde{u}(s) \int_{\alpha \in [0,1]} \phi(\alpha) s^{\alpha} \mathrm{d}\alpha - u(0) \frac{1}{s} \int_{\alpha \in [0,1]} \phi(\alpha) s^{\alpha} \mathrm{d}\alpha, \quad s \in \mathbb{C}_{+}. \qquad [2.103]$$

# PART 2

# Mechanical Systems

Chapter 3

# Restrictions Following from the Thermodynamics for Fractional Derivative Models of a Viscoelastic Body

In the study of motion of mechanical systems, we begin with constitutive models of a viscoelastic body containing fractional derivatives of stress and strain. Isothermal processes and spatially one-dimensional cases only will be considered. Constitutive models should satisfy restrictions that follow from the second law of thermodynamics, i.e. the entropy inequality. The entropy inequality in the case of isothermal processes reduces to the dissipation inequality (see [ATA 02b, BAG 86, FAB 92]).

We consider the Fourier and the Laplace transform in $\mathcal{S}'$ and $\mathcal{S}'_+$, respectively, or in $\mathcal{K}'_+$ in the case of the Laplace transform. Thus, our assumption is that the solution with all its derivatives is of the polynomial (exponential) growth, in the case of the tempered (exponentially bounded) distributions. Moreover, we are interested in solutions being integrable functions, equal to zero on $(-\infty, 0]$, so such assumptions are also included. If we work with locally integrable functions, equal to zero on $(-\infty, 0)$, with the polynomial or exponential growth at infinity, then we use the classical Laplace transform and the convolution.

This chapter includes results from [ATA 02a, ATA 02b, ATA 03, ATA 13a, ATA 11b, ATA 12c].

## 3.1. Method based on the Fourier transform

The restriction following from the dissipation inequality based on the Fourier transform is formulated as follows. We apply the Fourier transform to the constitutive

$(\sigma - \varepsilon)$ equation of linear viscoelastic body (see, for example, [3.3] and [3.11]). Then, the complex modulus is defined as

$$\hat{E}(\omega) = \frac{\hat{\sigma}(\omega)}{\hat{\varepsilon}(\omega)} = E'(\omega) + \mathrm{i}E''(\omega), \quad \omega \in [0, \infty). \qquad [3.1]$$

The dissipation inequality requires that the tangent of the mechanical loss angle $\delta$ and both $E'$ and $E''$ are non-negative for all values of $\omega \in [0, \infty)$

$$\tan\delta(\omega) = \frac{E''(\omega)}{E'(\omega)} \geq 0, \quad E'(\omega) \geq 0, \quad E''(\omega) \geq 0, \qquad [3.2]$$

see [BAG 86, p. 140]. We refer to [FAB 92] for the relation of conditions [3.2] with other forms of dissipation inequality.

### 3.1.1. *Linear fractional model*

Suppose that the stress $\sigma$ and its derivatives of real (not necessarily integer) order with respect to time at time instant $t$ depend linearly on the strain $\varepsilon$ and its derivatives of real order. Then, following [ATA 11b], we may write

$$\sum_{n=0}^{N} a_n \, {}_0\mathrm{D}_t^{\alpha_n} \sigma(t) = \sum_{m=0}^{M} b_m \, {}_0\mathrm{D}_t^{\beta_m} \varepsilon(t), \quad t > 0, \qquad [3.3]$$

where $a_0, \ldots, a_N$, $b_0, \ldots, b_M$, $\ldots$, $\alpha_0, \ldots, \alpha_N$ and $\beta_0, \ldots, \beta_M$ are real constants. Note that in [3.3], the coefficients $\{a_n\}_{n=0,\ldots,N}$, $\{b_m\}_{m=0,\ldots,M}$ have the physical meaning of relaxation times and are assumed to be given. Equation [3.3] is called the generalized linear fractional model of a viscoelastic body and includes many known models. For example, in the case when $a_0 = 1$, $a_i = 0$, $i = 1, \ldots, N$, $b_0 \neq b_1 \neq 0$, $b_j = 0$, $\beta_0 = 0$, $\beta_1 = 1$, $\beta_j = 0$, $j = 2, \ldots, M$, we obtain the Kelvin–Voigt model of viscoelastic body.

For the constitutive equation [3.3], we may formulate the restrictions in a specific form. First, for technical purposes, the orders of the fractional derivatives in [3.3] are assumed to satisfy

$$\begin{aligned} &0 \leq \alpha_0 < \alpha_1 < \ldots < \alpha_n < \ldots < \alpha_N \leq 1 \text{ and} \\ &0 \leq \beta_0 < \beta_1 < \ldots < \beta_m < \ldots < \beta_M \leq 1. \end{aligned} \qquad [3.4]$$

Our aim is to find restrictions on parameters of the model [3.3], i.e. on $\{\alpha_n\}_{n=0,\ldots,N}$, $\{\beta_m\}_{m=0,\ldots,M}$, $\{a_n\}_{n=0,\ldots,N}$, $\{b_m\}_{m=0,\ldots,M}$, so that the generalized linear fractional model of a viscoelastic body satisfies the requirements of the dissipation inequality.

Following the procedure analogous to the one presented in [BAG 86], we apply the Fourier transform to [3.3] and obtain

$$\hat{\sigma}(\omega)\sum_{n=0}^{N}a_n(\mathrm{i}\omega)^{\alpha_n}=\hat{\varepsilon}(\omega)\sum_{m=0}^{M}b_m(\mathrm{i}\omega)^{\beta_m},\quad \omega\in\mathbb{R}. \qquad [3.5]$$

Having in mind the restrictions that are the consequences of the dissipation inequality, we consider $\omega\in[0,\infty)$. Next, we write [3.5] in the form [3.1]

$$\hat{E}(\omega)=\frac{\sum_{m=0}^{M}b_m(\mathrm{i}\omega)^{\beta_m}}{\sum_{n=0}^{N}a_n(\mathrm{i}\omega)^{\alpha_n}},\quad \omega\in[0,\infty).$$

The restrictions on parameters $\alpha_n$, $\beta_m$, $a_n$ and $b_m$ follow from conditions [3.2] (see [BAG 86, FAB 92]). A straightforward calculation leads to

$$\hat{E}(\omega)=\frac{\sum_{m=0}^{M}b_m\omega^{\beta_m}\cos\frac{\beta_m\pi}{2}+\mathrm{i}\sum_{m=0}^{M}b_m\omega^{\beta_m}\sin\frac{\beta_m\pi}{2}}{\sum_{n=0}^{N}a_n\omega^{\alpha_n}\cos\frac{\alpha_n\pi}{2}+\mathrm{i}\sum_{n=0}^{N}a_n\omega^{\alpha_n}\sin\frac{\alpha_n\pi}{2}},\quad \omega\in[0,\infty).$$

Introducing $\hat{E}_{nom}$ as

$$\hat{E}_{nom}(\omega)=\hat{E}(\omega)\left|\sum_{n=0}^{N}a_n\,(\mathrm{i}\omega)^{\alpha_n}\right|^2=E'_{nom}(\omega)+\mathrm{i}E''_{nom}(\omega),\quad \omega\in[0,\infty),$$

we obtain ($\omega\in[0,\infty)$)

$$\begin{aligned}E'_{nom}(\omega)&=\left(\sum_{n=0}^{N}a_n\omega^{\alpha_n}\cos\frac{\alpha_n\pi}{2}\right)\left(\sum_{m=0}^{M}b_m\omega^{\beta_m}\cos\frac{\beta_m\pi}{2}\right)\\&+\left(\sum_{n=0}^{N}a_n\omega^{\alpha_n}\sin\frac{\alpha_n\pi}{2}\right)\left(\sum_{m=0}^{M}b_m\omega^{\beta_m}\sin\frac{\beta_m\pi}{2}\right),\\E''_{nom}(\omega)&=\left(\sum_{n=0}^{N}a_n\omega^{\alpha_n}\cos\frac{\alpha_n\pi}{2}\right)\left(\sum_{m=0}^{M}b_m\omega^{\beta_m}\sin\frac{\beta_m\pi}{2}\right)\\&-\left(\sum_{n=0}^{N}a_n\omega^{\alpha_n}\sin\frac{\alpha_n\pi}{2}\right)\left(\sum_{m=0}^{M}b_m\omega^{\beta_m}\cos\frac{\beta_m\pi}{2}\right).\end{aligned} \qquad [3.6]$$

We assume that $\alpha_n, \beta_m \in [0,1]$, $n = 0,1,\ldots,N$, $m = 0,1,\ldots,M$, so that $\frac{\alpha_n \pi}{2}, \frac{\beta_m \pi}{2} \in [0, \frac{\pi}{2}]$, and consequently, sine and cosine of those angles are positive. Therefore, assuming $a_n, b_m \geq 0$ we obtain that $E'_{nom}(\omega) \geq 0$, $\omega \in [0,\infty)$, and hence [3.2]$_2$ is satisfied. In the sequel, we will restrict our attention to this case (i.e. $a_n, b_m \geq 0$).

After a straightforward calculation from [3.6], we conclude that [3.2]$_3$ holds if and only if

$$E''_{nom}(\omega) = -\sum_{\substack{n\in\{0,1,\ldots,N\} \\ m\in\{0,1,\ldots,M\}}} \omega^{\alpha_n+\beta_m} \sin\frac{(\alpha_n-\beta_m)\pi}{2} a_n b_m \geq 0, \quad \omega \in [0,\infty). \quad [3.7]$$

LEMMA 3.1.– [ATA 11b] Let [3.4] hold and $a_n, b_m \geq 0$, $n = 0,1,\ldots,N$, $m = 0,1,\ldots,M$. Suppose that $\alpha_N \neq \beta_M$. Then, a necessary condition for [3.7] is that $\alpha_N < \beta_M$.

In other words, the highest order of fractional derivatives of stress in [3.3] could not be greater than the highest order of fractional derivatives of strain.

PROOF.– We observe that for large $\omega$, the sign of $E''_{nom}(\omega)$ coincides with the sign of the term in the sum on the right-hand side of [3.7] with the largest power of $\omega$. It follows from [3.4] that the latter is achieved for $\alpha_N$ and $\beta_M$. Therefore, in order that $E''_{nom}(\omega) > 0$, $\alpha_N$ has to be less than $\beta_M$, as claimed. ■

REMARK 3.1.–

i) As in lemma 3.1, we can prove that if $\alpha_N = \beta_M$, then a necessary condition becomes $\alpha_n < \beta_m$, for the largest $\alpha_n$ and $\beta_m$, which do not coincide.

ii) A similar conclusion for a particular problem with the constitutive equation of the form [3.3] has been obtained in [ATA 03].

An equation of the form: $a_0\sigma + a_\alpha \, {}_0\mathrm{D}_t^\alpha \sigma = b_0\varepsilon$, $0 < \alpha < 1$, cannot be a constitutive equation of a viscoelastic body that satisfies the dissipation inequality. Indeed, for the mentioned model, equation [3.7] reduces to $-\omega^\alpha a_\alpha b_0 \sin\frac{\alpha\pi}{2}$, which is strictly less than zero for all $\omega \geq 0$.

For practical purposes, it can happen that condition [3.7] is too general and thus hardly applicable to concrete problems. Therefore, it will be useful to extract from [3.7] particular conditions on parameters $a_n$, $b_m$, $\alpha_n$ and $\beta_m$, which guarantee that [3.2]$_3$ is satisfied. We separately consider several cases.

3.1.1.1. – *Case I*

Suppose that $\alpha_n \neq \beta_m$, for all $n, m$, i.e. there are $N+1$ and $M+1$ terms of different order in [3.3]. Since we assumed that $a_n, b_m \geq 0$, having in mind [3.4], by choosing that the orders of fractional derivatives satisfy

$$0 \leq \alpha_0 < \alpha_1 < \ldots < \alpha_n < \ldots < \alpha_N < \beta_0 < \beta_1 < \ldots < \beta_m < \ldots < \beta_M \leq 1, \quad [3.8]$$

we have that all $\sin \frac{(\alpha_n - \beta_m)\pi}{2}$ are negative, and consequently, $E''_{nom}(\omega) \geq 0$, $\omega \in [0, \infty)$. In other words, $a_n, b_m \geq 0$ and [3.8] are sufficient conditions for $[3.2]_3$.

REMARK 3.2.– Note that the fractional linear model of a fluid-like viscoelastic body

$$\left(1 + \frac{a}{b}\, {}_0\mathrm{D}_t^{\alpha-\beta}\right) \sigma(t) = E \left(a\, {}_0\mathrm{D}_t^{\alpha} + c\, {}_0\mathrm{D}_t^{\gamma} + \frac{ac}{b}\, {}_0\mathrm{D}_t^{\alpha+\gamma-\beta}\right) \varepsilon(t), \quad t > 0, \quad [3.9]$$

proposed in [SCH 00], belongs to this class. Indeed, by setting $N = 1$, $a_0 = 1$, $\alpha_0 = 0$, $a_1 = \frac{a}{b}$, $\alpha_1 = \alpha - \beta$, $M = 2$, $b_0 = Ea$, $\beta_0 = \alpha$, $b_1 = Ec$, $\beta_1 = \gamma$, $b_2 = E\frac{ac}{b}$, $\beta_2 = \alpha + \gamma - \beta$ in [3.3], we obtain [3.9]. Thermodynamical restrictions are satisfied since constants $E$, $a$, $b$, $c$ are positive, while [3.8] in this case becomes

$$0 \leq \alpha - \beta < \alpha < \gamma < \alpha + \gamma - \beta \leq 1,$$

and is satisfied according to assumption $0 < \beta < \alpha < \gamma < \frac{1}{2}$.

3.1.1.2. – *Case II*

Suppose that $M > N$ and $\alpha_i = \beta_i$, $i = 0, 1, \ldots, N$, i.e. there are $N+1$ first terms of the same order and $M - N$ terms left in [3.3]. Then, [3.7] reduces to

$$\begin{aligned} E''_{nom}(\omega) &= \sum_{i,j \in \{0,1,\ldots,N\}, i<j} \omega^{\alpha_i + \alpha_j} \sin \frac{(\alpha_j - \alpha_i)\pi}{2} (a_i b_j - a_j b_i) \\ &- \sum_{i \in \{0,1,\ldots,N\}, m>N} \omega^{\alpha_i + \beta_m} \sin \frac{(\alpha_i - \beta_m)\pi}{2} a_i b_m \geq 0, \quad \omega \in [0, \infty). \end{aligned}$$

We are interested in a special case when each term in the above sum is non-negative. Then, the following should hold:

$$\frac{a_i}{b_i} \geq \frac{a_j}{b_j}, \quad i, j \in \{0, 1, \ldots, N\}, \; i < j,$$

$$\alpha_i \leq \beta_m, \quad i \in \{0, 1, \ldots, N\}, \; m > N,$$

which implies that

$$\frac{a_0}{b_0} \geq \frac{a_1}{b_1} \geq \ldots \geq \frac{a_N}{b_N} \geq 0,$$
$$0 \leq \alpha_0 < \alpha_1 < \ldots < \alpha_N < \beta_{N+1} < \ldots < \beta_M \leq 1.$$

3.1.1.3. – *Case III*

Suppose that $N > M$ and $\alpha_{N-M+i} = \beta_i$, $i = 0, 1, \ldots, M$, i.e. there are $M+1$ last terms of the same order and $N - M$ terms left in [3.3]. Then, [3.7] reduces to

$$\begin{aligned} E''_{nom}(\omega) &= -\sum_{n<N-M, i\in\{0,1,\ldots,M\}} \omega^{\alpha_n+\alpha_{N-M+i}} \sin\frac{(\alpha_n - \alpha_{N-M+i})\pi}{2} a_n b_i \\ &+ \sum_{i,j\in\{0,1,\ldots,M\}, i<j} \omega^{\beta_i+\beta_j} \sin\frac{(\beta_j - \beta_i)\pi}{2} \\ &\times (a_{N-M+i} b_j - a_{N-M+j} b_i) \geq 0, \quad \omega \in [0,\infty). \end{aligned}$$

Again, we concentrate on a special case when each term in the above sum is non-negative. By assumption [3.4], it follows that all terms in the first part of the sum are positive. Thus,

$$\frac{a_{N-M+i}}{b_i} \geq \frac{a_{N-M+j}}{b_j}, \quad i, j \in \{0, 1, \ldots, M\}, \; i < j,$$

which implies

$$\frac{a_{N-M}}{b_0} \geq \frac{a_{N-M+1}}{b_1} \geq \ldots \geq \frac{a_N}{b_M} \geq 0.$$

3.1.1.4. – *Case IV*

Suppose that $N = M$ and $\alpha_i = \beta_i$, $i = 0, 1, \ldots, N$, i.e. there are $N+1$ terms of the same order on both sides of [3.3]. In this case, [3.7] becomes

$$E''_{nom}(\omega) = \sum_{i,j\in\{0,1,\ldots,N\}, i<j} \omega^{\alpha_i+\alpha_j} \sin\frac{(\alpha_j - \alpha_i)\pi}{2} (a_i b_j - a_j b_i) \geq 0, \quad \omega \in [0,\infty).$$

As in the previous cases, we look for those parameters for which all terms in the above sum are non-negative. Hence, we have to restrict our attention only to those parameters $a_n, b_m \geq 0$ that in addition satisfy

$$\frac{a_i}{b_i} \geq \frac{a_j}{b_j}, \quad i < j.$$

This implies that

$$\frac{a_0}{b_0} \geq \frac{a_1}{b_1} \geq \ldots \geq \frac{a_N}{b_N} \geq 0.$$

REMARK 3.3.– Note that the fractional Zener model

$$\left(1 + a\, {}_0\mathrm{D}_t^\alpha\right) \sigma(t) = E \left(1 + b\, {}_0\mathrm{D}_t^\alpha\right) \varepsilon(t), \quad t > 0 \qquad [3.10]$$

belongs to this class. Indeed, by setting $N = 1$, $a_0 = 1$, $\alpha_0 = 0$, $a_1 = a$, $\alpha_1 = \alpha$, $M = 1$, $b_0 = E$, $\beta_0 = 0$, $b_1 = Eb$, $\beta_1 = \alpha$ in [3.3], we obtain [3.9]. Thermodynamical restrictions are satisfied since constants $E$, $a$, $b$ are positive, while [3.8] in this case becomes $b \geq a$. For the experimental evidence supporting the validity of the fractional Zener model, see [NÄS 13] and references therein.

REMARK 3.4.– It can be shown that in other cases that are not listed above (e.g. when there are $K < \min\{N, M\}$ terms of the same order in [3.3], it is not possible to make all terms in [3.7] non-negative simultaneously, thus it is more difficult to find some particular conditions on parameters $a_n$, $b_m$, $\alpha_n$ and $\beta_m$, which implies [3.2].

### 3.1.2. *Distributed-order fractional model*

The distributed-order model is introduced as a generalization of [3.3] (see [BAG 00a, BAG 00b, CAP 01] and the references therein) that leads to the following type of $\sigma - \varepsilon$ relation, used in [ATA 02a, ATA 03],

$$\int_0^1 \phi_\sigma(\gamma)\, {}_0\mathrm{D}_t^\gamma \sigma(t)\, \mathrm{d}\gamma = E \int_0^1 \phi_\varepsilon(\gamma)\, {}_0\mathrm{D}_t^\gamma \varepsilon(t)\, \mathrm{d}\gamma, \quad t > 0, \qquad [3.11]$$

where $\phi_\sigma$ and $\phi_\varepsilon$ are constitutive functions, or distributions, which characterize the material under consideration and $E$ is the generalized Young modulus.

If the integrals in the constitutive equation [3.11] are interpreted in the distributional setting as

$$\left\langle \int_{\operatorname{supp}\phi} \phi(\alpha)\, {}_0\mathrm{D}_t^\alpha h(t)\, \mathrm{d}\alpha, \varphi(t) \right\rangle = \left\langle \phi(\alpha), \left\langle {}_0\mathrm{D}_t^\alpha h(t), \varphi(t) \right\rangle \right\rangle, \quad \varphi \in \mathcal{S}(\mathbb{R}),$$

then [3.11] takes the form

$$\left\langle \phi_\sigma(\gamma), \left\langle {}_0\mathrm{D}_t^\gamma \sigma\left(t\right), \varphi(t)\right\rangle\right\rangle = E\left\langle \phi_\varepsilon(\gamma), \left\langle {}_0\mathrm{D}_t^\gamma \varepsilon\left(t\right), \varphi(t)\right\rangle\right\rangle, \quad \varphi \in \mathcal{S}(\mathbb{R}),$$

see [ATA 09b]. We assume that $\phi_\sigma$ and $\phi_\varepsilon$ are positive integrable functions of $\gamma \in [0,1]$, or distributions with compact support in $[0,1]$, and $\varphi$ denotes a test function belonging to the space $\mathcal{D}(\mathbb{R})$.

Note that by choosing

$$\phi_\sigma(\gamma) = \sum_{n=0}^{N} a_n \delta(\gamma - \alpha_n) \quad \text{and} \quad \phi_\varepsilon(\gamma) = \sum_{m=0}^{M} b_m \delta(\gamma - \beta_m),$$

where $\delta$ is the Dirac distribution, we obtain [3.3]. In the special case when $\phi_\sigma = \delta$, the constitutive equation [3.11] becomes

$$\sigma\left(t\right) = E \int_0^1 \phi_\varepsilon\left(\gamma\right) {}_0\mathrm{D}_t^\gamma \varepsilon\left(t\right) \mathrm{d}\gamma, \quad t > 0, \qquad [3.12]$$

with $\phi_\varepsilon$ given. Of course, the stress $\sigma$ could depend on integer-order derivatives as well, so that equation [3.12] may become

$$\sigma\left(t\right) = \sum_{k=0}^{K} c_k \varepsilon^{(k)}\left(t\right) + E \int_0^1 \phi_\varepsilon\left(\gamma\right) {}_0\mathrm{D}_t^\gamma \varepsilon\left(t\right) \mathrm{d}\gamma, \quad t > 0, \qquad [3.13]$$

where $K \geq 1$.

To obtain the thermodynamical restrictions, we apply the Fourier transform to [3.11] and obtain

$$\hat{\sigma}\left(\omega\right) \int_0^1 \phi_\sigma\left(\gamma\right) \left(\mathrm{i}\omega\right)^\gamma \mathrm{d}\gamma = E\hat{\varepsilon}\left(\omega\right) \int_0^1 \phi_\varepsilon\left(\gamma\right) \left(\mathrm{i}\omega\right)^\gamma \mathrm{d}\gamma, \quad \omega \in [0,\infty). \qquad [3.14]$$

From [3.14], the complex modulus becomes

$$\hat{E}\left(\omega\right) = E \frac{\int_0^1 \phi_\varepsilon\left(\gamma\right) \left(\mathrm{i}\omega\right)^\gamma \mathrm{d}\gamma}{\int_0^1 \phi_\sigma\left(\gamma\right) \left(\mathrm{i}\omega\right)^\gamma \mathrm{d}\gamma}, \quad \omega \in [0,\infty). \qquad [3.15]$$

Writing

$$\hat{E}(\omega) = E(E'(\omega) + \mathrm{i}E''(\omega)), \quad \omega \in [0, \infty),$$

and using $(\mathrm{i}\omega)^\gamma = \omega^\gamma \left(\cos \frac{\gamma\pi}{2} + \mathrm{i} \sin \frac{\gamma\pi}{2}\right)$ in [3.15], it follows that

$$E' = \frac{C_1 C_2 + S_1 S_2}{C_2^2 + S_2^2} \quad \text{and} \quad E'' = \frac{S_1 C_2 - C_1 S_2}{C_2^2 + S_2^2}, \qquad [3.16]$$

where ($\omega \in [0, \infty)$)

$$C_1(\omega) = \int_0^1 \phi_\varepsilon(\gamma)\, \omega^\gamma \cos \frac{\gamma\pi}{2} \mathrm{d}\gamma, \quad C_2(\omega) = \int_0^1 \phi_\sigma(\gamma)\, \omega^\gamma \cos \frac{\gamma\pi}{2} \mathrm{d}\gamma,$$

$$S_1(\omega) = \int_0^1 \phi_\varepsilon(\gamma)\, \omega^\gamma \sin \frac{\gamma\pi}{2} \mathrm{d}\gamma, \quad S_2(\omega) = \int_0^1 \phi_\sigma(\gamma)\, \omega^\gamma \sin \frac{\gamma\pi}{2} \mathrm{d}\gamma. \qquad [3.17]$$

Therefore, the tangent of the mechanical loss angle $\delta$ is

$$\tan \delta = \frac{E''}{E'} = \frac{S_1 C_2 - C_1 S_2}{C_1 C_2 + S_1 S_2}. \qquad [3.18]$$

The thermodynamical restrictions [3.2] take the form

$$E > 0, \quad C_1(\omega) C_2(\omega) + S_1(\omega) S_2(\omega) \geq 0,$$
$$S_1(\omega) C_2(\omega) - C_1(\omega) S_2(\omega) \geq 0, \quad \omega \in [0, \infty). \qquad [3.19]$$

By using definitions of $S_1, \ldots, C_2$, we write conditions [3.19] in expanded form as

$$\left(\int_0^1 \phi_\varepsilon(\gamma)\, \omega^\gamma \cos \frac{\gamma\pi}{2} \mathrm{d}\gamma\right)\left(\int_0^1 \phi_\sigma(\gamma)\, \omega^\gamma \cos \frac{\gamma\pi}{2} \mathrm{d}\gamma\right)$$
$$+ \left(\int_0^1 \phi_\varepsilon(\gamma)\, \omega^\gamma \sin \frac{\gamma\pi}{2} \mathrm{d}\gamma\right)\left(\int_0^1 \phi_\sigma(\gamma)\, \omega^\gamma \sin \frac{\gamma\pi}{2} \mathrm{d}\gamma\right) \geq 0, \quad \omega \in [0, \infty),$$
$$\left(\int_0^1 \phi_\varepsilon(\gamma)\, \omega^\gamma \sin \frac{\gamma\pi}{2} \mathrm{d}\gamma\right)\left(\int_0^1 \phi_\sigma(\gamma)\, \omega^\gamma \cos \frac{\gamma\pi}{2} \mathrm{d}\gamma\right)$$
$$- \left(\int_0^1 \phi_\varepsilon(\gamma)\, \omega^\gamma \cos \frac{\gamma\pi}{2} \mathrm{d}\gamma\right)\left(\int_0^1 \phi_\sigma(\gamma)\, \omega^\gamma \sin \frac{\gamma\pi}{2} \mathrm{d}\gamma\right) \geq 0, \quad \omega \in [0, \infty).$$
$$[3.20]$$

Since the restrictions on functions, or distributions, $\phi_\sigma$ and $\phi_\varepsilon$ are difficult to be obtained in the general case from conditions [3.20], we present several special cases.

3.1.2.1. – *Case I*

Let the constitutive functions $\phi_\sigma$ and $\phi_\varepsilon$ in [3.11] be given by

$$\phi_\varepsilon(\gamma) = c\phi_\sigma(\gamma), \quad c > 0. \qquad [3.21]$$

Then, $\tan\delta = 0$ and we conclude that the body behaves as an elastic body.

3.1.2.2. – *Case II*

Let the constitutive functions $\phi_\sigma$ and $\phi_\varepsilon$ in [3.11] take the form

$$\phi_\sigma(\alpha) = a^\alpha \text{ and } \phi_\varepsilon(\alpha) = b^\alpha.$$

Note that this choice of $\phi_\sigma$ and $\phi_\varepsilon$ is the simplest one guaranteeing the dimensional homogeneity. We look for the restrictions on $a$ and $b$. Then, constitutive equation [3.11] becomes a distributed-order model with exponential-type constitutive functions

$$\int_0^1 a^\alpha \, {}_0\mathrm{D}_t^\alpha \sigma(t)\, d\alpha = E \int_0^1 b^\alpha \, {}_0\mathrm{D}_t^\alpha \varepsilon(t)\, d\alpha, \quad t > 0. \qquad [3.22]$$

Note that for $a = b$, we have $\sigma = E\varepsilon$, i.e. the constitutive model [3.22] reduces to Hooke's law. Interpreting these integrals as the Riemann sums, we obtain

$$\sum_{n=0}^{N} a^{\alpha_n} \, {}_0\mathrm{D}_t^{\alpha_n} \sigma(t)\, \Delta\alpha_n = E \sum_{n=0}^{N} b^{\alpha_n} \, {}_0\mathrm{D}_t^{\alpha_n} \varepsilon(t)\, \Delta\alpha_n, \quad t > 0,$$

where $N \to \infty$ and $\Delta\alpha_n \to 0$. Putting $a_n = a^{\alpha_n}\Delta\alpha_n$ and $b_n = b^{\alpha_n}\Delta\alpha_n$, we obtain [3.3] with $N = M$ and all terms of the same order. Taking $a, b > 0$, we have that $[3.2]_2$ holds, while, as in Case IV (with equal number of terms of the same order on both sides of [3.3]), condition

$$\left(\frac{a}{b}\right)^{\alpha_i} \geq \left(\frac{a}{b}\right)^{\alpha_j}, \quad i < j$$

provides validity of [3.3], along with $E > 0$. Since $\alpha_i \leq \alpha_j$, $i < j$, the latter is equivalent to

$$E > 0, \quad \frac{a}{b} \leq 1, \text{ i.e., } a \leq b. \qquad [3.23]$$

3.1.2.3. – *Case III*

Suppose that the constitutive functions $\phi_\sigma$ and $\phi_\varepsilon$ in [3.11] are given by

$$\phi_\sigma(\gamma) = \delta(\gamma) \text{ and } \phi_\varepsilon(\gamma) = \tau_0^\gamma, \qquad [3.24]$$

where $\tau_0$ is a constant. Thus, the constitutive equation [3.11] becomes

$$\sigma(t) = E\int_0^1 \tau_0^\gamma \, {}_0\mathrm{D}_t^\gamma \varepsilon(t)\, \mathrm{d}\gamma, \quad t > 0, \qquad [3.25]$$

which is of type [3.12] and represents the distributed-order model of the Kelvin–Voigt type. Condition [3.19] becomes

$$\int_0^1 (\tau_0\omega)^\gamma \sin\frac{\gamma\pi}{2}\,\mathrm{d}\gamma \geq 0, \quad \int_0^1 (\tau_0\omega)^\gamma \cos\frac{\gamma\pi}{2}\,\mathrm{d}\gamma \geq 0, \quad \omega \in [0,\infty), \qquad [3.26]$$

or

$$E > 0 \text{ and } \tau_0 > 0. \qquad [3.27]$$

3.1.2.4. – *Case IV*

Let the constitutive function $\phi_\sigma$ and constitutive distribution $\phi_\varepsilon$ in [3.11] be given by

$$\phi_\sigma(\gamma) = C\tau_1^\gamma \text{ and } \phi_\varepsilon(\gamma) = \delta(\gamma), \qquad [3.28]$$

where $C$ and $\tau_1$ are constants. The constitutive equation corresponding to [3.28] becomes

$$C\int_0^1 \tau_1^\gamma \, {}_0\mathrm{D}_t^\gamma \sigma(t)\, \mathrm{d}\gamma = E\varepsilon(t), \quad t > 0. \qquad [3.29]$$

Condition [3.19] leads to

$$-C\int_0^1 (\tau_1\omega)^\gamma \sin\frac{\gamma\pi}{2}\,\mathrm{d}\gamma \geq 0, \quad C\int_0^1 (\tau_1\omega)^\gamma \cos\frac{\gamma\pi}{2}\,\mathrm{d}\gamma \geq 0, \quad \omega \in [0,\infty). \qquad [3.30]$$

From [3.30], we conclude that the constitutive equation [3.29] violates the dissipation inequality for any value of $C \neq 0$. This is an interesting fact and it generalizes the results of [ATA 02a, BAG 86]. Namely, suppose that we chose the constitutive functions $\phi_\sigma$ and $\phi_\varepsilon$ in [3.11] as

$$\phi_\varepsilon(\gamma) = \delta(\gamma) \quad \text{and} \quad \phi_\sigma(\gamma) = \delta(\gamma) + a\,\delta(\gamma - \alpha).$$

The result [3.30] forbids the constitutive equation of the form

$$(1 + a\,{}_0\mathrm{D}_t^\alpha)\,\sigma(t) = E\varepsilon(t), \quad t > 0,$$

where $0 < \alpha < 1$, $a > 0$.

3.1.2.5. – *Case V*

Suppose that the constitutive functions $\phi_\sigma$ and $\phi_\varepsilon$ in [3.11] are of the form

$$\phi_\sigma(\gamma) = \delta(\gamma) + a\,\delta(\gamma - \alpha) \quad \text{and} \quad \phi_\varepsilon(\gamma) = \delta(\gamma) + b\,\delta(\gamma - \alpha), \qquad [3.31]$$

where $a$, $b$ and $0 < \alpha < 1$ are constants. This choice corresponds to the fractional Zener model

$$(1 + a\,{}_0\mathrm{D}_t^\alpha)\,\sigma(t) = E\,(1 + b\,{}_0\mathrm{D}_t^\alpha)\,\varepsilon(t), \quad t > 0, \qquad [3.32]$$

see [NÄS 13]. By substituting [3.31] into condition [3.19], we have ($\omega \in [0, \infty)$)

$$\begin{gathered} b\omega^\alpha \sin\frac{\alpha\pi}{2} - a\omega^\alpha \sin\frac{\alpha\pi}{2} \geq 0, \\ \left(1 + b\omega^\alpha \cos\frac{\alpha\pi}{2}\right)\left(1 + a\omega^\alpha \cos\frac{\alpha\pi}{2}\right) + \left(b\omega^\alpha \sin\frac{\alpha\pi}{2}\right)\left(a\omega^\alpha \sin\frac{\alpha\pi}{2}\right) \geq 0. \end{gathered} \qquad [3.33]$$

From [3.33], it follows that

$$E > 0 \quad \text{and} \quad b \geq a \geq 0, \qquad [3.34]$$

which is a well-known result (see [ATA 02a, BAG 86]).

3.1.2.6. – *Case VI*

Consider the case when $\phi_\sigma$ and $\phi_\varepsilon$ in [3.11] are given by

$$\begin{aligned}&\phi_\sigma(\gamma) = \delta(\gamma) + a\,\delta(\gamma-\alpha) \quad \text{and} \\ &\phi_\varepsilon(\gamma) = E\left(\delta(\gamma) + b\,\delta(\gamma-\alpha) + c\,\delta(\gamma-\beta)\right),\end{aligned} \qquad [3.35]$$

where $a$, $b$, $c$, $\alpha$ and $\beta$ are constants with $0 < \alpha < 1$, $0 < \beta < 1$. This choice corresponds to the seven-parameter model

$$\left(1 + a\,{}_0\mathrm{D}_t^\alpha\right)\sigma(t) = E\left(1 + b\,{}_0\mathrm{D}_t^\alpha + c\,{}_0\mathrm{D}_t^\beta\right)\varepsilon(t), \quad t > 0,$$

used in [PRI 03, ROS 01b, STA 04a]. As a matter of fact, in [PRI 03], a special case of [3.35] was used where $a = \tau^\alpha$, $b = \tau^\alpha$, $c = \tau^\beta$, with $\tau = const$ (see [PRI 03, eq. [3.37]]), while in [ROS 01b], it was assumed that $a = \tau_\varepsilon^\alpha$, $b = \tau_\sigma^\alpha$, $c = \tau_\sigma^\beta$ with $\tau_\sigma = const$, $\tau_\varepsilon = const$. From [3.35], we have that condition [3.20] becomes ($\omega \in [0,\infty)$)

$$1 + ab\omega^{2\alpha} + \omega^\alpha(a+b)\cos\frac{\alpha\pi}{2} + c\omega^\beta\cos\frac{\beta\pi}{2} + ac\omega^{\alpha+\beta}\cos\frac{(\beta-\alpha)\pi}{2} \geq 0,$$

$$\omega^\alpha(b-a)\sin\frac{\alpha\pi}{2} + c\omega^\beta\sin\frac{\alpha\pi}{2} + ac\omega^{\alpha+\beta}\sin\frac{(\beta-\alpha)\pi}{2} \geq 0, \qquad [3.36]$$

together with $E > 0$. Thus, the constants in [3.35] must satisfy

$$E > 0, \quad b \geq a > 0, \quad c > 0, \quad \alpha < \beta. \qquad [3.37]$$

Conditions [3.37] contain, as a special case, the results presented in [PRI 03, ROS 01b].

### 3.1.3. *Constitutive equations for rod bending*

In the mathematical theory of elastic rods, the bending moment $M$ is assumed to be an arbitrary function of the change in the curvature of the rod's axis $\kappa$, i.e. $M = M(\kappa)$, where $\kappa = \frac{1}{r} - \frac{1}{R}$ with $r$ and $R$ being radii of the curvature of the rod's axis in the present and initial state, respectively. In the case of viscoelastic rods, the bending moment is a function of the curvature and its time derivatives.

We postulate the constitutive moment–curvature relation for a viscoelastic rod

$$\int_0^1 \phi_M(\gamma)\,{}_0\mathrm{D}_t^\gamma M\,\mathrm{d}\gamma = \int_0^1 \phi_\kappa(\gamma)\,{}_0\mathrm{D}_t^\gamma \kappa\,\mathrm{d}\gamma, \qquad [3.38]$$

motivated by the stress–strain constitutive equation [3.11]. We used $\phi_M$ and $\phi_\kappa$ to denote the constitutive weight functions, or distributions. Moment–curvature relations may be used for the study of the motion and stability of viscoelastic rods (see [LI 01, STA 07, STA 01]).

Note that in the case of an elastic material, with the choice of $\phi_M = \delta$ and $\phi_\kappa = EI\delta$, where $\delta$ is the Dirac distribution, the constitutive moment–curvature relation [3.38] becomes

$$M = EI\kappa,$$

i.e. the moment–curvature relation of the classical Euler–Bernoulli rod theory. Here, $E$ is the modulus of elasticity and $I$ is the second moment of inertia of the cross-sections. If we choose $\phi_M(\gamma) = \delta(\gamma)$ and $\phi_\kappa(\gamma) = EI(\delta(\gamma) + b\,\delta(\gamma - \alpha))$, equation [3.38] becomes the fractional Kelvin–Voigt moment–curvature relation

$$M = EI\left(1 + a\,{}_0\mathrm{D}_t^\alpha\right)\kappa. \qquad [3.39]$$

The fractional Zener moment–curvature relation

$$\left(1 + a\,{}_0\mathrm{D}_t^\alpha\right)M = EI\left(1 + b\,{}_0\mathrm{D}_t^\alpha\right)\kappa \qquad [3.40]$$

is obtained with the choice: $\phi_M(\gamma) = \delta(\gamma) + a\,\delta(\gamma - \alpha)$ and $\phi_\kappa(\gamma) = EI(\delta(\gamma) + b\,\delta(\gamma - \alpha))$.

For the choice of the constitutive distributions as: $\phi_M(\gamma) = \delta(\gamma) + a\,\delta(\gamma - \alpha)$ and $\phi_\kappa(\gamma) = EI(\delta(\gamma) + b\,\delta(\gamma - \alpha) + c\,\delta(\gamma - \beta))$, we obtain

$$\left(1 + a\,{}_0\mathrm{D}_t^\alpha\right)M = EI\left(1 + b\,{}_0\mathrm{D}_t^\alpha + c\,{}_0\mathrm{D}_t^\beta\right)\kappa. \qquad [3.41]$$

The constitutive moment–curvature relations [3.39]–[3.41] may also be derived from the corresponding stress–strain constitutive equation of a viscoelastic body and the coefficients must satisfy the restrictions following from dissipation inequality.

The constitutive equations [3.39]–[3.41] were used in [ATA 02c, BAC 00, STA 01, STA 02, STA 04a].

### 3.1.4. *Stress relaxation and creep for two special cases of viscoelastic bodies*

Following [ATA 02a], we analyze the behavior in the stress relaxation and creep experiments of two models of the viscoelastic body: the distributed-order model of the exponential type [3.22] and of the Kelvin–Voigt type [3.25].

3.1.4.1. *Distributed-order model of the exponential type*

Consider the distributed-order model of the exponential type [3.22]

$$\int_0^1 a^\gamma {}_0\mathrm{D}_t^\gamma \sigma(t)\,\mathrm{d}\gamma = E\int_0^1 b^\gamma {}_0\mathrm{D}_t^\gamma \varepsilon(t)\,\mathrm{d}\gamma,\quad t>0 \qquad [3.42]$$

and suppose that $a$, $b$ and $E$ satisfy [3.23]. By applying Laplace transform to [3.42], we get

$$\tilde{\sigma}(s)\frac{as-1}{\ln(as)} = E\tilde{\varepsilon}(s)\frac{bs-1}{\ln(bs)},\quad \mathrm{Re}\, s>0. \qquad [3.43]$$

By solving [3.43] with respect to $\tilde{\sigma}$, we obtain

$$\tilde{\sigma}(s) = E\tilde{\varepsilon}(s)\frac{\ln(as)}{\ln(bs)}\frac{bs-1}{as-1},\quad \mathrm{Re}\, s>0. \qquad [3.44]$$

We write [3.44] as ($\mathrm{Re}\, s>0$)

$$\begin{aligned}\frac{\tilde{\sigma}(s)}{E} &= \left(\frac{bs}{\ln(bs)}\frac{\ln(as)}{as-1} - \frac{1}{\ln(bs)}\frac{\ln(as)}{as-1}\right)\tilde{\varepsilon}(s)\\ &= \frac{1}{a}\left(b\frac{1}{s-\frac{1}{a}}s\tilde{\varepsilon}(s) - \frac{1}{s-\frac{1}{a}}\tilde{\varepsilon}(s)\right.\\ &\quad \left. + \ln\frac{a}{b}\left(b\frac{1}{\ln(bs)}\frac{1}{s-\frac{1}{a}}s\tilde{\varepsilon}(s) - \frac{1}{\ln(bs)}\frac{1}{s-\frac{1}{a}}\tilde{\varepsilon}(s)\right)\right).\end{aligned} \qquad [3.45]$$

The inversion of [3.45] could be easily obtained if we use a function defined as

$$f(t) = \mathcal{L}^{-1}\left[\frac{1}{\ln s}\right](t) = \int_0^\infty \frac{t^{u-1}}{\Gamma(u)}\mathrm{d}u,\quad t>0, \qquad [3.46]$$

see [ERD 53], and the convolution theorem. The result of the inversion of [3.45] is

$$\sigma(t) = E\frac{b}{a}\left(\mathrm{e}^{\frac{t}{a}} * \left(\varepsilon^{(1)}(t) - \frac{1}{b}\varepsilon(t)\right) + \ln\frac{a}{b}F(t) * \left(\varepsilon^{(1)}(t) - \frac{1}{b}\varepsilon(t)\right)\right),\quad t>0, \qquad [3.47]$$

where $*$ denotes the convolution ($f(t) * g(t) = \int_0^t f(\tau)\, g(t-\tau)\, \mathrm{d}\tau$) and

$$F(t) = \frac{1}{b}\mathrm{e}^{\frac{t}{a}} * \left( \int_0^\infty \frac{1}{\Gamma(u)} \left(\frac{t}{b}\right)^{u-1} \mathrm{d}u \right), \quad t > 0.$$

Using the integration by parts and conditions $\varepsilon(0) = 0$, $F(0) = 0$, in [3.47], we obtain

$$\begin{aligned} \sigma(t) &= E\frac{b}{a}\Bigg( \varepsilon(t) + \left(\frac{1}{a} - \frac{1}{b}\right) \mathrm{e}^{\frac{t}{a}} * \varepsilon(t) \\ &\quad + \ln\frac{a}{b} \left( \left(\frac{1}{a} - \frac{1}{b}\right) F(t) + \frac{1}{b} \int_0^\infty \frac{1}{\Gamma(u)} \left(\frac{t}{b}\right)^{u-1} \mathrm{d}u \right) * \varepsilon(t) \Bigg), \quad t > 0. \end{aligned} \qquad [3.48]$$

By solving [3.43] with respect to $\tilde{\varepsilon}$, we get

$$\tilde{\varepsilon}(s) = \frac{\tilde{\sigma}(s)}{E} \frac{\ln(bs)}{\ln(as)} \frac{as-1}{bs-1}, \quad \operatorname{Re} s > 0. \qquad [3.49]$$

Using the same procedure as in the case of the inversion of [3.48], we obtain

$$\begin{aligned} \varepsilon(t) &= \frac{1}{E}\frac{a}{b}\Bigg( \sigma(t) + \left(\frac{1}{b} - \frac{1}{a}\right) \mathrm{e}^{\frac{t}{b}} * \sigma(\tau) \\ &\quad + \ln\frac{b}{a} \left( \left(\frac{1}{b} - \frac{1}{a}\right) F(t) + \frac{1}{b} \int_0^\infty \frac{1}{\Gamma(u)} \left(\frac{t}{a}\right)^{u-1} \mathrm{d}u \right) * \sigma(t) \Bigg), \quad t > 0. \end{aligned} \qquad [3.50]$$

Now we examine the asymptotic behavior of the solution to [3.48] and [3.50] for a special type of applied stress (strain). Suppose that

$$\sigma(t) = \sigma_0 H(t), \quad t > 0,$$

where $H$ is the Heaviside function. This corresponds to the creep test. Then, $\tilde{\sigma}(s) = \frac{\sigma_0}{s}$, so that [3.49] yields

$$\tilde{\varepsilon}(s) = \frac{\sigma_0}{E} \frac{1}{s} \frac{\ln(bs)}{\ln(as)} \frac{as-1}{bs-1}, \quad \operatorname{Re} s > 0. \qquad [3.51]$$

Now, if $\lim_{t\to\infty}\varepsilon(t)$ exists, it is given by $\lim_{t\to\infty}\varepsilon(t) = \lim\limits_{s\to 0}(s\tilde{\varepsilon}(s))$ (see [DOE 67]). Using this in [3.51], it follows that

$$\lim_{t\to\infty}\varepsilon(t) = \frac{\sigma_0}{E}. \qquad [3.52]$$

Also, if $\lim_{t\to 0^+}\varepsilon(t) = \varepsilon(0^+)$ exists, it is given by $\varepsilon(0^+) = \lim_{s\to\infty}(s\tilde{\varepsilon}(s))$. Therefore,

$$\lim_{t\to 0}\varepsilon(t) = \varepsilon(0^+) = \frac{\sigma_0}{E}\frac{a}{b} < \frac{\sigma_0}{E} = \lim_{t\to\infty}\varepsilon(t). \qquad [3.53]$$

Since $\mathcal{L}\left[\varepsilon^{(1)}(t)\right](s) = s\tilde{\varepsilon}(s) - \varepsilon(0^+)$, we have

$$\mathcal{L}\left[\varepsilon^{(1)}(t)\right](s) = \frac{\sigma_0}{E}\frac{\ln(bs)}{\ln(as)}\frac{as-1}{bs-1} - \frac{\sigma_0}{E}\frac{a}{b}, \quad \operatorname{Re} s > 0. \qquad [3.54]$$

From [3.54], it follows

$$\lim_{t\to\infty}\varepsilon^{(1)}(t) = \lim_{s\to 0}\left(s\mathcal{L}\left[\varepsilon^{(1)}(t)\right](s)\right) = 0 \text{ and}$$
$$\lim_{t\to 0^+}\varepsilon^{(1)}(t) = \lim_{s\to\infty}\left(s\mathcal{L}\left[\varepsilon^{(1)}(t)\right](s)\right) = \infty.$$

The function $\varepsilon$ is shown in Figure 3.1 on the basis of the estimates [3.52] and [3.53].

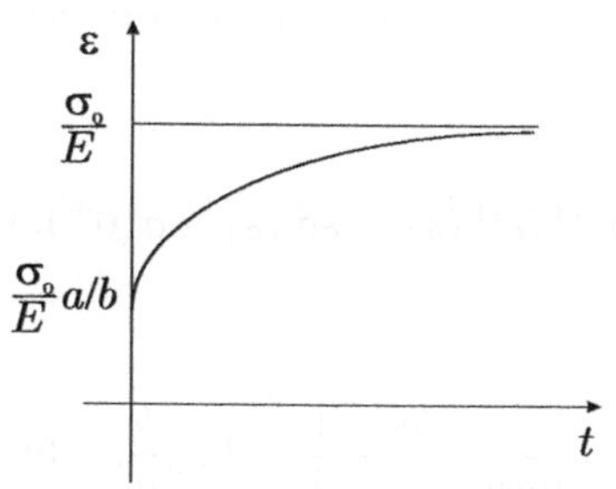


**Figure 3.1.** *Creep for the model [3.42]*

Next we assume that

$$\varepsilon(t) = \varepsilon_0 H(t), \quad t > 0. \qquad [3.55]$$

This corresponds to the stress relaxation test. From [3.44], we obtain

$$\tilde{\sigma}(s) = E\varepsilon_0 \frac{1}{s} \frac{\ln(as)}{\ln(bs)} \frac{bs-1}{as-1}, \quad \operatorname{Re} s > 0. \tag{3.56}$$

Again $\lim_{t\to\infty} \sigma(t) = \lim_{s\to 0} (s\tilde{\sigma}(s))$, so that

$$\lim_{t\to\infty} \sigma(t) = E\varepsilon_0.$$

Also, for $t \to 0^+$, we have

$$\sigma(0^+) = \lim_{t\to 0^+} \sigma(t) = \lim_{s\to\infty} (s\tilde{\sigma}(s)) = E\varepsilon_0 \frac{b}{a} > E\varepsilon_0 = \lim_{t\to\infty} \sigma(t). \tag{3.57}$$

In Figure 3.2, the function $\sigma$ is shown qualitatively using asymptotic estimates.

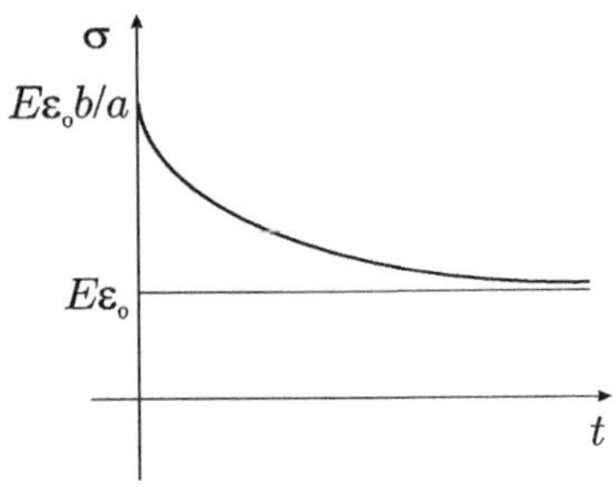


**Figure 3.2.** *Stress relaxation for the model [3.42]*

Since we have that $\mathcal{L}\left[\sigma^{(1)}(t)\right](s) = s\tilde{\sigma}(s) - \sigma(0^+)$, by using [3.56] and [3.57], we obtain

$$\mathcal{L}\left[\sigma^{(1)}(t)\right](s) = E\varepsilon_0 \frac{\ln(as)}{\ln(bs)} \frac{bs-1}{as-1} - E\varepsilon_0 \frac{b}{a}, \quad \operatorname{Re} s > 0.$$

Since

$$\lim_{s\to 0} \left(s\mathcal{L}\left[\sigma^{(1)}(t)\right](s)\right) = 0 \text{ and } \lim_{s\to\infty} \left(s\mathcal{L}\left[\sigma^{(1)}(t)\right](s)\right) = -\infty,$$

it follows that

$$\lim_{t\to 0+} \sigma^{(1)}(t) = -\infty \text{ and } \lim_{t\to\infty} \sigma^{(1)}(t) = 0.$$

In drawing $\sigma$ in Figure 3.2, we used these facts.

Note that as a consequence of the dissipation inequality (condition $a \leq b$), the stress relaxes ($\lim_{t\to\infty} \sigma(t) \leq \sigma(0^+)$), as expected since it is an inherent property of real materials. However, regarding [3.55], we should be aware that the inertial limitations of physical testing devices prevent instantaneous strain applications and very fast ramp times are intractable due to issues such as overshoot, vibration and poorly approximated strain histories (see [TRO 12]).

3.1.4.2. *Distributed-order model of the Kelvin–Voigt type*

Consider the distributed-order model of the Kelvin–Voigt type [3.25]

$$\sigma(t) = E \int_0^1 \tau_0^\gamma \, {}_0\mathrm{D}_t^\gamma \varepsilon(t) \, \mathrm{d}\gamma, \quad t > 0, \qquad [3.58]$$

along with the restrictions [3.27]. The constitutive equation [3.58] shows that the stress $\sigma$ depends on all time derivatives of the strain ${}_0\mathrm{D}_t^\gamma \varepsilon$, $0 < \gamma < 1$. It has a rheological model interpretation as a parallel connection of infinitely many spring-pot elements, i.e. elements whose stress–strain relation is determined by $\sigma_{sp} = E\tau_0^\gamma \, {}_0\mathrm{D}_t^\gamma \varepsilon$. Applying the Laplace transform to [3.58], we obtain

$$\tilde{\sigma}(s) = E\tilde{\varepsilon}(s) \frac{\tau_0 s - 1}{\ln(\tau_0 s)}, \quad \operatorname{Re} s > 0. \qquad [3.59]$$

The convolution theorem used in [3.59] leads to

$$\sigma(t) = \frac{E}{\tau_0} f\left(\frac{t}{\tau_0}\right) * \left(\tau_0 \varepsilon^{(1)}(t) - \varepsilon(t)\right), \quad t > 0,$$

with $f$ given by [3.46]. To determine $\varepsilon$ as a function of $\sigma$, we solve [3.59] for $\tilde{\varepsilon}$ so that

$$\tilde{\varepsilon}(s) = \frac{1}{E}\tilde{\sigma}(s) \frac{\ln(\tau_0 s)}{\tau_0 s - 1}, \quad \operatorname{Re} s > 0. \qquad [3.60]$$

The inversion of [3.60] leads to

$$\varepsilon(t) = \frac{1}{E\tau_0}\left(E_1\left(\frac{t}{\tau_0}\right)\mathrm{e}^{\frac{t}{\tau_0}}\right) * \sigma(t), \quad t > 0,$$

where $E_1(t) = \int_0^\infty \frac{1}{u}\mathrm{e}^{-u}\mathrm{d}u$ is the exponential integral (see [ABR 64]). The asymptotic behavior for the stress with the strain given by [3.55] is

$$\sigma(0^+) = \lim_{t\to 0+}\sigma(t) = \lim_{s\to\infty}(s\tilde{\sigma}(s)) = \infty \text{ and } \lim_{t\to\infty}\sigma(t) = \lim_{s\to 0}(s\tilde{\sigma}(s)) = 0.$$

Thus, the model [3.58] has complete stress relaxation, i.e. it has the fluid-like behavior.

### 3.1.5. *Variable-order fractional derivative: application to stress relaxation problem*

Suppose that a constitutive equation of a viscoelastic material is given as

$$\sigma(t) = E\,{}_0\mathrm{D}_t^{\alpha(t)}\varepsilon(t), \quad t > 0, \qquad [3.61]$$

where the operator of the variable-order fractional derivative ${}_0\mathrm{D}_t^{\alpha(t)}$ is given by [2.33] and $\alpha(t) \in (0,1)$, $t \geq 0$, is a given function, as stated in [ATA 13a]. The material corresponding to [3.61] changes its properties with time. For example, if $\alpha$ is a decreasing function, satisfying $\alpha(0)$ close to 1 and $\alpha(T)$, $T > 0$ close to 0, then [3.61] represents a solidifying material, changing its properties from those corresponding to a viscous liquid to those corresponding to a solid. In stress relaxation experiments, we take $\varepsilon = H$. As a test example we take $E = 1$, $\alpha(t) = B\mathrm{e}^{-ct}$, with $B = 0.9$ and $c = 0.6$. We used $N = M = 5$ in evaluating ${}_0\mathrm{D}_t^{\alpha(t)}\varepsilon$ in [3.61] by the expansion formula [2.35]. The results are shown in Figure 3.3.

As could be seen, unlike the case of the constant-order fractional derivative, here we have a change of sign in the values of stress due to the initial viscous behavior of the material.

It is an open question what restrictions the dissipation inequality imposes upon the constitutive equation [3.61]. The inequality that must be satisfied becomes

$$E\int_0^T \left({}_0\mathrm{D}_t^{\alpha(t)}\varepsilon(t)\right)\varepsilon^{(1)}(t)\,\mathrm{d}t \geq 0,$$

for arbitrary $\varepsilon$ and $T$.

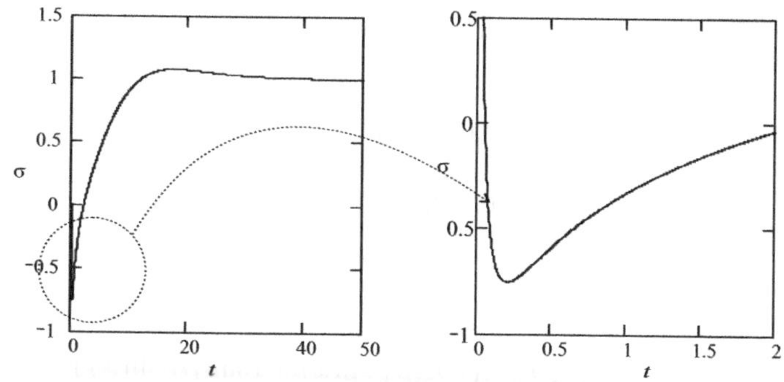

**Figure 3.3.** *Stress relaxation curve corresponding to [3.61]*

### 3.1.6. *Linear constitutive equation with fractional derivatives of complex order*

Consider the constitutive equation of the form

$$\sigma(t) = \left(b_1 \, {}_0\mathrm{D}_t^{\beta_1} + b_2 \, {}_0\mathrm{D}_t^{\beta_2}\right) \varepsilon(t), \quad t > 0, \qquad [3.62]$$

where $\beta_1, \beta_2 \in \mathbb{C}$ and $b_1, b_2 \in \mathbb{R}$. We examine the restrictions on coefficients in [3.62] so that it is the acceptable constitutive model. We follow the presentation of [ATA 12c]. Applying the Laplace transform to [3.62], we obtain

$$\tilde{\sigma}(s) = \left(b_1 \, s^{\beta_1} + b_2 \, s^{\beta_2}\right) \tilde{\varepsilon}(s), \quad \operatorname{Re} s > 0. \qquad [3.63]$$

For given $\varepsilon$, the stress $\sigma$ will be obtained from the inverse Laplace transform of the right-hand side in [3.63], i.e.

$$\sigma(t) = \mathcal{L}^{-1}\left[b_1 \, s^{\beta_1} + b_2 \, s^{\beta_2}\right](t) * \varepsilon(t), \quad t > 0.$$

Here, the question arises whether the stress $\sigma$, obtained in this way, will be a real-valued function of real argument, if $\varepsilon$ is a real-valued function of real argument. In other words, we have to determine restrictions on parameters in the constitutive equation [3.62] in such a way that the inverse Laplace transform of the complex function $b_1 \, s^{\beta_1} + b_2 \, s^{\beta_2}$ is a real-valued function. Afterwards, we will look for additional restrictions on those parameters that follow from the dissipation inequality.

The key role in deciding whether $\sigma$ will be a real-valued function has the following theorem (see [DOE 55, volume I, p. 293, theorem 2]): let $f(s) = \mathcal{L}[F](s)$. If the function $f$ is real valued for all real $s$ in the half-plane of convergence right from some real $x_0$, then the function $F$ is also real valued almost everywhere. Alternatively, it is enough that $f$ is real valued at a sequence of equidistant points on the real axis in order that the function $F$ is real valued almost everywhere.

Therefore, we have to find restrictions on $b_1, b_2, \beta_1, \beta_2$, which provide that $b_1\, s^{\beta_1} + b_2\, s^{\beta_2}$ belongs to $\mathbb{R}$ for all real $s > x_0$, for some $x_0 \in \mathbb{R}$. Without loss of generality, we can assume $x_0 > 0$. For that purpose, set $s = x \in \mathbb{R}$, $\beta_i = \beta_i^{\mathrm{re}} + \mathrm{i}\beta_i^{\mathrm{im}}$, $\beta_i^{\mathrm{re}}, \beta_i^{\mathrm{im}} \in \mathbb{R}$, $i = 1, 2$. Then, we have

$$\begin{aligned} b_1\, x^{\beta_1} + b_2\, x^{\beta_2} &= b_1\, e^{\beta_1 \ln x} + b_2\, e^{\beta_2 \ln x} \\ &= b_1\, x^{\beta_1^{\mathrm{re}}} \left(\cos(\beta_1^{\mathrm{im}} \ln x) + \mathrm{i} \sin(\beta_1^{\mathrm{im}} \ln x)\right) \\ &\quad + b_2\, x^{\beta_2^{\mathrm{re}}} \left(\cos(\beta_2^{\mathrm{im}} \ln x) + \mathrm{i} \sin(\beta_2^{\mathrm{im}} \ln x)\right). \end{aligned}$$

If $b_1\, x^{\beta_1} + b_2\, x^{\beta_2}$ has to be real valued for all $x > x_0$, its imaginary part has to vanish for all $x > x_0$, and we obtain

$$b_1\, x^{\beta_1^{\mathrm{re}}} \sin(\beta_1^{\mathrm{im}} \ln x) + b_2\, x^{\beta_2^{\mathrm{re}}} \sin(\beta_2^{\mathrm{im}} \ln x) = 0, \quad x > x_0. \qquad [3.64]$$

We immediately come to the conclusion that the following set of conditions guarantees equality [3.64]:

$$b_1 = b_2, \quad \beta_1^{\mathrm{re}} = \beta_2^{\mathrm{re}}, \quad \beta_1^{\mathrm{im}} = -\beta_1^{\mathrm{im}}.$$

Thus, we have obtained a sufficient condition for the well-posedness of the linear fractional complex-order constitutive equation [3.62].

THEOREM 3.1.– Let $b_1, b_2 \in \mathbb{R}$ and $\beta_1, \beta_2 \in \mathbb{C}$, $\beta_i = \beta_i^{\mathrm{re}} + \mathrm{i}\beta_i^{\mathrm{im}}$, $\beta_i^{\mathrm{re}}, \beta_i^{\mathrm{im}} \in \mathbb{R}$, $i = 1, 2$. Suppose that $b_1 = b_2$, $\beta_1^{\mathrm{re}} = \beta_2^{\mathrm{re}}$ and $\beta_1^{\mathrm{im}} = -\beta_1^{\mathrm{im}}$. Then, the inverse Laplace transform of $b_1\, s^{\beta_1} + b_2\, s^{\beta_2}$ is a real-valued function almost everywhere.

PROOF.– The claim follows from [DOE 55, volume I, p. 293, Satz 2] and [3.64]. ■

REMARK 3.5.– As a result of theorem 3.1, we have that if the constitutive equation contains a sum of the fractional derivatives of the strain whose orders are complex conjugated numbers, then the corresponding model is physically acceptable. Moreover, according to the above consideration, we can take an arbitrary linear combination of such sums. Even more, we can also allow the same type of fractional derivatives of the strain. Thus, we can consider the following constitutive equation:

$$\sigma(t) + \sum_{i=1}^{N} a_i \left({}_0\mathrm{D}_t^{\alpha_i} + {}_0\mathrm{D}_t^{\bar{\alpha}_i}\right) \sigma(t) = \varepsilon(t) + \sum_{j=1}^{M} b_j \left({}_0\mathrm{D}_t^{\beta_j} + {}_0\mathrm{D}_t^{\bar{\beta}_j}\right) \varepsilon(t), \quad t > 0,$$

where $a_i, b_j \in \mathbb{R}$ and $\alpha_i, \beta_j \in \mathbb{C}$, $i = 1, \ldots, N$, $j = 1, \ldots, M$.

In fact, we will prove in the following that the set of conditions in theorem 3.1 is the only one that implies the claim. Indeed, suppose that $b_1, b_2 \in \mathbb{R}$ and $\beta_i^{\text{re}}, \beta_i^{\text{im}} \in \mathbb{R}$, $i = 1, 2$, are arbitrary. We will show that if the parameters in [3.64] do not satisfy theorem 3.1, then [3.64] is violated for some $x > x_0$. First assume that $b_1 \neq b_2$ and $b_1, b_2 > 0$ (the cases $b_1 \neq b_2$ and $b_1, b_2 < 0$ are analogous). Since sine is a periodic function, it follows that $\sin(\beta_1^{\text{im}} \ln x)$ and $\sin(\beta_2^{\text{im}} \ln x)$ will be of the same sign for some $x$ large enough. But then for such $x$, the sum in [3.64] does not vanish. On the other hand, if $b_1$ is positive and $b_2$ is negative (or vice versa), again for some large $x$, $\sin(\beta_1^{\text{im}} \ln x)$ and $\sin(\beta_2^{\text{im}} \ln x)$ will take different signs, and we will sum two positive or two negative numbers, which in both cases differ from zero.

Let us consider the following constitutive equation:

$$\sum_{i=0}^{N} a_i \left( {}_0\mathrm{D}_t^{\alpha_i} + {}_0\mathrm{D}_t^{\bar{\alpha}_i} \right) \sigma(t) = \sum_{j=0}^{M} b_j \left( {}_0\mathrm{D}_t^{\beta_j} + {}_0\mathrm{D}_t^{\bar{\beta}_j} \right) \varepsilon(t), \quad t > 0, \qquad [3.65]$$

with $a_i, b_j \in \mathbb{R}$ and $\alpha_i, \beta_j \in \mathbb{C}$, $i = 0, \ldots, N$, $j = 0, \ldots, M$.

Equation [3.65] also generalizes the well-known constitutive equations in viscoelasticity:

1) The classical Zener model is obtained for $a_0 = b_0 = \frac{1}{2}$, $a_2 = \frac{a}{2}$, $b_2 = \frac{b}{2}$, $\alpha_0 = \beta_0 = 0$, $\alpha_1 = \beta_1 = 1$, and all other parameters vanishing.

2) The fractional Zener model is obtained for $a_0 = b_0 = \frac{1}{2}$, $a_1 = \frac{a}{2}$, $b_1 = \frac{b}{2}$, $\alpha_0 = \beta_0 = 0$, $\alpha_1 = \beta_1 = \alpha$, with $\alpha \in (0, 1)$, and all other parameters vanishing.

As we showed earlier, in both classical and fractional Zener models, the only restriction on coefficients in the constitutive equation is $a \leq b$, which follows from the Clausius–Duhem inequality applied for isothermal conditions.

For the sake of simplicity, we will start by considering the simplest case with both sums in [3.65] containing only one term, i.e.

$$a \left( {}_0\mathrm{D}_t^{\alpha} + {}_0\mathrm{D}_t^{\bar{\alpha}} \right) \sigma(t) = b \left( {}_0\mathrm{D}_t^{\beta} + {}_0\mathrm{D}_t^{\bar{\beta}} \right) \varepsilon(t), \quad t > 0, \qquad [3.66]$$

with $\alpha, \beta \in \mathbb{C} \backslash \mathbb{R}$ and $a, b \in \mathbb{R}$. The case $\alpha, \beta \in \mathbb{R}$ belongs to the type of problems studied earlier.

We apply the Fourier transform with respect to $t$ to [3.66] and obtain

$$a \left( (\mathrm{i}\omega)^{\alpha} + (\mathrm{i}\omega)^{\bar{\alpha}} \right) \hat{\sigma}(\omega) = b \left( (\mathrm{i}\omega)^{\beta} + (\mathrm{i}\omega)^{\bar{\beta}} \right) \hat{\varepsilon}(\omega), \quad \omega \in [0, \infty).$$

The complex modulus $\hat{E}$ becomes

$$\hat{E}(\omega) = \frac{b\left((\mathrm{i}\omega)^{\beta} + (\mathrm{i}\omega)^{\bar{\beta}}\right)}{a\left((\mathrm{i}\omega)^{\alpha} + (\mathrm{i}\omega)^{\bar{\alpha}}\right)}, \quad \omega \in [0, \infty).$$

The dissipation inequality requires that condition [3.2] should be satisfied. For that purpose, introduce

$$\hat{E}_{nom}(\omega) = E'_{nom}(\omega) + \mathrm{i}E''_{nom}(\omega) = \hat{E}(\omega)\left|a\left((\mathrm{i}\omega)^{\alpha} + (\mathrm{i}\omega)^{\bar{\alpha}}\right)\right|^{2}, \quad \omega \in [0, \infty).$$

Then, the sign of the real and imaginary parts of $\hat{E}$ and $\hat{E}_{nom}$ coincides, and we will examine $\hat{E}_{nom}$ instead of $\hat{E}$. After a simple calculation, we obtain (by setting $\beta = A + \mathrm{i}B$ and $\alpha = C + \mathrm{i}D$, $A, B, C, D \in \mathbb{R}$)

$$\begin{aligned} E'_{nom}(\omega) &= 2\cos\left((B - D)\ln\omega + (A - C)\frac{\pi}{2}\right)\cosh\left((B + D)\frac{\pi}{2}\right) \\ &\quad + 2\cos\left((B + D)\ln\omega + (A + C)\frac{\pi}{2}\right)\cosh\left((B - D)\frac{\pi}{2}\right), \end{aligned}$$

which should be non-negative for all $\omega \geq 0$. A similar calculation can be performed for the imaginary part of $\hat{E}_{nom}$.

We note the following: since cosine is a periodic function and $\ln\omega$ ranges through all of $\mathbb{R}$ for $\omega \geq 0$, a necessary condition for $E'_{nom}$ to remain greater than or equal to zero is that $B - D = B + D = 0$, i.e. $B = D = 0$. However, this condition implies that the imaginary parts of $\alpha$ and $\beta$ vanish, or in other words, that the orders of fractional derivatives in the constitutive equation [3.66] have to be real and could not be of a complex order.

As a conclusion, we may state that according to theorem 3.1, a linear fractional complex-order viscoelastic constitutive equation should contain pairwise complex conjugated fractional derivatives of stress and strain, but on the other hand, for [3.66], the second law of thermodynamics implies that such an equation is not acceptable.

One possibility for overcoming this problem could be in relaxing condition [3.2], and requiring that both inequalities hold for $\omega$ in some interval of positive numbers instead of in all of $\mathbb{R}_{+}$. For a different approach to complex-order constitutive equations of linear viscoelasticity, see [MAK 91, MAK 92, MAK 93a, MAK 93b].

## 3.2. Thermodynamical restrictions via the internal variable theory

We present another approach to the problem of obtaining restrictions on the parameters in constitutive equations with fractional derivatives, following the presentation of [ATA 02b]. The method is based on the internal variable theory and we will apply it to the fractional Zener model [3.32].

First, we review some notions of the internal variable theory for the constitutive equation corresponding to a standard linear viscoelastic body (Zener model) that under uniaxial, isothermal deformation has a stress–strain relation of the form

$$\sigma(t) + a\sigma^{(1)}(t) = E\left(\varepsilon(t) + b\varepsilon^{(1)}(t)\right), \quad t > 0. \qquad [3.67]$$

The second law of thermodynamics implies that the following restrictions on the constants must be satisfied in [3.67]

$$E > 0, \quad b \geq a > 0. \qquad [3.68]$$

We recall the description of the constitutive relation [3.67] and of the thermodynamic stability conditions [3.68] in the context of an internal variable theory.

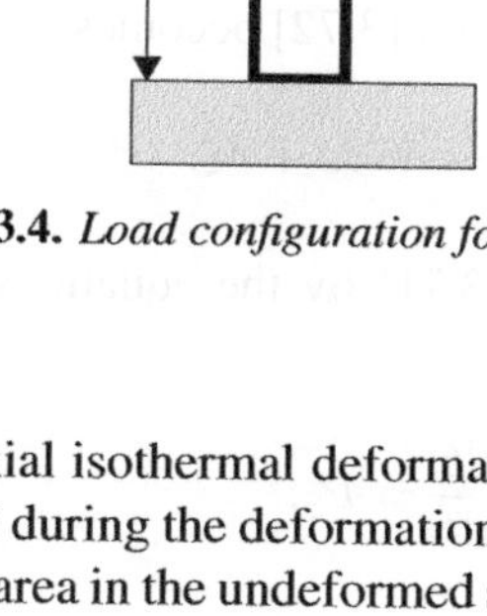


**Figure 3.4.** *Load configuration for a rod*

Consider a rod in the uniaxial isothermal deformation (Figure 3.4). The length is $L$ in the undeformed state and $l$ during the deformation. The rod is loaded by the force $P$, and $A$ is the cross-sectional area in the undeformed state. Stress and strain are given by

$$\sigma = \frac{P}{A} \text{ and } \varepsilon = \frac{l}{L} - 1. \qquad [3.69]$$

We describe a state of the rod by two variables: strain $\varepsilon$ and internal variable $\xi$. The equilibrium state of the unloaded body corresponds to

$$\varepsilon = 0, \quad \xi = 0. \tag{3.70}$$

The internal energy $U$, the entropy $S$ and the free energy $U - TS$ are all functions of $\varepsilon$ and $\xi$ and we may write the Gibbs equation for the free energy $U - TS$ in the form

$$\frac{\mathrm{d}}{\mathrm{d}t}(U - TS) = \sigma V \frac{\mathrm{d}\varepsilon}{\mathrm{d}t} - \Theta \frac{\mathrm{d}\xi}{\mathrm{d}t}, \tag{3.71}$$

where $T$ is the temperature, assumed to be constant, $\Theta$ is the "force" associated with the internal variable $\xi$ so that $\Theta \frac{\mathrm{d}\xi}{\mathrm{d}t}$ is the power of the force $\Theta$ and $V$ is the volume of the body, which is assumed to be constant. We assume that both $\sigma$ and $\Theta$ are linear in the variables $\varepsilon$ and $\xi$ so that

$$\sigma = E_\infty \varepsilon + \beta\xi \;\text{ and }\; \Theta = \gamma\varepsilon + \delta\xi, \tag{3.72}$$

where $E_\infty$, $\beta$, $\gamma$ and $\delta$ are constants. From [3.72], we conclude that $\xi$ is proportional to the difference between the instantaneous and equilibrium stress, denoted by $E_\infty \varepsilon$. Note that with [3.72], the force $P$ is given as

$$P = A\left(E_\infty \varepsilon + \beta\xi\right). \tag{3.73}$$

The integrability of the free energy ($\frac{\partial}{\partial \xi} V\sigma = \frac{\partial}{\partial \varepsilon}\Theta$) requires

$$V\beta = -\gamma. \tag{3.74}$$

Therefore, with [3.74], equation [3.72] becomes

$$\sigma = E_\infty \varepsilon + \beta\xi \;\text{ and }\; \Theta = -V\beta\varepsilon + \delta\xi. \tag{3.75}$$

We replace $\frac{\mathrm{d}U}{\mathrm{d}t}$ and $\frac{\mathrm{d}\xi}{\mathrm{d}t}$ in [3.71] by the equations of balance of energy and of internal variable

$$\frac{\mathrm{d}U}{\mathrm{d}t} = \dot{Q} + \sigma V \frac{\mathrm{d}\varepsilon}{\mathrm{d}t} \;\text{ and }\; \frac{\mathrm{d}\xi}{\mathrm{d}t} = P_\xi, \tag{3.76}$$

where $\dot{Q}$ is the heating and $P_\xi$ is the production of the internal variable $\xi$. Thus, from [3.74], we obtain an equation of balance of entropy in the form

$$\frac{\mathrm{d}S}{\mathrm{d}t} - \frac{\dot{Q}}{T} = \frac{\Theta}{T} P_\xi \geq 0, \tag{3.77}$$

where we have indicated that the entropy production is non-negative. The inequality [3.77] may be satisfied by setting $P_\xi = \gamma\Theta$ with a non-negative coefficient $\gamma$. Therefore, we assume

$$\xi^{(1)} = \gamma\Theta, \quad \gamma \geq 0. \qquad [3.78]$$

Elimination of $\xi$ and $\Theta$ between [3.75] and [3.78] leads to

$$\sigma + \frac{1}{-\delta\gamma}\sigma^{(1)} = \left(E_\infty + V\frac{\beta^2}{\delta}\right)\varepsilon + \frac{E_\infty}{-\gamma\delta}\varepsilon^{(1)}. \qquad [3.79]$$

which is of the form [3.67]. We identify the coefficients in [3.67] with the coefficients of the internal variable theory as follows

$$a = \frac{1}{-\delta\gamma}, \qquad E = E_\infty + V\frac{\beta^2}{\delta}, \qquad b = \frac{E_\infty}{-\gamma\delta}\frac{1}{E_\infty + \frac{V\beta^2}{\delta}}. \qquad [3.80]$$

Thus, the viscoelastic constitutive equation [3.67] may be viewed as a consequence of the internal variable theory. It results upon the elimination of all explicit reference to the internal variable field.

The internal variable theory provides itself for an easy evaluation of thermodynamic stability conditions. Indeed, elimination of $\dot{Q}$ between [3.76] and [3.77] leads to inequality

$$\frac{\mathrm{d}}{\mathrm{d}t}(U - TS - Pl) \leq l\frac{\mathrm{d}P}{\mathrm{d}t}. \qquad [3.81]$$

Thus, for a constant force ($\frac{\mathrm{d}P}{\mathrm{d}t} = 0$), the Gibbs free energy $U - TS - Pl$ tends to a minimum and it assumes that the minimum is in the equilibrium. The matrix of the second derivatives of the Gibbs free energy must be a positive definite. By [3.71] and [3.75], we must have

$$\begin{bmatrix} E_\infty & \beta V \\ \beta V & -\delta \end{bmatrix} \text{ positive definite, or } E_\infty > 0, \quad -E_\infty\delta - \beta^2 V > 0, \qquad [3.82]$$

so that $E_\infty > 0$ and $\delta < 0$ or $E_\infty + \frac{\beta^2 V}{\delta} > 0$. By using this in [3.80], we obtain conditions [3.68].

We apply the procedure to two specific examples.

### 3.2.1. *Case I*

Consider a constitutive equation

$$\left(1 + a\,{}_0\mathrm{D}_t^{\alpha}\right)\sigma\left(t\right) = E\left(1 + b\,{}_0\mathrm{D}_t^{\alpha}\right)\varepsilon\left(t\right) + g\left(\varepsilon\left(t\right)\right), \quad t > 0, \qquad [3.83]$$

where $g$ is the given functional. If $g \equiv 0$, then the constitutive equation [3.83] reduces to the fractional Zener model [3.32]. Our aim is to include this type of equation into the internal variable framework. We repeat the relevant system of equations [3.75]–[3.77]

$$\begin{aligned} \sigma &= E_{\infty}\varepsilon + \beta\xi, \quad \Theta = -V\beta\varepsilon + \delta\xi, \\ \xi^{(1)} &= P_{\xi}, \quad \frac{\mathrm{d}S}{\mathrm{d}t} - \frac{1}{T}\frac{\mathrm{d}Q}{\mathrm{d}t} = \frac{\Theta}{T}P_{\xi} \geq 0, \end{aligned} \qquad [3.84]$$

so that relation between $\sigma$ and $\varepsilon$, given by [3.83], must be satisfied. Suppose that the $\alpha$th derivative of $\xi$ is given by a constitutive equation for the internal variable as

$${}_0\mathrm{D}_t^{\alpha}\xi\left(t\right) = \gamma\Theta\left(t\right) + X\left(t\right), \quad t > 0, \qquad [3.85]$$

where $X$ is a functional that will be specified later. Equation [3.85] is a generalization of [3.78] and it is of central importance in the analysis that follows. Equation [3.78] is obtained if we take $\alpha = 1$ and $X = 0$. The special case of [3.85] with $X = 0$, $\alpha \neq 1$ was used for an internal variable description of the fractional derivative model in [PAP 97][1]. Our intention is to choose $X$ so that $[3.84]_4$ is satisfied. From $[3.75]_2$ and [3.85], it follows

$${}_0\mathrm{D}_t^{\alpha}\xi\left(t\right) + \left(-\alpha\delta\right)\xi\left(t\right) = K\varepsilon\left(t\right) + X\left(t\right), \quad t > 0, \qquad [3.86]$$

where $K = -\gamma V\beta$. Suppose that $X$ is given as

$$X\left(t\right) = {}_0\mathrm{D}_t^{\alpha}x\left(t\right) + \left(-\alpha\delta\right)x\left(t\right), \quad t > 0, \qquad [3.87]$$

where $x$ is another function. With [3.87], equation [3.86] could be written as

$${}_0\mathrm{D}_t^{\alpha}\left(\xi\left(t\right) - x\left(t\right)\right) + \left(-\alpha\delta\right)\left(\xi\left(t\right) - x\left(t\right)\right) = K\varepsilon\left(t\right), \quad t > 0. \qquad [3.88]$$

1 In [PAP 97], equation [3.85] with $X = 0$ is used in the context of finite deformations of a viscoelastic body.

The solution to fractional-order differential equation [3.88], for $t > 0$, becomes

$$\begin{aligned}\xi(t) - x(t) &= \frac{K}{\Gamma(1-\alpha)} \frac{E_{\alpha,\alpha}(\alpha\delta t^{\alpha})}{t^{1-\alpha}} \left[\int_0^t \frac{\xi(\tau) + x(\tau)}{(t-\tau)^{\alpha}} \mathrm{d}\tau\right]_{t=0} \\ &\quad + K \frac{E_{\alpha,\alpha}(\alpha\delta t^{\alpha})}{t^{1-\alpha}} * \varepsilon(t), \end{aligned} \qquad [3.89]$$

see [SAM 93, p. 837] or [POD 99], where $E_{\alpha,\beta}$ is the two-parameter Mittag-Leffler function. Using the two-parameter Mittag-Leffler-type function $e_{\alpha,\beta}$ and the fact that at $t = 0$ both $\xi$ and $x$ are equal to zero, we have that [3.89] becomes

$$\xi(t) = K e_{\alpha,\alpha}(t, -\alpha\delta) * \varepsilon(t) + x(t), \quad t > 0. \qquad [3.90]$$

We specify $x$ in the form

$$x(t) = K \mathrm{e}^{-\lambda t} * \left(\varepsilon(t) + e_{\alpha,\alpha}(t, -\alpha\delta) * \left(\Lambda\varepsilon(t) - \varepsilon^{(1)}(t)\right)\right), \quad t > 0, \qquad [3.91]$$

where $\lambda$ and $\Lambda$ are constants. Combining [3.90] and [3.91], it follows ($t > 0$)

$$\begin{aligned}\xi(t) &= K\left(e_{\alpha,\alpha}(t, -\alpha\delta) * \varepsilon(t) + \mathrm{e}^{-\lambda t} * (\varepsilon(t) + e_{\alpha,\alpha}(t, -\alpha\delta)\right. \\ &\quad \left. * \left(\Lambda\varepsilon(t) - \varepsilon^{(1)}(t)\right)\right)\bigg). \end{aligned} \qquad [3.92]$$

We use [3.92] to determine $\xi^{(1)}$, and thus $P_{\xi}$ (see $[3.84]_3$). The first derivative of $\xi$ becomes

$$\begin{aligned}\xi^{(1)}(t) &= K\left(\varepsilon(0) e_{\alpha,\alpha}(t, -\alpha\delta) + \varepsilon(t) + \Lambda e_{\alpha,\alpha}(t, -\alpha\delta) * \varepsilon(t)\right. \\ &\quad \left. - \lambda \mathrm{e}^{-\lambda t} * \left(\varepsilon(t) + e_{\alpha,\alpha}(t, -\alpha\delta) * \left(\Lambda\varepsilon(t) - \varepsilon^{(1)}(t)\right)\right)\right), \quad t > 0. \end{aligned} \qquad [3.93]$$

Since $\varepsilon(0) = 0$, with $K = -\gamma\beta V$, $\gamma > 0$, we have

$$\begin{aligned}\xi^{(1)}(t) &= \gamma\left(-\beta V\varepsilon(t) - \beta V\Lambda e_{\alpha,\alpha}(t, -\alpha\delta) * \varepsilon(t)\right. \\ &\quad \left. + \beta V\lambda \mathrm{e}^{-\lambda t} * \left(\varepsilon(t) + e_{\alpha,\alpha}(t, -\alpha\delta) * \left(\Lambda\varepsilon(t) - \varepsilon^{(1)}(t)\right)\right)\right), \quad t > 0. \end{aligned} \qquad [3.94]$$

Suppose that $\Lambda$ and $\lambda$ are selected so that (see [3.84])

$$-\beta V\Lambda = K\delta = -\gamma V\beta\delta, \quad \beta V\lambda = K\delta = -\gamma V\beta\delta, \tag{3.95}$$

or

$$\Lambda = \gamma\delta, \quad \lambda = -\gamma\delta. \tag{3.96}$$

Then, by using $\Theta = -V\beta\varepsilon + \delta\xi$ and [3.92], we have that [3.94] becomes ($t > 0$)

$$\begin{aligned}\xi^{(1)}(t) &= \gamma\left(-\beta V\varepsilon(t) + \delta K\left(e_{\alpha,\alpha}(t,-\alpha\delta) * \varepsilon(t) + \mathrm{e}^{-\lambda t} * \varepsilon(t)\right.\right.\\ &\quad \left.\left.+\, \mathrm{e}^{-\lambda t} * e_{\alpha,\alpha}(t,-\alpha\delta) * \left(\Lambda\varepsilon(t) - \varepsilon^{(1)}(t)\right)\right)\right)\\ &= \gamma(-\beta V\varepsilon(t) + \delta\xi(t)) = \gamma\Theta(t).\end{aligned} \tag{3.97}$$

Thus, with $\gamma > 0$, condition $[3.84]_4$ is satisfied. Note that [3.97] guarantees that the constitutive equation for the internal variable $\xi$ is local in time, since the right-hand side of [3.97] is a function of $\varepsilon$ and $\xi$.

We determine the form of the constitutive equation ($\sigma - \varepsilon$ relation) that follows from [3.86]. Combining $[3.84]_1$, [3.86], [3.87] and [3.91], for $t > 0$, we have

$$\begin{aligned}&{}_0\mathrm{D}_t^\alpha\left(\frac{\sigma(t) - E_\infty\varepsilon(t)}{\beta}\right) + (-\alpha\delta)\left(\frac{\sigma(t) - E_\infty\varepsilon(t)}{\beta}\right)(t)\\ &\quad = (-\alpha V\beta)\,\varepsilon(t) + {}_0\mathrm{D}_t^\alpha x(t) + (-\alpha\delta)\,x(t),\end{aligned} \tag{3.98}$$

where

$$x(t) = -\alpha V\beta\mathrm{e}^{-\lambda t} * \left(\varepsilon(t) + e_{\alpha,\alpha}(t,-\alpha\delta) * \left(\alpha\delta\varepsilon(t) - \varepsilon^{(1)}(t)\right)\right), \quad t > 0. \tag{3.99}$$

From [3.98], for $t > 0$, we obtain

$$\begin{aligned}\sigma(t) + \frac{1}{-\alpha\delta}\,{}_0\mathrm{D}_t^\alpha\sigma(t) &= \left(E_\infty + \frac{V\beta^2}{\delta}\right)\varepsilon(t) + \frac{E_\infty}{-\alpha\delta}\,{}_0\mathrm{D}_t^\alpha\varepsilon(t)\\ &\quad + \frac{\beta}{-\alpha\delta}\,{}_0\mathrm{D}_t^\alpha x(t) + \beta x(t),\end{aligned} \tag{3.100}$$

so that with

$$a = \frac{1}{-\alpha\delta}, \quad E = E_\infty + \frac{V\beta^2}{\delta}, \quad b = \frac{E_\infty}{-\alpha\delta}\frac{1}{E} = a\frac{E_\infty}{E}, \tag{3.101}$$

[3.100] becomes ($t > 0$)

$$(1 + a\,{}_0\mathrm{D}_t^\alpha)\,\sigma(t) = E\,(1 + b\,{}_0\mathrm{D}_t^\alpha)\,\varepsilon(t) + E\left(1 - \frac{b}{a}\right)\left(\frac{1}{a} + {}_0\mathrm{D}_t^\alpha\right) d(t), \tag{3.102}$$

where

$$d(t) = \mathrm{e}^{-\frac{t}{a}} * \left(\varepsilon(t) - e_{\alpha,\alpha}\left(t, \frac{1}{a}\right) * \left(\frac{1}{a}\varepsilon(t) + \varepsilon^{(1)}(t)\right)\right), \quad t > 0. \tag{3.103}$$

Note that for $\alpha = 1$, we have $e_{1,1}\left(t, \frac{1}{a}\right) = \mathrm{e}^{-\frac{t}{a}}$, $t > 0$, so that

$$\begin{aligned} d(t) &= \mathrm{e}^{-\frac{t}{a}} * \left(\varepsilon(t) - \mathrm{e}^{-\frac{t}{a}} * \left(\frac{1}{a}\varepsilon(t) + \varepsilon^{(1)}(t)\right)\right) \\ &= \mathrm{e}^{-\frac{t}{a}} * \left(\varepsilon(t) + \int_0^t \frac{\mathrm{d}}{\mathrm{d}u}\left(\mathrm{e}^{-\frac{u}{a}}\varepsilon(t-u)\right) \mathrm{d}u\right) \\ &= \mathrm{e}^{-\frac{t}{a}} * (\varepsilon(t) - \varepsilon(t)) = 0, \quad t > 0, \end{aligned} \tag{3.104}$$

and [3.102] reduces to [3.67].

### 3.2.2. *Case II*

Another generalization of the procedure with the internal variables may be formulated as follows. Instead of [3.75], we assume the following relations:

$$\sigma(t) = E_\infty\varepsilon(t) + \beta\,{}_0\mathrm{D}_t^{1-\alpha}\xi(t), \quad \Theta(t) = -V\beta\varepsilon(t) + \delta\,{}_0\mathrm{D}_t^{1-\alpha}\xi(t), \quad t > 0. \tag{3.105}$$

Also, we take [3.78] in the same form, i.e.

$$\xi^{(1)} = \gamma\Theta, \quad \gamma \geq 0. \tag{3.106}$$

We apply the operator ${}_0\mathrm{D}_t^\alpha$ to $[3.105]_1$. Note that for $0 < \alpha < 1$, $0 < \beta < 1$, $\alpha + \beta < 1$,

$${}_0\mathrm{D}_t^\beta\,{}_0\mathrm{D}_t^\alpha\xi(t) = {}_0\mathrm{D}_t^{\alpha+\beta}\xi(t) - \frac{t^{-1-\alpha}}{\Gamma(-\alpha)}\left[{}_0\mathrm{I}_t^{1-\beta}\xi(t)\right]_{t=0}, \quad t > 0, \tag{3.107}$$

see [KIL 06, p. 75] or [2.19]. Suppose that $\xi \in C^1([0,\infty))$ . Then, $\left[{}_0\mathrm{I}_t^{1-\beta}\xi(t)\right]_{t=0} = 0$, so that

$$ {}_0\mathrm{D}_t^{\beta}\,{}_0\mathrm{D}_t^{\alpha}\xi(t) = {}_0\mathrm{D}_t^{\alpha+\beta}\xi(t), \quad t>0. $$

Further, let $\beta = 1-\alpha$. Since $\xi \in C^1([0,\infty))$

$$ \lim_{\beta\to 1-\alpha} {}_0\mathrm{D}_t^{\alpha+\beta}\xi(t) = \xi^{(1)}(t), \quad t>0, $$

we have that [3.106] holds true. By applying ${}_0\mathrm{D}_t^{\alpha}$ to $[3.105]_1$ and by using [3.106], we obtain ($t>0$)

$$ \begin{aligned} \Theta(t) &= \frac{1}{\gamma\beta}\left({}_0\mathrm{D}_t^{\alpha}\sigma(t) - E_\infty\,{}_0\mathrm{D}_t^{\alpha}\varepsilon(t)\right), \\ {}_0\mathrm{D}_t^{1-\alpha}\xi(t) &= \frac{1}{\gamma\beta\delta}\left({}_0\mathrm{D}_t^{\alpha}\sigma(t) - E_\infty\,{}_0\mathrm{D}_t^{\alpha}\varepsilon(t)\right) + \frac{V\beta}{\delta}\varepsilon(t). \end{aligned} $$

Therefore, $[3.105]_1$ becomes

$$ \sigma(t) + \frac{1}{-\delta\gamma}\,{}_0\mathrm{D}_t^{\alpha}\sigma(t) = \left(E_\infty + \frac{V\beta^2}{\delta}\right)\varepsilon(t) + \frac{E_\infty}{-\gamma\delta}\,{}_0\mathrm{D}_t^{\alpha}\varepsilon(t), \quad t>0, \qquad [3.108] $$

which is of the form [3.32], if we identify the coefficients in [3.108] as in [3.80]. Condition [3.82] then leads to

$$ \left(1 + a\,{}_0\mathrm{D}_t^{\alpha}\right)\sigma(t) = E\left(1 + b\,{}_0\mathrm{D}_t^{\alpha}\right)\varepsilon(t), \quad t>0, \quad E>0, \;\; a \le b, \qquad [3.109] $$

as before. Thus, [3.105] corresponds to the fractional Zener model [3.109], when the internal variable $\xi$ and its corresponding "force" $\Theta$ are eliminated.

# Chapter 4

# Vibrations with Fractional Dissipation

In this chapter, we discuss linear and nonlinear vibrations of a single-degree-of-freedom mechanical system with fractional derivative type dissipation. Since we continue to use the Laplace transform in the framework of $\mathcal{S}'_+$ and $\mathcal{K}'_+$, we assume that all functions and distributions belong to $\mathcal{K}'_+$, so that their Laplace transforms are analytic functions in $\operatorname{Re} s > s_0$, for some $s_0 \geq 0$. Recall that in order for a function $f \in L^1_{loc}([0,\infty))$ to determine an exponentially bounded distribution (element of $\mathcal{K}'_+$), it is sufficient that $f$ is of the exponential growth. Note, for locally integrable and exponentially bounded functions, we work with the classical Laplace transform, i.e. the distributional and classical Laplace transforms coincide.

This chapter includes results from [ATA 11a, ATA 07a, ATA 10c, ATA 12d, ATA 13e, BAC 00, POP 13].

## 4.1. Linear vibrations with fractional dissipation

### 4.1.1. *Linear vibrations with the single fractional dissipation term*

Consider the equation of linear vibrations with the fractional dissipation term of order $\alpha \in (0,1)$

$$f^{(2)}(t) + \mu\, {}_0\mathrm{D}_t^{\alpha} f(t) + \omega^2 f(t) = 0, \quad t > 0, \qquad f(0) = f_0, \qquad f^{(1)}(0) = f_1, \qquad [4.1]$$

where $f_0$, $f_1$ and $\omega, \mu > 0$ are constants. The solutions to [4.1] corresponding to $\alpha = 0$ and $\alpha = 1$ become

$$f(t) = \begin{cases} f_0 \cos\left(t\sqrt{\omega^2+\mu}\right) + f_1 \frac{\sin\left(t\sqrt{\omega^2+\mu}\right)}{\sqrt{\omega^2+\mu}}, \\ \qquad \text{for } \alpha = 0,\ t > 0, \\ f_0 \cos\left(t\sqrt{\omega^2+\frac{\mu^2}{4}}\right) \mathrm{e}^{-\frac{\mu t}{2}} + \left(f_1 + \frac{\mu}{2} f_0\right) \frac{\sin\left(t\sqrt{\omega^2+\frac{\mu^2}{4}}\right)}{\sqrt{\omega^2+\frac{\mu^2}{4}}} \mathrm{e}^{-\frac{\mu t}{2}}, \\ \qquad \text{for } \alpha = 1,\ t > 0. \end{cases} \quad [4.2]$$

Following [BAC 00], we solve [4.1] by the use of the Laplace transform method. Applying the Laplace transform to [4.1], we obtain

$$\tilde{f}(s) = \frac{s f_0 + f_1}{s^2 + \mu s^\alpha + \omega^2}, \quad \operatorname{Re} s > 0, \quad [4.3]$$

where we used the fact that $\left[{}_0\mathrm{I}_t^{1-\alpha} f(t)\right]_{t=0^+} = 0$, since $f$ is bounded at zero.

4.1.1.1. *Case* $\omega \neq 0$

Then, there exists $s_0 > 0$, such that for $\operatorname{Re} s > s_0$, there holds $\left|\frac{\mu}{s^2}\left(s^\alpha + \frac{\omega^2}{\mu}\right)\right| < 1$, so that from [4.3], it follows

$$\begin{aligned} \tilde{f}(s) &= \frac{s f_0}{s^2 + \mu s^\alpha + \omega^2} + \frac{f_1}{s^2 + \mu s^\alpha + \omega^2} \\ &= f_0 \frac{1}{s} \frac{1}{1 + \frac{\mu}{s^2}\left(s^\alpha + \frac{\omega^2}{\mu}\right)} + f_1 \frac{1}{s^2} \frac{1}{1 + \frac{\mu}{s^2}\left(s^\alpha + \frac{\omega^2}{\mu}\right)} \\ &= f_0 \sum_{k=0}^{\infty} \frac{(-1)^k \mu^k}{s^{2k+1}} \left(s^\alpha + \frac{\omega^2}{\mu}\right)^k + f_1 \sum_{k=0}^{\infty} \frac{(-1)^k \mu^k}{s^{2k+2}} \left(s^\alpha + \frac{\omega^2}{\mu}\right)^k \\ &= f_0 \sum_{k=0}^{\infty} \sum_{j=0}^{k} (-1)^k \binom{k}{j} \frac{\mu^j \omega^{2(k-j)}}{s^{2k+1-\alpha j}} \\ &\quad + f_1 \sum_{k=0}^{\infty} \sum_{j=0}^{k} (-1)^k \binom{k}{j} \frac{\mu^j \omega^{2(k-j)}}{s^{2(k+1)-\alpha j}}, \quad \operatorname{Re} s > s_0. \end{aligned} \quad [4.4]$$

The inverse Laplace transform of [4.4] in term-by-term steps is based on [DOE 67, Theorem 29.2], and yields the solution of [4.1], for $\alpha \in (0,1)$, in the form

$$\begin{aligned} f(t) &= f_0 \sum_{k=0}^{\infty} \sum_{j=0}^{k} (-1)^k \binom{k}{j} \frac{\mu^j \omega^{2(k-j)} t^{2k-\alpha j}}{\Gamma(2k+1-\alpha j)} \\ &\quad + f_1 \sum_{k=0}^{\infty} \sum_{j=0}^{k} (-1)^k \binom{k}{j} \frac{\mu^j \omega^{2(k-j)} t^{2k+1-\alpha j}}{\Gamma(2k+2-\alpha j)}. \end{aligned} \quad [4.5]$$

We can prove that the series [4.5] are convergent for $t > 0$.

REMARK 4.1.– In the case of the forced vibration, instead of [4.1], we have

$$\frac{d^2}{dt^2} f(t) + \mu\, {}_0D_t^{\alpha} f(t) + \omega^2 f(t) = h(t), \quad t > 0, \quad [4.6]$$

where $h$ is given and initial conditions are $[4.1]_{2,3}$. The solution to [4.6] and $[4.1]_{2,3}$ is

$$\begin{aligned} f(t) &= f_0 \sum_{k=0}^{\infty} \sum_{j=0}^{k} (-1)^k \binom{k}{j} \frac{\mu^j \omega^{2(k-j)} t^{2k-\alpha j}}{\Gamma(2k+1-\alpha j)} \\ &\quad + f_1 \sum_{k=0}^{\infty} \sum_{j=0}^{k} (-1)^k \binom{k}{j} \frac{\mu^j \omega^{2(k-j)} t^{2k+1-\alpha j}}{\Gamma(2k+2-\alpha j)} \\ &\quad + h(t) * \left( \sum_{k=0}^{\infty} \sum_{j=0}^{k} (-1)^k \binom{k}{j} \frac{\mu^j \omega^{2(k-j)} t^{2k+1-\alpha j}}{\Gamma(2k+2-\alpha j)} \right), \quad t > 0. \end{aligned} \quad [4.7]$$

4.1.1.2. *Case $\omega = 0$*

In this case, [4.1] becomes

$$f^{(2)}(t) + \mu\, {}_0D_t^{\alpha} f(t) = 0, \quad t > 0, \qquad f(t) = f_0, \qquad f^{(1)}(0) = f_1. \quad [4.8]$$

Two special cases of the solution to [4.8] are

$$f(t) = \begin{cases} f_0 \cos\left(t\sqrt{\mu}\right) + f_1 \frac{1}{\sqrt{\mu}} \sin\left(t\sqrt{\mu}\right), & \text{for } \alpha = 0,\ t > 0, \\ f_0 e^{-\mu t} + f_1 \frac{1}{\mu}\left(1 - e^{-\mu t}\right), & \text{for } \alpha = 1,\ t > 0. \end{cases} \quad [4.9]$$

For $\alpha \in (0,1)$ and $\omega = 0$, expression [4.3] becomes

$$\tilde{f}(s) = \frac{sf_0}{s^2 + \mu s^\alpha} + \frac{f_1}{s^2 + \mu s^\alpha} = f_0 \sum_{k=0}^{\infty} \frac{(-\mu)^k}{s^{(2-\alpha)k+1}} + f_1 \sum_{k=0}^{\infty} \frac{(-\mu)^k}{s^{(2-\alpha)k+2}}, \qquad [4.10]$$

where the series converges for $|\mu s^{\alpha-2}| < 1$ implying $\operatorname{Re} s > |\mu|^{\frac{1}{2-\alpha}}$. Taking the inverse transform of [4.10], we obtain the general solution of [4.1] for the case $\omega = 0$ as

$$f(t) = f_0 \sum_{k=0}^{\infty} \frac{(-\mu)^k t^{(2-\alpha)k}}{\Gamma((2-\alpha)k+1)} + f_1 \sum_{k=0}^{\infty} \frac{(-\mu)^k t^{(2-\alpha)k+1}}{\Gamma((2-\alpha)k+2)}, \quad t > 0. \qquad [4.11]$$

The solution [4.10] to [4.8] may also be written as

$$f(t) = f_0 E_{2-\alpha,1}\left(-\mu t^{2-\alpha}\right) + f_1 t E_{2-\alpha,2}\left(-\mu t^{2-\alpha}\right), \quad t > 0,$$

(see [KIL 06, p. 302]), where $E_{\alpha,\beta}$ is the two-parameter Mittag-Leffler function.

### 4.1.2. *Fractional derivative-type creeping motion*

We consider a special case of [4.1] when the inertial forces can be neglected. Then, with $\alpha \in (0,1)$, we have

$${}_0\mathrm{D}_t^\alpha f(t) + Bf(t) = C, \quad t > 0, \qquad [4.12]$$

where $B$ and $C$ are constants.

Three special cases of the solution [4.12] are

$$f(t) = \begin{cases} \frac{C}{1+B}, & \text{for } \alpha = 0,\ t > 0, \\ \frac{C}{B}\left(1 - \mathrm{e}^{B^2 t}\left(1 + \operatorname{erf}\left(-B\sqrt{t}\right)\right)\right), & \text{for } \alpha = \frac{1}{2},\ t > 0, \\ f_0 \mathrm{e}^{-Bt} + \frac{C}{B}\left(1 - \mathrm{e}^{-Bt}\right), & \text{for } \alpha = 1,\ t > 0. \end{cases}$$

To solve [4.12], we take the Laplace transform and obtain

$$(s^\alpha + B)\tilde{f}(s) = \frac{C}{s} + \left[{}_0\mathrm{I}_t^{1-\alpha} f(t)\right]_{t=0^+}. \qquad [4.13]$$

As before, we assume that $\left[{}_0\mathrm{I}_t^{1-\alpha}f(t)\right]_{t=0^+} = 0$. If we suppose that $B \neq 0$, we obtain

$$\tilde{f}(s) = \frac{C}{s\left(s^\alpha + B\right)} = C\frac{1}{s^{\alpha+1}\left(1+\frac{B}{s^\alpha}\right)}, \quad \operatorname{Re} s > |B|^{\frac{1}{\alpha}}. \qquad [4.14]$$

The inverse Laplace transform of [4.14] becomes

$$f(t) = C\sum_{k=0}^{\infty}\frac{(-B)^k t^{\alpha(k+1)}}{\Gamma\left(\alpha(k+1)+1\right)}, \quad t > 0. \qquad [4.15]$$

In writing [4.15], we used term-by-term inversion, since the series expansion of [4.14] converges for $\operatorname{Re} s > |B|^{\frac{1}{\alpha}}$.

The general solution [4.12] may be written in the form

$$f(t) = C \cdot H(t) * \left(t^{\alpha-1}E_{\alpha,\alpha}\left(-Bt^\alpha\right)\right) + C \cdot t^{\alpha-1}E_{\alpha,\alpha}\left(-Bt^\alpha\right), \quad t > 0,$$

(see [KIL 06, p. 302]), where $E_{\alpha,\beta}$ is the two-parameter Mittag-Leffler function.

### 4.1.3. *Linear vibrations with the multiterm fractional dissipation*

We consider an equation representing a generalization of [4.1] given by

$$\sum_{i=0}^{m} A_i\, {}_0\mathrm{D}_t^{\alpha_i} y(t) = f(t), \quad t > 0, \qquad [4.16]$$

where $\alpha_i = n_i - 1 + \gamma_i,\ i = 0, ..., m,\ n_i \in \mathbb{N},\ \gamma_i \in [0,1)\,;\ 0 \leq \alpha_0 < \alpha_1 < ... < \alpha_m \leq n_m;\ A_i \neq 0,\ i = 0, ..., m.$

Equation [4.16] has been treated by different methods: Laplace or Mellin integral transforms (see [KLI 01]), numerical methods (see [DIE 04]) or other methods suitable for some special form of [4.16]. For example, sequential linear equations of fractional order of type [4.16] are treated by special methods (see [KIL 06, S 7.1]). The equations of the type [4.16], when the fractional order is any real number, are elaborated often for some special cases.

For the general case of [4.16], we mention [DIE 04, HAD 96, LUC 99]. In [DIE 04, Section 3], authors analyze equation [4.16] with the Caputo fractional derivatives. They show that the solutions to [4.16] may be approximated arbitrarily closely by a system of linear fractional differential equations of rational order. This allows us to obtain an arbitrarily good approximation for the unknown solution to [4.16]. All results on [4.16] are given with the limitations: $\alpha_i - \alpha_{i-1} \leq 1,\ i = 1, ..., m$. In [HAD 96], the authors used a modification of the Mikusiński operational calculus (see [MIK 87]) to obtain an operator form of solutions to fractional linear equations of real orders. The fractional derivatives are of the Riemann–Liouville or of the Caputo form. Let us note that the convergence structure in the Mikusiski operational calculus is still an open problem and in [HAD 96] the convergence of series has not been explained.

If we apply the Laplace transform to [4.16], we have

$$\tilde{y}(s) = \frac{1}{\sum_{i=0}^{m} A_i s^{\alpha_i}} \left( \tilde{f}(s) + \sum_{i=0}^{m} \sum_{j=0}^{n_i} s^{n_i - j - 1} \left[ \left( {}_0\mathrm{I}_t^{1-\alpha_i} y\left(t\right) \right)^{(j)} \right]_{t=0^+} \right), \quad \mathrm{Re}\, s > 0. \tag{4.17}$$

THEOREM 4.1.– [ATA 12d] Suppose that $n_m \geq 2, \alpha_m - \alpha_0 > 1, A_m = 1, f = 0$. Let $0 \leq \alpha_0 < ... < \alpha_{m-1} < \alpha_m = n_m \in \mathbb{N}$. Then, equation [4.16] has a unique solution $y$ in the space $AC_{loc}^{n_m}([0, \infty))$, which satisfies the initial condition $y^{(j)}(0^+) = 0,\ j = 0, ..., n_m - 2$ and $y^{(n_m - 1)}(0^+) = 1$.

Additionally, let $\alpha_m - \alpha_{m-1} = k + \varepsilon$, $k \in \mathbb{N}$, $\varepsilon \in (0, 1)$ and let a solution satisfy $y^{(n_m + k)} \in L_{loc}^1([0, \infty))$. Then, $y_q = y^{(q-1)}$, $q = 1, ..., k + 1$, represents a linearly independent system of $k + 1$ solutions to [4.16], with $f = 0$, having the properties:

1) for $q = 1, ..., k$: $y_q \in AC_{loc}^{n_m}([0, \infty))$, $y_k^{(n_m)} \in C([0, b])$; $y_q^{(n_m - q)} = 1$; $y_q^{(i)}(0^+) = 0, i = 0, ..., n_m, i \neq n_m - q$;

2) for $q = k + 1$: $y_{k+1}^{(i)}(0^+) = 0, i = 0, ..., n_m - 1, i \neq n_m - (k + 1)$.

If the assumption $y^{(n_m + k)} \in L_{loc}^1([0, \infty))$ is not satisfied, we can assert that only $k$ linearly independent solutions exist on $[0, \infty)$; $y_{k+1}$ is a solution on $[0, \infty)$. The solution $y$ is given by

$$y(t) = \frac{1}{A_m} \frac{t^{\alpha_m - 1}}{\Gamma(\alpha_m)} + \left( \frac{1}{A_m} \frac{\tau^{\alpha_m - 1}}{\Gamma(\alpha_m)} * \phi(\tau) \right)(t),\ t \geq 0, \tag{4.18}$$

with

$$\phi(t) = \sum_{r=1}^{\infty}(-1)^r \phi_r(t), \ \ \phi_r(t) = \left( \sum_{i=0}^{m-1} \frac{A_i}{A_m} \frac{1}{\Gamma(\alpha_m - \alpha_i)} t^{\alpha_m - \alpha_i - 1} \right)^{*r}, \ \ t \geq 0, \tag{4.19}$$

where $(\cdot)^{*r}$ means $r$-times convolution, $(\cdot) * \ldots * (\cdot)$ if $r \geq 2$ and $(\cdot)^{*0} = 1$, $(\cdot)^{*1} \equiv (\cdot)$.

As a first example, we consider equation [4.1]. In this case, we have $m = 2$, $n_2 = 2$, $A_0 = \omega^2$, $A_1 = \mu$, $A_2 = 1$, $f = 0$, $\alpha_2 = 2$, $\alpha_1 = \alpha$. Thus, [4.1] has two linearly independent solutions $y$ and $y^{(1)}$. Comparing two terms in [4.5], we conclude that this relation holds. Theorem 4.1 gives further properties of the solution.

As a second example, we consider equation [4.1] with $\alpha = \frac{3}{2}$, i.e.

$$\frac{\mathrm{d}^2}{\mathrm{d}t^2} y(t) + A\,{}_0\mathrm{D}_t^{\frac{3}{2}} y(t) + A_0\, y(t) = 0, \ \ t \geq 0, \tag{4.20}$$

$$y(0^+) = 0, \quad y^{(1)}(0^+) = C. \tag{4.21}$$

The unique solution $y$ in the space $AC^2_{loc}([0,\infty))$ is given by [4.18] with $m = 2$, $A_2 = 1$, $\alpha_2 = 2$, $A_1 = A$, $\alpha_1 = \frac{3}{2}$

$$y(t) = Ct + C\,(t * \phi(t)), \ \ t \geq 0,$$

where

$$\begin{aligned} \phi(t) &= \sum_{r=1}^{\infty}(-1)^r \phi_r(t), \ \ t \geq 0 \ \text{ and} \\ \phi_r(t) &= \left( \sum_{i=0}^{2} A_i \frac{t^{1-\alpha_i}}{\Gamma(2-\alpha_i)} \right)^{*r} = \left( A \frac{t^{-\frac{1}{2}}}{\Gamma\left(2 - \frac{3}{2}\right)} + A_0 t \right)^{*r}, \ \ t \geq 0. \end{aligned}$$

Then, $\phi_1(t) = A\frac{1}{\Gamma(\frac{1}{2})} t^{-\frac{1}{2}} + A_0 t$. Since $(t \geq 0)$

$$\frac{t^{a-1}}{\Gamma(a)} * \frac{t^{b-1}}{\Gamma(b)} = \frac{t^{a+b-1}}{\Gamma(a+b)} \quad \text{and}$$

$$\left( \frac{t^{a-1}}{\Gamma(a)} * \frac{t^{b-1}}{\Gamma(b)} \right)^{*r} = \sum_{j=0}^{r} \binom{r}{j} \left( \frac{t^{a-1}}{\Gamma(a)} \right)^{*r-j} * \left( \frac{t^{b-1}}{\Gamma(b)} \right)^{*j},$$

for $b = \frac{1}{2}$ and $a = 2$, we obtain

$$\begin{aligned}
\phi_r(t) &= \sum_{j=0}^{r} \binom{r}{j} A^j A_0^{r-j} \left( \frac{t^{-\frac{1}{2}}}{\Gamma\left(\frac{1}{2}\right)} \right)^{*(r-j)} * \left( \frac{t}{\Gamma(2)} \right)^{*j} \\
&= \sum_{j=0}^{r} \binom{r}{j} A^j A_0^{r-j} \frac{t^{\frac{r-j}{2}-1}}{\Gamma\left(\frac{r-j}{2}\right)} * \frac{t^{2j-1}}{\Gamma(2j)} \\
&= \sum_{j=0}^{r} \binom{r}{j} A^j A_0^{r-j} \frac{t^{\frac{r-j}{2}+2j-1}}{\Gamma\left(\frac{r-j}{2}+2j\right)}, \quad t \geq 0, \\
t * \phi_r(t) &= t * \left( \sum_{r=1}^{\infty} (-1)^r \phi_r(t) \right) = \sum_{r=1}^{\infty} (-1)^r \left( t * \phi_r(t) \right) \\
&= \sum_{r=1}^{\infty} (-1)^r \left( t * \left( \sum_{j=0}^{r} \binom{r}{j} A^j A_0^{r-j} \frac{t^{\frac{r-j}{2}+2j-1}}{\Gamma\left(\frac{r-j}{2}+2j\right)} \right) \right) \\
&= \sum_{r=1}^{\infty} (-1)^r \sum_{j=0}^{r} \binom{r}{j} A^j A_0^{r-j} \frac{t^{\frac{r-j}{2}+2j+1}}{\Gamma\left(\frac{r-j}{2}+2j+2\right)}, \quad t \geq 0.
\end{aligned}$$

Therefore

$$y(t) = Ct + C \sum_{r=1}^{\infty} (-1)^r \sum_{j=0}^{r} \binom{r}{j} A^j A_0^{r-j} \frac{t^{\frac{r-j}{2}+2j+1}}{\Gamma\left(\frac{r-j}{2}+2j+2\right)}, \quad t \geq 0, \qquad [4.22]$$

is the solution to [4.20] and [4.21].

Equation [4.16] may also be considered in the space of tempered distributions $\mathcal{S}'_+$.

THEOREM 4.2.– [ATA 10c] Consider

$$\sum_{i=0}^{m} A_i \, {}_0\mathrm{D}_t^{\alpha_i} y(t) = g(t), \quad \text{in } \mathcal{S}'_+, \qquad [4.23]$$

where $A_i \neq 0$, $0 \leq \alpha_0 < \alpha_1 < ... < \alpha_m$. Then, [4.23] has a unique solution $\mathcal{S}'_+$ if and only if

$$\sum_{i=0}^{m} A_i s^{\alpha_i} \neq 0, \quad \mathrm{Re}\, s > 0. \qquad [4.24]$$

Also, let

$$l(t) = \mathcal{L}^{-1}\left[\left(\sum_{i=0}^{m} A_i s^{\alpha_i}\right)^{-1}\right](t), \quad t > 0$$

and $l(t) = 0$, $t < 0$. Then, the solution to [4.23] is given as

$$y(t) = l(t) * g(t), \quad t > 0. \qquad [4.25]$$

Moreover, if $0 \le \alpha_0 < \alpha_1 < ... < \alpha_m < 2$, then $l$ is locally integrable in $\mathbb{R}$. If, in addition, $\alpha_m - \alpha_0 > 1$, then $l$ is absolutely continuous in $\mathbb{R}$.

REMARK 4.2.– The violation of the condition $\sum_{i=0}^{m} A_i s^{\alpha_i} \neq 0$, $\operatorname{Re} s > 0$, leads to the violation of the dissipation inequality if [4.23] is interpreted as the constitutive equation of linear viscoelasticity. To show this, we suppose that [4.23] represents a constitutive equation

$$\sigma(t) = \sum_{i=0}^{m} A_i \, {}_0\mathrm{D}_t^{\alpha_i} \varepsilon(t).$$

Let

$$\sigma(t) = H(t) - H(t-\tau), \quad t > 0.$$

This corresponds to application of stress at $t = 0$ and unloading at $t = \tau$. Then

$$\begin{aligned} \tilde{\varepsilon}(s) &= \frac{1}{\sum_{i=0}^{m} A_i s^{\alpha_i}} \left(\frac{1}{s} - \frac{\mathrm{e}^{-\tau s}}{s}\right) \quad \operatorname{Re} s > 0, \quad \text{or} \\ \varepsilon^{(1)}(t) &= l(t) H(t) - l(t-\tau) H(t-\tau). \end{aligned}$$

The dissipation of mechanical energy is

$$A_d = \int_0^{\infty} \sigma(t) \varepsilon^{(1)}(t) \, \mathrm{d}t = \int_0^{\tau} l(t) \, \mathrm{d}t.$$

If $\sum_{i=0}^{m} A_i s_0^{\alpha_i} = 0$, $s_0 = u + iv$, $u > 0$, then $l$ has a term that shows oscillation with increasing amplitude (see [ATA 09b]), so that by choosing appropriate $\tau$ we can always violate the condition $A_d > 0$.

#### 4.1.4. *Linear fractional two-compartmental model with fractional derivatives of different order*

The contribution of fractional calculus to evolution problems is not only confined to mechanical systems. The same tool can be applied to pharmacokinetics. As an example, consider a two-compartmental biological system. This is a system that, for specific choice of parameters, can exhibit both oscillatory and non-oscillatory behaviors. The first compartment represents the place where the drug is applied and the second compartment represents the plasma and other body regions where the kinetics of the drug is uniform. These two compartments are described by a system of differential equations with the Caputo fractional derivatives, as done in [POP 13]

$$
\begin{aligned}
{}_{0}^{C}\mathrm{D}_{t}^{\alpha_1} q_1(t) &= -K_1 q_1(t), \quad t > 0, \\
{}_{0}^{C}\mathrm{D}_{t}^{\alpha_2} q_2(t) &= \mathcal{K}\left(q_1(t)\right) - K_2 q_2(t), \quad t > 0, \qquad [4.26] \\
q_1(0) &= q_{10}, \quad q_2(0) = q_{20}, \qquad [4.27]
\end{aligned}
$$

where $0 < \alpha_1 < 1$, $0 < \alpha_2 < 1$, $q_1$ and $q_2$ are concentrations that are to be determined, and $K_1$ and $K_2$ are given constants. The term $\mathcal{K}\left(q_1\right)$ will be specified later. System [4.26] is a generalization of the integer-order compartmental system

$$
q_1^{(1)}(t) = -K_1 q_1(t), \quad q_2^{(1)}(t) = \mathcal{K}\left(q_1(t)\right) - K_2 q_2(t), \quad t > 0, \qquad [4.28]
$$

see [RES 04].

The balance of mass (the mass leaving the first compartment must be equal to the mass entering the second compartment from the first one) for the case of integer-order system [4.28] requires that $K_1 = \mathcal{K}$. The same holds if $\alpha_1 = \alpha_2$ in [4.26]. For the fractional-order dynamics, with $\alpha_1 \neq \alpha_2$ in [4.26], the relation between $K_1$ and $\mathcal{K}$ must be determined so that the balance of mass is satisfied. The problem is interesting and leads the authors in [DOK 10] to state that the mass conservation "... is possible only in the special case where a common fractional order is considered for all ordinary differential equations". To find the relation between $K_1$ and $\mathcal{K}$ when $\alpha_1 \neq \alpha_2$, we start with the Taylor formula for the fractional derivatives [2.81].

The mass leaving the first compartment may be expressed in two ways as

$$
\begin{aligned}
\mathrm{d}q_1(t) &= q_1(t+\Delta t) - q_1(t) = \frac{{}_{0}^{C}\mathrm{D}_{t}^{\alpha_1} q_1(t)}{\Gamma(\alpha_1+1)}(\Delta t)^{\alpha_1} + O\left((\Delta t)^{2\alpha_1}\right) \\
&= \frac{{}_{0}^{C}\mathrm{D}_{t}^{\alpha_2} q_1(t)}{\Gamma(\alpha_2+1)}(\Delta t)^{\alpha_2} + O((\Delta t)^{\alpha_2}), \quad t > 0,
\end{aligned}
$$

where $O$ is the Landau symbol "big $O$". Recall that $a(x) = O(|x|^{\alpha})$ means $|a(x)| \leq M|x|^{\alpha}$, as $|x| \to 0$, or $|x| \to \infty$. Functions $q_1$ and $q_2$ are assumed to satisfy all the necessary assumptions, so that the Taylor formula can be applied.

The change of mass in the second compartment is

$$\begin{aligned} dq_2(t) &= q_2(t+\Delta t) - q_2(t) = \frac{{}_0^C\mathrm{D}_t^{\alpha_2} q_2(t)}{\Gamma(1+\alpha_2)} (\Delta t)^{\alpha_2} + O\left((\Delta t)^{2\alpha_2}\right) \\ &= \frac{1}{\Gamma(1+\alpha_2)} \left[\mathcal{K}(q_1(t))(\Delta t)^{\alpha_2} - K_2 q_2 (\Delta t)^{\alpha_2}\right], \quad t > 0. \end{aligned}$$

The mass entering the second compartment from the first is

$$dq_1(t) = \frac{{}_0^C\mathrm{D}_t^{\alpha_2} q_1(t)}{\Gamma(\alpha_2+1)} (\Delta t)^{\alpha_2} = -\frac{1}{\Gamma(1+\alpha_2)} \left[\mathcal{K}(q_1(t))(\Delta t)^{\alpha_2}\right], \quad t > 0,$$

or

$$\mathcal{K}(q_1(t)) = -{}_0^C\mathrm{D}_t^{\alpha_2} q_1(t), \quad t > 0.$$

System [4.26], when the mass is preserved, becomes

$$\begin{aligned} {}_0^C\mathrm{D}_t^{\alpha_1} q_1(t) &= -K_1 q_1(t), \quad t > 0, \\ {}_0^C\mathrm{D}_t^{\alpha_2} q_2(t) &= -{}_0^C\mathrm{D}_t^{\alpha_2} q_1(t) - K_2 q_2(t), \quad t > 0. \end{aligned} \qquad [4.29]$$

Note that for the case when there is no mass leaving the second compartment into surroundings ($K_2 = 0$), we have

$${}_0^C\mathrm{D}_t^{\alpha_2} (q_2(t) + q_1(t)) = 0, \quad \text{or} \quad q_1(t) + q_2(t) = const, \quad t > 0,$$

i.e. $q_1(t) + q_2(t) = const.$ is a conservation law for [4.29]. This example shows that fractionalization of an integer-order system must be performed so that the constraints coming from physics are taken into account.

Finally, we give a solution to [4.29] subject to [4.27] as ($t > 0$)

$$\begin{aligned} q_1(t) &= q_{10} E_{\alpha_1}(-K_1 t^{\alpha_1}), \\ q_2(t) &= (q_{10} + q_{20}) E_{\alpha_2}(-K_2 t^{\alpha_2}) \\ &\quad + K_2 q_{10} \left( t^{\alpha_2 - 1} E_{\alpha_2,\alpha_2} \left( -K_2 t^{\alpha_2} \right) \right) * E_{\alpha_1} \left( -K_1 t^{\alpha_1} \right) - q_1(t), \end{aligned} \quad [4.30]$$

where $E_\alpha$ and $E_{\alpha,\beta}$ are one- and two-parameter Mittag-Leffler functions, respectively.

In Figure 4.1, we show concentrations $q_1$ and $q_2$ as functions of time, obtained according to [4.30] with $K_1 = K_2 = q_{01} = 1$ and $q_{02} = 0$.

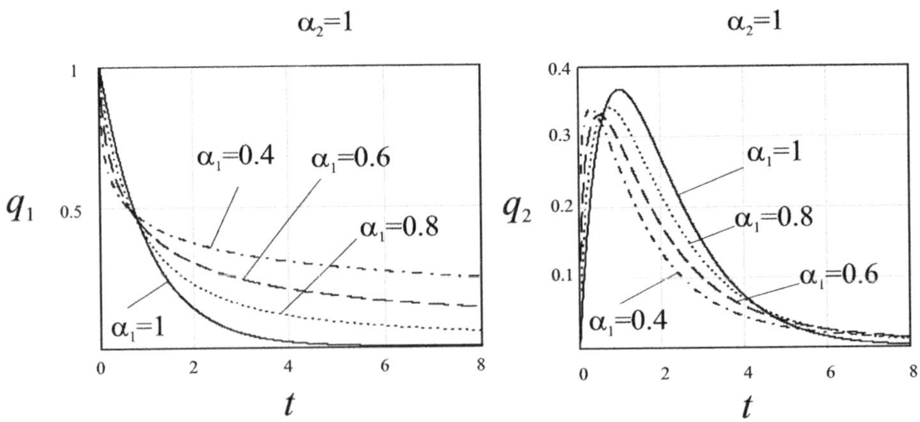


**Figure 4.1.** *Solutions [4.30] of [4.29]*

## 4.2. Bagley–Torvik equation

An equation of the type [4.1] arises in another important mechanical problem. Since the seminal paper of Torvik and Bagley [TOR 84], many authors have studied the fractional differential equation of the form

$$Ay^{(2)}(t) + B\, {}_0\mathrm{D}_t^{\alpha} y(t) + Cy(t) = f(t), \quad t > 0, \ \alpha \in [1, 2), \quad [4.31]$$

$$y(0) = y_0, \quad y^{(1)}(0) = v_0. \quad [4.32]$$

In the Bagley–Torvik equation [4.31], $A$, $B$ and $C$ are given constants and the forcing term $f$ is a known function. In the work by Torvik and Bagley, $\alpha = \frac{3}{2}$.

Equation [4.31] represents a model of motion of a rigid plate immersed in a Newtonian fluid. In [POD 99], the Bagley–Torvik equation [4.31] was studied both analytically and numerically. In [LI 10, RAY 12], the Haar wavelet method was developed for a class of equations to which [4.31] belongs, while in [RAJ 11] a stochastic technique was developed for the solution of a fractional-order system represented by [4.31]. A collocation method was used in [QAS 10] to solve [4.31], and in [WAN 10], a method based on linear combination of fundamental solutions that are expressed in terms of $\alpha$-exponential functions is presented. The homotopy perturbation method was developed in [ZOL 09]. Finally, we mention [RAY 05] where the Adomian decomposition method was applied to recover the solution obtained in [POD 99]. The important questions of existence and numerical solution of [4.31] were discussed in [DIE 10, DIE 02b].

Following [ATA 13e], we transform the Bagley–Torvik equation [4.31] as follows. Note that for $\alpha \in [1,2)$

$$
\begin{aligned}
{}_0\mathrm{D}_t^{\alpha} y(t) &= \frac{1}{\Gamma(2-\alpha)} \frac{\mathrm{d}^2}{\mathrm{d}t^2} \int_0^t \frac{y(\tau)}{(t-\tau)^{\alpha-1}} \mathrm{d}\tau \\
&= \frac{\mathrm{d}}{\mathrm{d}t} \left( \frac{\mathrm{d}}{\mathrm{d}t} \frac{1}{\Gamma(1-\beta)} \int_0^t \frac{y(\tau)}{(t-\tau)^{\beta}} \mathrm{d}\tau \right) = \frac{\mathrm{d}}{\mathrm{d}t} {}_0\mathrm{D}_t^{\beta} y(t), \quad t > 0,
\end{aligned}
$$

where $\alpha = 1+\beta$, $\beta \in [0,1)$. Then, [4.31] is equivalent to

$$
z^{(1)}(t) + Cy(t) = f(t), \quad z(t) = Ay^{(1)}(t) + B\,{}_0\mathrm{D}_t^{\beta} y(t), \quad t > 0. \tag{4.33}
$$

To solve [4.33], we will make use of the expansion formula for fractional derivative

$$
{}_0\mathrm{D}_t^{\beta} f(t) = \mathcal{A}(N) \frac{f(t)}{t^{\beta}} + \mathcal{B}(N) f^{(1)}(t) t^{1-\beta} + \mathcal{C}(N, V_p(t)) + R_N(t), \quad t \in (0,T], \tag{4.34}
$$

where

$$
\mathcal{A}(N) = \frac{\sin(\beta\pi)}{\beta\pi} \frac{\Gamma(N+\beta)}{\Gamma(N)}, \quad \mathcal{B}(N) = \frac{\Gamma(N+\beta)}{\Gamma(\beta)\Gamma(2-\beta)\Gamma(N+1)},
$$

$$
\mathcal{C}(N, V_p(t)) = \frac{1}{\Gamma(2-\beta)\Gamma(\beta-1)} \sum_{p=2}^{N} \frac{\Gamma(p-1+\beta)}{(p-2)!} \frac{V_p(t)}{t^{p-1+\beta}}, \quad t \in (0,T],
$$

$$
V_p^{(1)}(t) = t^{p-2} f(t), \quad t \in (0,T], \quad V_p(0) = 0, \ p = 2,3,\ldots, \tag{4.35}
$$

see [ATA 08b] and [2.31]. Remainder $R_N$ is given by

$$R_N(t) = \frac{t^{1-\beta}}{\Gamma(\beta-1)\Gamma(2-\beta)} \int_0^t f^{(2)}(\tau) \left( \sum_{p=N+1}^{\infty} \frac{\Gamma(p-1+\beta)}{p!} \left(\frac{\tau}{t}\right)^p \mathrm{d}\tau \right), \quad t \in (0,T].$$

Since $f \in C^2((0,T])$, we have

$$\begin{aligned} |R_N(t)| &= \frac{t^{1-\beta}}{\Gamma(2-\beta)\Gamma(\beta-1)} \left| \int_0^t f^{(2)}(\tau) \left( \sum_{p=N+1}^{\infty} \frac{\Gamma(p-1+\beta)}{p!} \left(\frac{\tau}{t}\right)^p \mathrm{d}\tau \right) \right| \\ &\leq \frac{t^{1-\beta}}{\Gamma(\beta-1)\Gamma(2-\beta)} \sum_{p=N+1}^{\infty} \frac{\Gamma(p-1+\beta)}{p!t^p} \int_0^t \left| f^{(2)}(\tau) \right| \tau^p \mathrm{d}\tau \\ &\leq M \frac{t^{2-\beta}}{\Gamma(\beta-1)\Gamma(2-\beta)} \sum_{p=N+1}^{\infty} \frac{\Gamma(p-1+\beta)}{(p+1)!} \\ &= \frac{M\Gamma(N+\beta)t^{2-\beta}}{\Gamma(3-\beta)\Gamma(\beta-1)(N+1)!}, \end{aligned} \quad [4.36]$$

where $M = \sup_{t \in (0,T]} \left| f^{(2)}(t) \right|$. Thus, if we take approximation of the fractional derivative

$${}_0\mathrm{D}_t^\beta f(t) \approx \mathcal{A}(N) \frac{f(t)}{t^\beta} + \mathcal{B}(N) f^{(1)}(t) t^{1-\beta} + \mathcal{C}(N, V_p(t)), \quad t \in (0,T], \quad [4.37]$$

with $\mathcal{A}$, $\mathcal{B}$, and $\mathcal{C}$ given by [4.35], the remainder $R_N$ can be made arbitrarily small by choosing $N$ sufficiently large.

### 4.2.1. *Solution procedure*

We use [4.31] with homogeneous initial conditions, i.e.

$$Ay^{(2)}(t) + B\,{}_0\mathrm{D}_t^\alpha y(t) + Cy(t) = f(t), \quad t > 0, \quad [4.38]$$

$$y(0) = 0, \quad y^{(1)}(0) = 0. \quad [4.39]$$

If $y(0) = y_0 \neq 0$, $y^{(1)}(0) = v_0 \neq 0$ in [4.39], the change of variables $y = y_0 + v_0 t + Y$ and the new definition of $f$ leads to system of the forms [4.38] and [4.39]. We rewrite [4.38] as

$$z^{(1)}(t) + Cy(t) = f(t), \quad z(t) = Ay^{(1)}(t) + B\,{}_0\mathrm{D}_t^\beta y(t), \quad t > 0. \quad [4.40]$$

Finally, from [4.34], [4.35] and $[4.40]_2$, we obtain

$$\begin{aligned} y^{(1)}(t) &= \frac{z(t) - B\left(\mathcal{A}(N)\frac{y(t)}{t^\beta} + \mathcal{C}(N, V_p(t)) + R_N(t)\right)}{A + B\,\mathcal{B}(N)t^{1-\beta}}, \quad t \in (0, T], \\ z^{(1)}(t) &= f(t) - Cy(t), \ \ t \in (0, T], \\ V_p^{(1)}(t) &= t^{p-2}y(t), \ \ t \in (0, T], \ p = 2, 3, \ldots, N, \end{aligned} \qquad [4.41]$$

subject to

$$y(0) = 0, \quad z(0) = 0, \quad V_p(0) = 0, \ \ p = 2, 3, \ldots, N. \qquad [4.42]$$

Thus, the Cauchy problem for the fractional differential equations [4.38] and [4.39] is transformed to a Cauchy problem for the system of $N$ first-order differential equations [4.41] and [4.42]. This is the central point of our analysis. We note that in [YUA 98, YUA 02], a different method of transforming fractional differential equations to the system of first-order equations is used. In applications, we will neglect the term $R_N$ in [4.41].

For the case of a nonlinear system

$$Ay^{(2)}(t) + B\,{}_0\mathrm{D}_t^\alpha y(t) + Cy(t) = F(t, y(t)), \ \ t > 0, \qquad [4.43]$$

$$y(0) = 0, \quad y^{(1)}(0) = 0, \qquad [4.44]$$

with $1 \le \alpha = 1 + \beta < 2$, the corresponding system of the first-order equations becomes

$$\begin{aligned} y^{(1)}(t) &= \frac{z(t) - B\left(\mathcal{A}(N)\frac{y(t)}{t^\beta} + \mathcal{C}(N, V_p(t)) + R_N(t)\right)}{A + B\,\mathcal{B}(N)t^{1-\beta}}, \quad t \in (0, T], \\ z^{(1)}(t) &= F(t, y(t)) - Cy(t), \ \ t \in (0, T], \\ V_p^{(1)}(t) &= t^{p-2}y(t), \ \ t \in (0, T], \ p = 2, 3, \ldots, N, \end{aligned} \qquad [4.45]$$

subject to

$$y(0) = 0, \quad z(0) = 0, \quad V_p(0) = 0, \ \ p = 2, 3, \ldots, N. \qquad [4.46]$$

Systems [4.43] and [4.44] correspond to a motion of a plate connected to a nonlinear spring and immersed in a Newtonian fluid.

Note that in expressions [4.41] and [4.45], $\mathcal{A}(N)\frac{y(t)}{t^\beta}$ satisfies $\lim_{t\to 0}\mathcal{A}(N)\frac{y(t)}{t^\beta} = 0$ so that at $t = 0$ it is not singular.

Systems [4.38] and [4.39] may be solved by the use of the Laplace transform. The solution is given in [POD 99, p. 231] and it becomes

$$y(t) = \frac{1}{A} f(t) * \left( \sum_{k=0}^{\infty} \frac{(-1)^k}{k!} \left(\frac{C}{A}\right)^k t^{2k+1} E^{(k)}_{\frac{1}{2},2+\frac{3k}{2}} \left(-\frac{B}{A}\sqrt{t}\right)\right), \quad t > 0, \qquad [4.47]$$

$E^{(k)}_{\frac{1}{2},2+\frac{3k}{2}}$ denotes the $k$th derivative of the two-parameter Mittag-Leffler function.

Another form of the solution to [4.38] and [4.39] is obtained by the inversion of the Laplace transform along the Hankel contour. It is given by ($t > 0$)

$$\begin{aligned} y(t) &= f(t) * K(t), \\ K(t) &= \frac{1}{\pi} \int_0^\infty \frac{Bq^\alpha \sin(\alpha\pi)}{(Aq^2 + Bq^\alpha \cos(\alpha\pi) + C)^2 + (Bq^\alpha \sin(\alpha\pi))^2} \mathrm{d}q \\ &\quad + 2\,\mathrm{Re}\left( \left. \frac{\mathrm{e}^{st}}{2As + \alpha B s^{\alpha-1}} \right|_{s=s_0} \right), \end{aligned}$$

where $s_0$ is a solution to

$$As^2 + Bs^\alpha + C = 0, \quad s \in \mathbb{C}.$$

We also cite the analytical solution for [4.38] with $f = 0$, i.e.

$$Ay^{(2)}(t) + B\, {}_0\mathrm{D}_t^\alpha y(t) + Cy(t) = 0, \quad t > 0, \qquad [4.48]$$

$$y(0) = 0, \quad y^{(1)}(0) = V, \qquad [4.49]$$

obtained in [ATA 13b] by a different method (see [4.22]). The solution to [4.48] and [4.49] is

$$y(t) = Vt + V \sum_{r=1}^{\infty} (-1)^r \sum_{j=0}^{r} \binom{r}{j} \left(\frac{B}{A}\right)^j \left(\frac{C}{A}\right)^{r-j} \frac{t^{\frac{r-j}{2}+2j+1}}{\Gamma(\frac{r-j}{2} + 2j + 2)}, \quad t > 0. [4.50]$$

We discuss several special cases of [4.38] and [4.39].

### 4.2.2. *Numerical examples*

4.2.2.1. *Case of linear restoring force*

Suppose that

$$\alpha = 1.5, \quad A = 1, \quad B = 0.5, \quad C = 0.5, \quad f(t) = \begin{cases} 8, & 0 \le t \le 1, \\ 0, & t > 1. \end{cases} \qquad [4.51]$$

This is test problem for many procedures. We solved it using [4.41] and [4.42] with $N = 30$. The results are shown in Figure 4.2. Our results for $y$ correspond well with the results obtained by other methods. In [RAY 12], the result for $t = 19.8438$ was given as $y_{SSR}(19.8438) = -1.44734$. With our method, we have a more precise result: $y_{our}(19.8438) = -1.58917$, since the exact solution, given by [4.47], is $y_{exact}(19.8438) = -1.52284$. Also in [ENE 10], the generalized Taylor collocation method was used to solve [4.38], [4.39] and [4.51]. The value of $y(1)$ was given as $y_{T-co}(1) = 2.954070000$. Our result is $y_{our}(1) = 2.95179355033523$ while the exact value is $y_{exact}(1) = 2.952584099$.

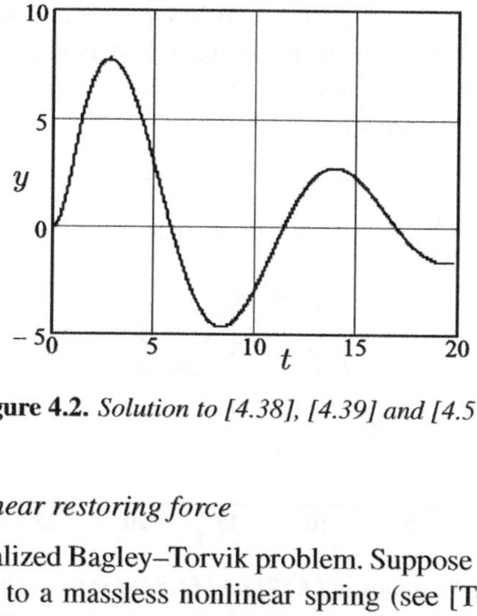


**Figure 4.2.** *Solution to [4.38], [4.39] and [4.51]*

4.2.2.2. *Case of nonlinear restoring force*

Consider the generalized Bagley–Torvik problem. Suppose that the plate immersed in a fluid is connected to a massless nonlinear spring (see [TOR 84, Figure 2]). We assume that the force $F$ is related to the displacement $y$ of the spring by $F = Cy + c_3 y^3$. The differential equation of motion of the plate then becomes

$$Ay^{(2)}(t) + B\,{}_0\mathrm{D}_t^{1.5} y(t) + Cy(t) + c_3 y^3(t) = 0, \quad t > 0, \qquad [4.52]$$

$$y(0) = 0, \quad y^{(1)}(0) = v_0. \qquad [4.53]$$

By using $y = v_0 t + Y$, we obtain

$$AY^{(2)}(t) + B\,{}_0\mathrm{D}_t^{1.5}(Y(t) + v_0 t) = -C(v_0 t + Y) - c_3(v_0 t + Y)^3, \quad t > 0, \qquad [4.54]$$

subject to $Y(0) = 0$, $Y^{(1)}(0) = 0$. Systems [4.45] and [4.46] become

$$\begin{aligned} Y^{(1)}(t) &\approx \frac{z(t) - B\frac{t^{0.5}}{\Gamma(1.5)} - B\left(\mathcal{A}(N)\frac{y(t)}{t^{0.5}} + \mathcal{C}(N, V_p(t))\right)}{A + B\,\mathcal{B}(N)t^{0.5}}, \quad t \in (0, T], \\ z^{(1)}(t) &= -C(v_0 t + Y) - c_3(v_0 t + Y)^3, \quad t \in (0, T], \\ V_p^{(1)}(t) &= t^{p-2}y(t), \quad t \in (0, T], \ p = 2, 3, \ldots, N, \end{aligned} \qquad [4.55]$$

with

$$y(0) = 0, \quad z(0) = 0, \quad V_p(0) = 0, \quad p = 2, 3, \ldots, N. \qquad [4.56]$$

We use $A = 1$, $B = C = 0.5$, $c_3 = 0.5$, $v_0 = 1$ and solve [4.55] and [4.56] for $N = 30$. The results are shown in Figure 4.3. It is seen that, due to the larger rigidity of the spring to which the plate is attached, in the nonlinear case $c_3 \neq 0$, the amplitudes of the oscillation are smaller.

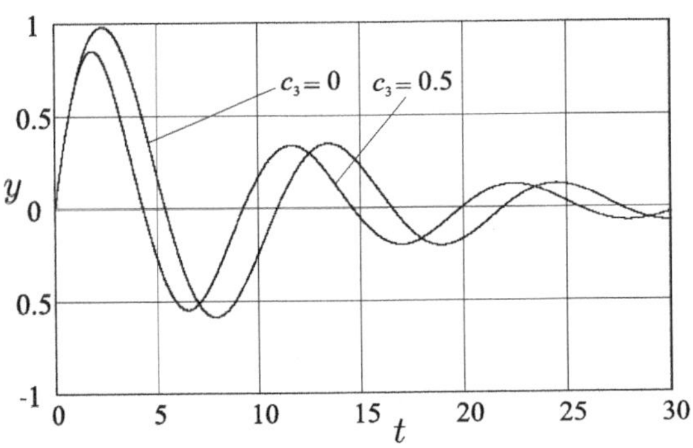


**Figure 4.3.** *Solutions of [4.52] and [4.53] for two values of* $c_3$

## 4.3. Nonlinear vibrations with symmetrized fractional dissipation

We consider an equation of the type [4.1], but with a different type of dissipation, following [ATA 11a]. Namely, we use the symmetric and antisymmetric Caputo fractional derivatives of order $0 < \alpha < 1$

$$
{}_0^-\mathcal{E}_T^\alpha u(t) = \frac{1}{2}\left({}_0^C\mathrm{D}_t^\alpha - {}_t^C\mathrm{D}_T^\alpha\right) u(t) \text{ and}
$$

$$
{}_0^+\mathcal{E}_T^\alpha u(t) = \frac{1}{2}\left({}_0^C\mathrm{D}_t^\alpha + {}_t^C\mathrm{D}_T^\alpha\right) u(t), \quad t \in [0, T], \tag{4.57}
$$

respectively. These Caputo-type fractional derivatives are also called the Riesz–Caputo fractional derivatives in [FRE 10]. For both of the above fractional derivatives, we can define the corresponding distributed-order fractional derivative. For example, $\int_0^1 {}_0^\pm\mathcal{E}_T^\alpha u(t)\phi(\alpha)\mathrm{d}\alpha$ are distributed-order Caputo-type fractional derivatives of $u$, where $\phi$ is a positive integrable function or a distribution of the form $\sum_{i=0}^n c_{\alpha_i}\,\delta(\alpha - \alpha_i)$, $0 \leq \alpha_0 < \alpha_1 < \ldots < \alpha_n \leq 1$, $c_{\alpha_i} \geq 0$, $i = 0, 1, \ldots, n$. The order of the fractional derivative is determined numerically from the set of measured data. This procedure usually leads to several values of the order of derivative that belong to a certain range. Distributed-order fractional derivatives model such situations more appropriately than single-value fractional derivatives.

We assume that an integrable function $u$ satisfies conditions which imply that its fractional derivative under consideration is again an integrable function in $[0, T]$, for every $T > 0$. For example, ${}_0\mathrm{D}_t^\alpha u$, $0 < \alpha < 1$, is integrable whenever $u \in AC([0, T])$.

We intend to study solvability, dissipativity and stability for the following two problems (one with ${}_0^-\mathcal{E}_T^\alpha u$, and other with ${}_0^+\mathcal{E}_T^\alpha u$) in $C^2([0, T])$, $T > 0$:

$$
u^{(2)}(t) + b\int_0^1 {}_0^\pm\mathcal{E}_T^\alpha u(t)\phi(\alpha)\mathrm{d}\alpha + F(u(t)) = 0, \quad t \in (0, T], \quad u(0) = u_0,
$$

$$
u^{(1)}(0) = v_0, \tag{4.58}
$$

where $b$ is a positive constant and $F$ is a locally Lipschitz continuous force function on $\mathbb{R}$. We assume that $F(0) = 0$ so that, with $u_0 = v_0 = 0$, $u = 0$ is a trivial solution to [4.58].

Equation [4.58] is a generalization of different equations that have been analyzed before. We refer to [SER 00] for a number of mathematical models describing oscillators with fractional damping of the form $u^{(2)}(t) + b \cdot {}_0\mathrm{D}_t^\alpha u(t) + F(u(t)) = 0$, where $b$ and $F$ are as above. Special cases are: duffing oscillator with $V(u) = -\frac{Ku^2}{2} + \frac{Au^4}{4}$, $A, K > 0$, a nonlinear pendulum with $V(u) = K(1 - \cos u)$, $K > 0$, and a linear oscillator $V(u) = \frac{\omega^2 u^2}{2}$ ($V$ is a potential). Another possible application of equation [4.58] is in non-local elasticity where the independent variable $t$ is replaced with the spatial variable $x$.

Let $V(u) = \int_0^u F(\lambda)\mathrm{d}\lambda$ be the potential of $F(u)$, $u \in \mathbb{R}$. In studying dissipation and stability of solutions to [4.58], the total mechanical energy of the system

$$E(u(t)) = \frac{\left(u^{(1)}(t)\right)^2}{2} + V(u(t)), \quad t \in [0,T], \; T > 0, \qquad [4.59]$$

plays the crucial role (see [SER 00]). Multiplying [4.58] with $u^{(1)}$ and then integrating, we obtain

$$\frac{\left(u^{(1)}(t)\right)^2}{2} + V(u(t)) - \frac{v_0^2}{2} = -b \int_0^t \left( \int_0^1 {}_0^{\pm}\mathcal{E}_T^{\alpha} u(s)\phi(\alpha)\mathrm{d}\alpha \right) \cdot u^{(1)}(s)\,\mathrm{d}s,$$
$$t \in [0,T], \; T > 0, \qquad [4.60]$$

where, without loss of generality, we have assumed that $V(u_0) = 0$.

Since $b > 0$, the dissipation of the mechanical energy in [4.60] depends on the sign of the expression

$${}_d^{\pm}\mathcal{D}_t(u) = \int_0^t \left( \int_0^1 {}_0^{\pm}\mathcal{E}_T^{\alpha} u(s)\phi(\alpha)\mathrm{d}\alpha \right) \cdot u^{(1)}(s)\,\mathrm{d}s, \quad t \in [0,T], \; T > 0.$$

The temporarily weak form of the dissipation inequality is satisfied if ${}_d^{\pm}\mathcal{D}_t(u) \geq 0$, $t \in [0,T]$, $T > 0$, for all $u$ within a certain class of functions. The latter inequality was proved in [ENE 99] for the left Riemann–Liouville derivative instead of the distributed-order Caputo-type fractional derivative, with initial conditions $u(0) = 0$, $u^{(1)}(0) = 0$. In [SER 00], it is proved that

$$\int_0^t {}_0^C\mathrm{D}_s^{\alpha} u(s) \cdot u^{(1)}(s)\,\mathrm{d}s \geq 0, \quad t \in [0,T], \; T > 0,$$

for function $u$ having a locally integrable derivative. Moreover, it was stated in [SER 00] that the dissipation inequality does not hold for the Riemann–Liouville fractional derivative, in general. In [STA 04b], it is proved that

$$\int_0^t {}_0\mathrm{D}_s^{\alpha} u(s) \cdot u(s)\,\mathrm{d}s \geq 0, \quad t \in [0,T], \; T > 0, \qquad [4.61]$$

for $0 < \alpha < 1$, provided $u$ satisfies suitable conditions.

### 4.3.1. *Solvability and dissipativity of [4.58]*

Equation [4.58], in the special case when $\phi(\alpha) = \delta(\alpha)$, is uniquely solvable in $C^2([0,T])$, for every $T > 0$, as stated in [SER 00, Theorem 1]. We examine solvability of [4.58] in the case when $\phi$ is continuous in $[0,1]$.

Recall that we assume $F$ to be a locally Lipschitz continuous function on $\mathbb{R}$. We consider the system

$$u^{(1)} = f_1(u,v), \quad v^{(1)} = f_2(u,v), \quad u(0) = u_0, \quad v(0) = v_0, \qquad [4.62]$$

where $f_1(u,v) = v$ and

$$f_2(u,v) = -b\int_0^1 \frac{1}{\Gamma(1-\alpha)}\left(\int_0^t \frac{v(\tau)}{(t-\tau)^\alpha}\mathrm{d}\tau \pm \int_t^T \frac{v(\tau)}{(\tau-t)^\alpha}\mathrm{d}\tau\right)\phi(\alpha)\mathrm{d}\alpha - F(u).$$

Let $H(u,v) = (f_1(u,v), f_2(u,v))$. Clearly, it is a continuous mapping from $C([0,T]) \times C([0,T])$ into $C([0,T])$. Then

$$\begin{aligned}
&|H(u(t)+u_1(t), v(t)+v_1(t)) - H(u(t),v(t))| \\
&\quad \le |f_1(u(t)+u_1(t), v(t)+v_1(t)) - f_1(u(t),v(t)))| \\
&\qquad + |f_2(u(t)+u_1(t), v(t)+v_1(t)) - f_2(u(t),v(t))| \\
&\quad \le |v_1(t)| + b\int_0^1 \frac{1}{\Gamma(1-\alpha)}\left(\int_0^t \frac{|v_1(\tau)|}{(t-\tau)^\alpha}\mathrm{d}\tau + \int_t^T \frac{|v_1(\tau)|}{(\tau-t)^\alpha}\mathrm{d}\tau\right)\phi(\alpha)\mathrm{d}\alpha \\
&\qquad + |F(u(t)+u_1(t)) - F(u(t))| \\
&\quad \le C \max_{t\in[0,T]}(|u_1(t)| + |v_1(t)|),
\end{aligned}$$

where we used local Lipschitz continuity of $F$. Now, we have

$$\sup_{t\in[0,T]} |H(u(t)+u_1(t), v(t)+v_1(t)) - H(u(t),v(t))| \le C(||u_1||_{C([0,T])} + ||v_1||_{C([0,T])}).$$

Thus, $H$ is Lipschitz continuous. Since [4.58] is equivalent to [4.62] in $C^2([0,T])$, by [HAL 93, Theorem 2.1] and the improved version given in [SER 00, Theorem 1], we have the following theorem.

THEOREM 4.3.– Problem [4.58] has a unique solution in $C^2([0,T])$, for every $T > 0$.

In the following, we turn our attention to the study of dissipativity of the solution to [4.58]. This will be done by the analysis of

$$^{\pm}\mathcal{D}_t(u) = \int_0^t {}_0^{\pm}\mathcal{E}_T^{\alpha} u(s) \cdot u^{(1)}(s)\mathrm{d}s = \frac{1}{2}\int_0^t \left({}_0^C\mathrm{D}_s^{\alpha} u(s) \pm {}_s^C\mathrm{D}_T^{\alpha} u(s)\right) u^{(1)}(s)\mathrm{d}s.$$

THEOREM 4.4.– The solution $u \in C^2([0,T])$, $T > 0$, of the initial value problem [4.58] is dissipative, i.e. for every $T > 0$

$$\frac{\left(u^{(1)}(t)\right)^2}{2} + V(u(T)) - \frac{v_0^2}{2} \leq 0. \qquad [4.63]$$

To obtain the estimate and prove theorem 4.4, we need the following lemma.

LEMMA 4.1.– Suppose that $u \in C^1(0,T)$, $T > 0$. Then

$$\int_0^T \left({}_0^C\mathrm{D}_t^{\alpha} u(t) - {}_t^C\mathrm{D}_T^{\alpha} u(t)\right) u^{(1)}(t)\mathrm{d}t \geq 0,$$

$$\int_0^T \left({}_0^C\mathrm{D}_t^{\alpha} u(t) + {}_t^C\mathrm{D}_T^{\alpha} u(t)\right) u^{(1)}(t)\mathrm{d}t = 0. \qquad [4.64]$$

PROOF.– In the following, we use the family of distributions $\{f_\beta\}_{\beta\in\mathbb{R}}$, given by [1.6]. With $\beta = 1-\alpha$, $\alpha \in (0,1)$, in [1.6], we define $\tilde{\phi}_{1-\alpha}(t) = f_{1-\alpha}(|t|)$, $t \in \mathbb{R}$. By [KIL 06, p. 16], the Fourier transform of $\tilde{\phi}_{1-\alpha}$ is

$$\mathcal{F}\left[\tilde{\phi}_{1-\alpha}(t)\right](y) = \int_{-\infty}^{\infty} \tilde{\phi}_{1-\alpha}(t)\mathrm{e}^{ity}\mathrm{d}t = 2\sin\frac{\alpha\pi}{2}|y|^{1-\alpha}, \quad y \in \mathbb{R}.$$

Thus, by the Bochner–Schwartz theorem (see [VLA 73, p. 141]), the regular distribution defined by $\tilde{\phi}_{1-\alpha}$ is positive definite, i.e. $\langle\tilde{\phi}_{1-\alpha}, \varphi * \check{\varphi}\rangle \geq 0$, $\varphi \in \mathcal{D}(\mathbb{R})$ ($\check{\varphi}(x) = \varphi(-x)$). In our case

$$I(\varphi) = \int_{\mathbb{R}}\int_{\mathbb{R}} \tilde{\phi}_{1-\alpha}(x-y)\varphi(x)\varphi(y)\mathrm{d}x\mathrm{d}y \geq 0, \qquad [4.65]$$

for every $\varphi \in \mathcal{D}(\mathbb{R})$. Let $\omega \in L^1_{loc}(\mathbb{R})$, $\omega = 0$ in $(-\infty, 0)$. Taking $\{\varphi_n\}_{n\in\mathbb{N}}$ to be a sequence in $\mathcal{D}(\mathbb{R})$, which converges to $\omega|_{[0,T]}$ in $L^1([0,T])$, [4.65] and the Lebesgue dominated convergence theorem imply that

$$I(\omega) = \int_0^T \int_0^T \phi_{1-\alpha}(|t-\tau|)\omega(t)\omega(\tau)\mathrm{d}t\mathrm{d}\tau \geq 0. \qquad [4.66]$$

Next, we write [4.66] as

$$\begin{aligned} I(\omega) &= \int_0^T \left( \int_0^t \phi_{1-\alpha}(|t-\tau|)\omega(t)\omega(\tau)\mathrm{d}\tau \right) \mathrm{d}t \\ &\quad + \int_0^T \left( \int_t^T \phi_{1-\alpha}(|\tau-t|)\omega(t)\omega(\tau)\mathrm{d}\tau \right) \mathrm{d}t, \end{aligned}$$

and interchange the order of integration in the second integral so that

$$\begin{aligned} &\int_0^T \left( \int_t^T \phi_{1-\alpha}(|\tau-t|)\omega(t)\omega(\tau)\mathrm{d}\tau \right) \mathrm{d}t \\ &\quad = \int_0^T \left( \int_0^t \phi_{1-\alpha}(|t-\tau|)\omega(t)\omega(\tau)\mathrm{d}\tau \right) \mathrm{d}t. \end{aligned} \qquad [4.67]$$

This implies

$$I(\omega) = 2\int_0^T \omega(t) \left( \int_0^t \frac{(t-\tau)^{-\alpha}}{\Gamma(1-\alpha)}\omega(\tau)\mathrm{d}\tau \right) \mathrm{d}t = 2\int_0^T \omega(t)\cdot {}_0\mathrm{I}_t^{1-\alpha}\omega(t)\mathrm{d}t \geq 0. [4.68]$$

Similarly, substituting $\int_0^T \left( \int_0^t \phi_{1-\alpha}(|t-\tau|)\omega(t)\omega(\tau)\mathrm{d}\tau \right) \mathrm{d}t$ in [4.66], with the term on the left-hand side of [4.67], we have

$$I(\omega) = 2\int_0^T \omega(t) \left( \int_t^T \frac{(\tau-t)^{-\alpha}}{\Gamma(1-\alpha)}\omega(\tau)\mathrm{d}\tau \right) \mathrm{d}t = 2\int_0^T \omega(t)\cdot {}_t\mathrm{I}_T^{\gamma}\omega(t)\mathrm{d}t \geq 0. [4.69]$$

Now, put $\omega = u^{(1)}$ and $\alpha = 1-\gamma$. This implies that $\omega$ is continuous and that

$${}_0\mathrm{I}_t^{\gamma}\omega = {}_0\mathrm{I}_t^{\gamma}u^{(1)} = {}_0^C\mathrm{D}_t^{\alpha}u \text{ and } {}_t\mathrm{I}_T^{\gamma}\omega = {}_t\mathrm{I}_T^{\gamma}u^{(1)} = {}_t^C\mathrm{D}_T^{\alpha}u, \qquad [4.70]$$

see [KIL 06, p. 92], or [2.2 and 2.43]. Now, putting [4.70] into [4.68] and [4.69], it follows that

$$I(u) = 2\int_0^T u^{(1)}(t) \cdot {}_0^C\mathrm{D}_t^\alpha u(t)\mathrm{d}t = -2\int_0^T u^{(1)}(t) \cdot {}_t^C\mathrm{D}_T^\alpha u(t)\mathrm{d}t \geq 0. \quad [4.71]$$

Finally, by adding and subtracting integrals in [4.71], we obtain [4.64]. ■

PROOF.– [Proof of Theorem 4.4] First, we consider the case $\phi(\cdot) = \delta(\cdot - \alpha)$, when $\int_0^1 {}_0^\pm\mathcal{E}_T^\alpha u(t)\phi(\alpha)\mathrm{d}\alpha$ reduces to ${}_0^\pm\mathcal{E}_T^\alpha u(t)$. The case $\phi(\alpha) = \sum_{i=0}^n c_{\alpha_i}\delta(\alpha - \alpha_i)$, $0 \leq \alpha_0 < \alpha_1 < \ldots < \alpha_n \leq 1$, $c_{\alpha_i} \geq 0$, $i = 0, 1, \ldots, n$, can be treated in a similar way. Equation [4.58] then becomes

$$u^{(2)}(t) + b \cdot {}_0^\pm\mathcal{E}_T^\alpha u(t) + F(u(t)) = 0, \quad t \in [0,T],\ T > 0. \quad [4.72]$$

The dissipativity relation of the total mechanical energy [4.60] (with $V(u_0) = 0$) becomes

$$\frac{\left(u^{(1)}(t)\right)^2}{2} + V(u(t)) - \frac{v_0^2}{2} = -b\int_0^t {}_0^\pm\mathcal{E}_T^\alpha u(s) \cdot u^{(1)}(s)\mathrm{d}s, t \in [0,T], T > 0. \quad [4.73]$$

From [4.64], it follows that both ${}_0^+\mathcal{E}_T^\alpha u$ and ${}_0^-\mathcal{E}_T^\alpha u$ satisfy a weak dissipation inequality ${}_d^\pm\mathcal{D}_t(u) \geq 0$ for $t = T$, $T > 0$, and hence [4.63] holds in the case $\phi(\cdot) = \delta(\cdot - \alpha)$. ■

To prove [4.63] in the case when $\phi$ is a positive integrable function, we need the next lemma. Since every positive integrable function is the limit of a sequence of positive continuous functions, this proves the theorem.

LEMMA 4.2.– Suppose that $u \in C^1([0,T])$, $\phi \in C([0,1])$ and $\phi(\alpha) \geq 0$ for all $\alpha \in [0,1]$. Then

$$\int_0^T \left(\int_0^1 {}_0^\pm\mathcal{E}_T^\alpha u(t)\phi(\alpha)\mathrm{d}\alpha\right) u^{(1)}(t)\mathrm{d}t \geq 0. \quad [4.74]$$

PROOF.– Since the Lebesgue and Riemann integrals coincide for continuous functions, we have

$$\int_0^1 {}_0^\pm\mathcal{E}_T^\alpha u(t)\phi(\alpha)\mathrm{d}\alpha = \lim_{n\to\infty}\sum_{k=1}^n {}_0^\pm\mathcal{E}_T^{\alpha_k} u(t)\phi(\alpha_k)\Delta\alpha_k,$$

where on the right-hand side of the previous expression we have the Riemann sums and $0 = \alpha_1 < \ldots < \alpha_n = 1$ is a partition of $[0, 1]$, with $\max_{n\to\infty} \Delta\alpha_k \to 0$. Since $\phi \geq 0$ on $(0, 1)$ and $\int_0^T {}^{\pm}_{0}\mathcal{E}^{\alpha_k}_{T} u(t)u^{(1)}(t)dt \geq 0$, by lemma 4.1, it follows that

$$\sum_{k=1}^{n} \phi(\alpha_k)\Delta\alpha_k \int_0^T {}^{\pm}_{0}\mathcal{E}^{\alpha_k}_{T} u(t)u^{(1)}(t)dt$$
$$= \int_0^T \left( \sum_{k=1}^{n} \phi(\alpha_k){}^{\pm}_{0}\mathcal{E}^{\alpha_k}_{T} u(t)\Delta\alpha_k \right) u^{(1)}(t)dt \geq 0. \qquad [4.75]$$

Since $|\sum_{k=1}^{n} \phi(\alpha_k){}^{\pm}_{0}\mathcal{E}^{\alpha_k}_{T} u(t)\Delta\alpha_k| \leq C$, for some $C > 0$, we can apply the Lebesgue dominated convergence theorem in [4.75] to obtain [4.74]. ■

REMARK 4.3.– Note that [4.68] and [4.69] imply

$$\int_0^T \omega(t) \cdot {}_0\mathrm{I}^{\alpha}_{t}\omega(t)dt = \int_0^T \omega(t) \cdot {}_t\mathrm{I}^{\alpha}_{T}\omega(t)dt \geq 0. \qquad [4.76]$$

The equality between integrals in [4.76] is a special case of the integration by parts formula for fractional derivatives $\int_0^T \phi(t) \cdot {}_0\mathrm{I}^{\alpha}_{t}\xi(t)dt = \int_0^T \xi(t) \cdot {}_t\mathrm{I}^{\alpha}_{T}\phi(t)dt$ (see [2.21]) when both functions are equal, i.e. $\phi = \xi = w$. However, the fact that both integrals in this case are non-negative seems to be new.

### 4.3.2. *Stability of the solution*

We study the stability of solutions $u$ to [4.58], assuming the coercivity of $V$ and $V(u_0) = 0$, as before. The Liapunov function is taken to be the total mechanical energy $E$ given by [4.59] and we assume that potential energy is continuous coercive function, i.e. there exists $c > 0$ such that

$$V(u) \geq c\|u\| = \sup_{t\in[0,T]} |u(t)|, \quad u \in AC^2[0, T],\ T > 0. \qquad [4.77]$$

THEOREM 4.5.– The solution $u$ of [4.58] and its first derivative $u^{(1)}$ are bounded with respect to the sup-norm on $[0, T]$, for every $T > 0$.

PROOF.– First assume that $\phi(\cdot) = \delta(\cdot - \alpha)$. Using [4.59] and the assumption $V(u_0) = 0$, we have

$$E(T) = \frac{\left(u^{(1)}(T)\right)^2}{2} + V(u(T)) - \frac{v_0^2}{2} = -b \int_0^T {}^{\pm}_{0}\mathcal{E}^{\alpha}_{T} u(t) \cdot u^{(1)}(t)dt.$$

Since the integral is non-negative (see [4.64]), it follows that

$$\frac{(u^{(1)}(T))^2}{2} + V(u(T)) \le \frac{v_0^2}{2}, \quad T > 0.$$

Using [4.77], we obtain that for every $T > 0$

$$\frac{\left(u^{(1)}(T)\right)^2}{2} + c|u(T)| \le \frac{v_0^2}{2}. \qquad [4.78]$$

Now, assume that $\phi$ is a positive and continuous function. To show that [4.78] is satisfied, we have to apply estimate [4.74] obtained in lemma 4.2 instead of [4.64] from lemma 4.1.

The claim in the general case when $\phi$ is positive and integrable is again obtained considering $\phi$ as the limit of a sequence of positive continuous functions. ■

## 4.4. Nonlinear vibrations with distributed-order fractional dissipation

Following [ATA 07a], we propose the generalization of the nonlinear equation of the mathematical pendulum that includes viscoinertial effects

$$\frac{\mathrm{d}^2}{\mathrm{d}t^2}\varphi(t) + \frac{\lambda}{ml^2}\int_0^2 \phi(\gamma)\, {}_0\mathrm{D}_t^{\gamma}\varphi(t)\,\mathrm{d}\gamma = -\frac{g}{l}\sin\varphi(t). \qquad [4.79]$$

We consider a class of equations of the form

$$\frac{\mathrm{d}}{\mathrm{d}t}\left(g(t)\frac{\mathrm{d}}{\mathrm{d}t}y(t)\right) + \tilde{\lambda}\int_0^2 \phi(\gamma)\, {}_0\mathrm{D}_t^{\gamma}y(t)\,\mathrm{d}\gamma = f(t, y(t)), \quad t \in (0, b), \qquad [4.80]$$

that includes [4.79] as a special case. In [4.80], functions $g$ and $\phi$ are given and $\tilde{\lambda}$ is a constant. Many of previously studied equations can be obtained from [4.80]. The solvability and the uniqueness of a solution of [4.80] in the case $f(t, y(t)) = h(t)$, in the space of tempered distributions with support in $[0, \infty)$, were studied in [ATA 05b], and the case $g(t) \equiv 0$, $\phi(\gamma) = \delta(\gamma - \gamma_0)$, $0 < \gamma_0 < 1$, was discussed in [DEL 96]. A similar problem was studied in [ATA 05a] in the space of exponentially bounded distributions with support $[0, \infty)$.

Equation [4.80] is of distributed-order type. In [ATA 02a] (see also [3.25]), a model of linear viscoelastic body was proposed in which, in a linear state, stress $\sigma$ depends on all time derivatives of the strain $\varepsilon$, so that

$$\sigma(t) = \lambda \int_0^1 \phi(\gamma)\, {}_0\mathrm{D}_t^\gamma \varepsilon(t)\, \mathrm{d}\gamma, \quad t \in [0, b], \qquad [4.81]$$

where a constant $\lambda$ and a function $\phi$ characterize a material under consideration and they are the so-called constitutive quantities. A slight generalization of [4.81] such that the viscoinertial bodies are included (see [HAR 03]) becomes

$$\sigma(t) = \lambda \int_c^d \phi(\gamma)\, {}_0\mathrm{D}_t^\gamma \varepsilon(t)\, \mathrm{d}\gamma, \quad t \in [a, b], \qquad [4.82]$$

where $0 \leq c \leq d \leq 2$, $\lambda$ and $\phi$ are determined experimentally. Constitutive equations of the type [4.81] or [4.82] must satisfy restrictions that follow from the entropy inequality so that $\lambda$ and $\phi$ must satisfy

$$\lambda \int_c^d \phi(\gamma)\, \omega^\gamma \sin\frac{\gamma\pi}{2} \mathrm{d}\gamma \geq 0 \quad \text{and}$$

$$\lambda \int_c^d \phi(\gamma)\, \omega^\gamma \cos\frac{\gamma\pi}{2} \mathrm{d}\gamma \geq 0, \quad 0 \leq \omega \leq \infty, \qquad [4.83]$$

see [3.26]. For model [4.82], the ambient damping coefficients depend on a vibrational mode (see [ROS 01b]). Thus, for the mathematical modeling of such a phenomenon, fractional derivatives in the constitutive equation offer a procedure that is in agreement with experiments.

### 4.4.1. *Existence of solutions*

If $g = r$ is a constant in [4.80], then we have

$$r \frac{\mathrm{d}^2}{\mathrm{d}t^2} y(t) + \tilde{\lambda} \int_0^2 \phi(\gamma)\, {}_0\mathrm{D}_t^\gamma y(t)\, \mathrm{d}\gamma = \tilde{f}(t, y(t)), \quad t > 0, \qquad [4.84]$$

where $\phi$ is a distribution supported by $[0, 2)$, $r$, $\tilde{\lambda}$ are real constants and $\tilde{f}$ is a suitable function. Above, the integration is done over the support of $\phi$ contained in $[0, d_1]$, for some $d_1 < 2$. If $r = 0$ and $\tilde{\lambda} \neq 0$, we have

$$\int_0^2 \phi(\gamma)\, {}_0\mathrm{D}_t^\gamma y(t)\, \mathrm{d}\gamma = f(t, y(t)),$$

and if $r \neq 0$ and $\tilde{\lambda} = 0$, then [4.84] reduces to

$$\frac{\mathrm{d}^2}{\mathrm{d}t^2} y(t) = f(t, y(t)).$$

If $r = \lambda = 0$, then we have to use the implicit function theorem to solve $f(t, y(t)) = 0$.

4.4.1.1. *Mild solution*

We consider equation [4.84] in the form ($\tilde{\lambda} = -\lambda$)

$$\frac{\mathrm{d}^2}{\mathrm{d}t^2} y(t) = \lambda \int_0^2 \phi(\gamma)\, {}_0\mathrm{D}_t^\gamma y(t)\, \mathrm{d}\gamma + f(t, y(t)), \quad t > 0, \qquad [4.85]$$

in $L^1_{loc}(\mathbb{R}) \cap C^1([0, \infty))$ with the assumptions

$$y(0) = q \text{ and } \frac{\mathrm{d}}{\mathrm{d}t} y(0) = p.$$

In this section, we assume that $\phi$ is continuous and that $\operatorname{supp}\phi \subset [0, d_1]$, $d_1 < 2$. We will use the following condition:

(A)$_l$ $f(t, y), (t, y) \in [0, \infty) \times \mathbb{R}$ is locally integrable and for some $a > 0$ there exist $\alpha > 0$ and $h \in L^1_{loc}([0, a])$ such that

$$|f(t, y)| \leq h(t) |y|^\alpha, \quad y \in \mathbb{R}, \ t \in [0, a]. \qquad [4.86]$$

Integrating [4.85] twice, we adjoin to [4.85]

$$y(t) = \lambda \int_0^2 \phi(\gamma)\, {}_0\mathrm{I}_t^{\gamma - 2} y(t)\, \mathrm{d}\gamma + {}_0\mathrm{I}_t^2 f(t, y(t)) + pt + q, \quad t > 0, \qquad [4.87]$$

where we used ${}_0\mathrm{I}_t^2 f(t) = \int_0^t \left( \int_0^u f(\xi)\, \mathrm{d}\xi \right) \mathrm{d}u$. A solution to [4.87] is called the mild solution of [4.85]. Clearly, a classical $C^2$-solution of [4.85] is the solution to [4.87], but a continuous solution of [4.87] need not be the solution of [4.85]. It is the case if we additionally have a solution of [4.87] which has a continuous second derivative.

In the following, we will first prove the existence of a mild solution (theorem 4.6) and then of a classical solution (theorem 4.7).

We consider [4.87] and [4.85] in the spaces $C^i([0,a])$, $i = 1,2$, consisting of functions that are continuous on $[0,a]$ as well as their derivatives up to order $i$ with the norm

$$\|y\|_i = \sup_{t\in[0,a]} \{|y(t)| \mid \left|y^{(j)}(t)\right|, \ j \le i\}, \quad i = 1,2.$$

THEOREM 4.6.– Assume $(\mathrm{A})_l$. Then, there exist $\delta > 0$ and $y \in C^1([0,\delta])$, which satisfies [4.87] on $[0,\delta]$.

PROOF.– We will show that $T : C^1([0,a]) \to C^1([0,a])$ defined by

$$Ty(t) = \lambda \int_0^2 \phi(\gamma)\ {}_0\mathrm{I}_t^{\gamma-2} y(t)\,\mathrm{d}\gamma + {}_0\mathrm{I}_t^2 f(t, y(t)) + pt + q, \quad t \ge 0,$$

is a compact operator and that it maps $B_r = \{y \in C^1[0,\delta] : \|y\|_1 \le r\}$ into itself, where $r > \max\{|p|,|q|\}$ and $\delta$ is determined with respect to $r$ ($a$ is noted in $(\mathrm{A})_l$). Then, the Schauder fixed point theorem assures that $T$ has at least one fixed point. This is the mild solution.

Let

$$\begin{aligned} Jy(t) &= \lambda \int_0^2 \phi(\gamma)\ {}_0\mathrm{I}_t^{\gamma-2} y(t)\,\mathrm{d}\gamma \\ &= \lambda \int_0^2 \frac{\phi(\gamma)}{\Gamma(2-\gamma)} \int_0^t \frac{y(x)}{(t-x)^{\gamma-1}}\mathrm{d}x\mathrm{d}\gamma, \quad t \ge 0, \\ Gy(t) &= \int_0^t (t-x)\, f(x, y(x))\,\mathrm{d}x, \quad t \ge 0. \end{aligned}$$

Obviously, $Ty = Jy + Gy + pt + q$.

We show that $t \mapsto Ty(t)$ has a continuous derivative. Since $\frac{\mathrm{d}}{\mathrm{d}t}y = y^{(1)}$ is continuous, integration by parts leads to

$$Jy(t) = \lambda \int_0^2 \frac{\phi(\gamma)}{(2-\gamma)\Gamma(2-\gamma)} \int_0^t (t-x)^{2-\gamma} y^{(1)}(x)\,\mathrm{d}x\mathrm{d}\gamma.$$

This implies

$$\frac{\mathrm{d}}{\mathrm{d}t}(Jy(t)) = \lambda \int_0^2 \frac{\phi(t)}{\Gamma(2-\gamma)} \int_0^t (t-x)^{1-\gamma} y^{(1)}(x)\,\mathrm{d}x\mathrm{d}\gamma, \quad t \geq 0.$$

Also, for $t \geq 0$

$$\frac{\mathrm{d}}{\mathrm{d}t}(Gy(t)) = \int_0^t f(x, y(x))\,\mathrm{d}x,$$

so $T : C^1([0,a]) \to C^1([0,a])$.

Next, we show that $T$ is compact. First, we prove that $J : C^1([0,a]) \to C^1([0,a])$ is compact. For $i = 0, 1$, we have

$$\begin{aligned}\frac{\mathrm{d}^i}{\mathrm{d}t^i}(Jy(t)) &= \lambda \int_0^{d_1} \phi(\gamma)\, {}_0\mathrm{I}_t^{\gamma-2} y(t)\,\mathrm{d}\gamma \\ &= \int_0^{d_1} \frac{\phi(\gamma)}{\Gamma(2-\gamma)} \int_0^t \frac{y^{(i)}(x)}{(t-x)^{\gamma-1}} \mathrm{d}x\mathrm{d}\gamma.\end{aligned}$$

Then, in the sense of $L(C^1([0,a]), C^1([0,a]))$ convergence,

$$\begin{aligned}Jy(t) &= \lambda \lim_{N\to\infty} \sum_{n=1}^{N} \phi(\gamma_n)\, {}_0\mathrm{I}_t^{\gamma_n-2} y(t)\,\Delta\gamma_n \\ &= \lambda \lim_{N\to\infty} \sum_{n=1}^{N} \frac{\phi(\gamma_n)}{\Gamma(2-\gamma_n)} \int_0^t \frac{y(x)}{(t-x)^{\gamma_n-1}} \mathrm{d}x\Delta\gamma_n,\end{aligned}$$

where $\gamma_n$ are points of interval $[0, d_1]$ in the usual definition of the Riemann sum defined for the integral. Indeed, with $i = 0, 1$,

$$\begin{aligned}&\left| \sum_{n=1}^{N} \frac{\phi(\gamma_n)}{\Gamma(2-\gamma_n)} \int_0^t \frac{y^{(i)}(x)}{(t-x)^{\gamma_n-1}} \mathrm{d}x\Delta\gamma_n - \int_0^{d_1} \frac{\phi(\gamma)}{\Gamma(2-\gamma)} \int_0^t \frac{y^{(i)}(x)}{(t-x)^{\gamma-1}} \mathrm{d}x\mathrm{d}\gamma \right| \\ &\quad \leq \sup_{t\in[0,a]} \left| y^{(i)}(t) \right| \left| \sum_{n=1}^{N} \frac{\phi(\gamma_n)}{\Gamma(2-\gamma_n)} \int_0^t \frac{1}{(t-x)^{\gamma_n-1}} \mathrm{d}x\Delta\gamma_n \right. \\ &\quad \left. - \int_0^{d_1} \frac{\phi(\gamma)}{\Gamma(2-\gamma)} \int_0^t \frac{1}{(t-x)^{\gamma-1}} \mathrm{d}x\mathrm{d}\gamma \right|,\end{aligned}$$

and when $N \to \infty$, the expression goes to zero. For each $n \in \mathbb{N}$, the operator ${}_0\mathrm{I}_t^{\gamma_n-2}$ : $C^1([0,a]) \to C^1([0,a]$ is compact since $\gamma_n < 2$ and therefore ${}_0\mathrm{I}_t^{\gamma_n-2}$ is the operator of fractional integration, which is compact. Thus, $J$ is compact as a limit of compact operators. Operator $G : C^1([0,a]) \to C^1([0,a]$,

$$y(t) \mapsto Gy(t) = {}_0\mathrm{I}_t^2 f(t, y(t)), \quad t \geq 0,$$

is compact since

$$\frac{\mathrm{d}^i}{\mathrm{d}t^i}(Gy(t)) = \int_0^t (t-x)^{1-i} y(x)\,\mathrm{d}x, \quad i = 0, 1, \; t \geq 0.$$

Thus, if $(z_n)_n$ is bounded in $C^1([0,a])$, then $Tz_n(t) = Jz_n(t) + Gz_n(t) + pt + q$, $n \in \mathbb{N}$, has a subsequence that is convergent in $C^1([0,a])$. Therefore, $T$ is also compact.

Let $B_r = \{y \in C^1[0,\delta] : \|y\| \leq r\}$, where $r > \max\{|p|, |q|\}$ and $\delta$ will be defined later. Let $y \in B_r$. Then

$$\begin{aligned} |Ty(t)| &\leq |\lambda| \int_0^2 \frac{|\phi(\gamma)|}{\Gamma(2-\gamma)} \int_0^t \frac{|y(x)|}{(t-x)^{\gamma-1}} \mathrm{d}x\mathrm{d}\gamma \\ &\quad + \int_0^t (t-x)\,|f(x, y(x))|\,\mathrm{d}x + |p|t + |q| \\ &\leq |\lambda|\, M_t \sup_{t\in[0,\delta]} |y(t)| + C_t \sup_{t\in[0,\delta]} |y(t)|^\alpha + |p|t + |q|, \quad t \in [0,\delta], \end{aligned}$$

where

$$\begin{aligned} M_t &= \int_0^2 \frac{|\phi(\gamma)|}{\Gamma(2-\gamma)} \int_0^t (t-x)^{1-\gamma}\,\mathrm{d}x\mathrm{d}\gamma = \int_0^2 \frac{|\phi(\gamma)|}{\Gamma(3-\gamma)} t^{2-\gamma}\mathrm{d}\gamma, \\ C_t &= \int_0^t h(x)(t-x)\,\mathrm{d}x, \quad t \in [0,\delta]. \end{aligned}$$

Thus

$$|Ty(t)| \leq |\lambda|\, M_\delta r + C_\delta r^\alpha + |p|\delta + |q|, \quad t \in [0,\delta].$$

Since $r > |q|$, we can choose $\delta$ such that the right-hand side is smaller than $r$ and obtain

$$|Ty(t)| \leq r, \quad t \in [0, \delta].$$

Similarly, by using $r > |p|$, we can shrink $\delta$ so that

$$\left| \frac{\mathrm{d}}{\mathrm{d}t}(Ty(t)) \right| \leq r, \quad t \in [0, \delta].$$

Therefore, we have $\|Ty\|_1 \leq r$. Thus, there exists $y \in C^1([0, \delta])$ such that it is the fixed point for $T$. It solves [4.87] and completes the proof. ∎

PROPOSITION 4.1.– The mild solution is the weak solution.

PROOF.– Since the mild solution of [4.85] is the fixed point for the operator $T$ (see theorem 4.6), it follows that for every $\theta \in C_0^\infty((0, \delta))$, $y$ satisfies

$$\left\langle \frac{\mathrm{d}^2}{\mathrm{d}t^2} y(t), \theta(t) \right\rangle = \lambda \int_0^2 \phi(\gamma) \langle {}_0\mathrm{D}_t^\gamma y(t), \theta(t) \rangle \, \mathrm{d}\gamma + \langle f(t, y(t)), \theta(t) \rangle,$$

$$y(0) = q, \quad \frac{\mathrm{d}}{\mathrm{d}t} y(0) = p.$$

Indeed

$$\begin{aligned}
\left\langle \frac{\mathrm{d}^2}{\mathrm{d}t^2} y, \theta \right\rangle &= \left\langle y, \frac{\mathrm{d}^2}{\mathrm{d}t^2} \theta \right\rangle \\
&= \left\langle \lambda \int_0^2 \phi(\gamma)\, {}_0\mathrm{I}_t^{\gamma-2} y(t) \, \mathrm{d}\gamma + {}_0\mathrm{I}_t^2 f(t, y(t)) + pt + q, \frac{\mathrm{d}^2}{\mathrm{d}t^2} \theta(t) \right\rangle \\
&= \lambda \left\langle \int_0^2 \phi(\gamma)\, {}_0\mathrm{I}_t^{\gamma-2} y(t) \, \mathrm{d}\gamma, \frac{\mathrm{d}^2}{\mathrm{d}t^2} \theta(t) \right\rangle + \left\langle {}_0\mathrm{I}_t^2 f(t, y(t)), \frac{\mathrm{d}^2}{\mathrm{d}t^2} \theta(t) \right\rangle \\
&\quad + \left\langle pt + q, \frac{\mathrm{d}^2}{\mathrm{d}t^2} \theta(t) \right\rangle \\
&= \lambda \left\langle \frac{\mathrm{d}^2}{\mathrm{d}t^2} \int_0^2 \phi(\gamma)\, {}_0\mathrm{I}_t^{\gamma-2} y(t) \, \mathrm{d}\gamma, \theta(t) \right\rangle + \langle f(t, y(t)), \theta(t) \rangle \\
&= \lambda \int_0^2 \phi(\gamma) \langle {}_0\mathrm{D}_t^\gamma y, \theta \rangle \, \mathrm{d}\gamma + \langle f(t, y(t)), \theta(t) \rangle. \quad ∎
\end{aligned}$$

4.4.1.2. *Classical (strong) solution*

Now, we will prove the existence of a classical solution for [4.85]. We need the following condition:

$(A)_s$ $f(t,y), (t,y) \in [0,\infty) \times \mathbb{R}$ is continuous and for some $a > 0$ there exists $h \in C([0,a])$ such that

$$|f(t,y) - f(t,z)| \leq h(t)\,|y-z|, \quad y,z \in \mathbb{R},\ t \in [0,a]. \qquad [4.88]$$

THEOREM 4.7.– Assume $(A)_s$. Then, there exist $\delta > 0$ and $y \in C^2([0,\delta])$, which satisfies [4.85] on $[0,\delta]$.

PROOF.– We will show that $T : C^2([0,a]) \to C^2([0,a])$, defined by

$$Ty(t) = \lambda \int_0^2 \phi(\gamma)\, {}_0\mathrm{I}_t^{\gamma-2} y(t)\, \mathrm{d}\gamma + {}_0\mathrm{I}_t^2 f(t, y(t)) + pt + q, \quad t \geq 0,$$

is a compact operator and that it maps $B_r = \{y \in C^2[0,\delta] : \|y\|_2 \leq r\}$ into itself, where $r > \max\{|p|,|q|\}$ and $\delta < a$ is determined with respect to $r$. Then, the Schauder fixed point theorem assures that $T$ has at least one fixed point. This is the classical solution.

Again, we put $Ty = Jy + Gy + pt + q$, where

$$\begin{aligned} Jy(t) &= \lambda \int_0^2 \phi(\gamma)\, {}_0\mathrm{I}_t^{\gamma-2} y(t)\, \mathrm{d}\gamma \\ &= \lambda \int_0^2 \frac{\phi(\gamma)}{\Gamma(2-\gamma)} \int_0^t \frac{y(x)}{(t-x)^{\gamma-1}} \mathrm{d}x\mathrm{d}\gamma, \quad t \geq 0, \\ Gy(t) &= \int_0^t (t-x) f(x, y(x))\, \mathrm{d}x, \quad t \geq 0. \end{aligned}$$

As in theorem 4.6, we have $t \mapsto Ty(t)$ which has a continuous first derivative. For the continuity of the second derivative $\frac{\mathrm{d}^2}{\mathrm{d}t^2}(T(y))$, we note

$$\frac{\mathrm{d}^2}{\mathrm{d}t^2}(Ty(t)) = \lambda \int_0^2 \frac{\phi(t)}{\Gamma(2-\gamma)} \int_0^t (t-x)^{1-\gamma} y^{(2)}(x)\, \mathrm{d}x\mathrm{d}\gamma, \quad t \geq 0,$$

and $\frac{\mathrm{d}^2}{\mathrm{d}t^2}(Gy(t)) = f(x, y(x)),\ t \geq 0$. Thus, $T : C^2([0,a]) \to C^2([0,a])$.

Now, we prove the compactness of $T$ in the ball $B(0,r)$ of $C^2([0,a])$, where $r > \max\{|p|,|q|\}$. We have the same proof for the operator $J$ as in the proof of theorem 4.6 since we have to take $i = 0,1,2$. Consider operator $G$. From the Arzelà–Ascoli theorem, it follows that for a given bounded sequence $(y_n)_n$, in $C^2([0,a])$, and

$$z_n(t) = {}_0\mathrm{I}_t^2 f(t, y_n(t)), \quad t \in [0,a], \ n \in \mathbb{N},$$

there exists a subsequence, again denoted by $(z_n)_n$, such that it converges to some continuous function $z$ and that $(\frac{\mathrm{d}}{\mathrm{d}t} z_n)_n$ converges to $\frac{\mathrm{d}}{\mathrm{d}t} z$, which is also continuous in $[0,a]$. We will prove that there is a subsequence of $(z_n)_n$, again denoted in the same way, such that $\frac{\mathrm{d}^2}{\mathrm{d}t^2} z_n = f(\cdot, y_n)$, $n \in \mathbb{N}$ converges to $\frac{\mathrm{d}^2}{\mathrm{d}t^2} z$, which is also continuous on $[0,a]$. This will complete the proof that $G$ is compact in $C^2([0,a])$. From $(\mathrm{A})_s$, for $t \in [0,a]$,

$$\left| \frac{\mathrm{d}^2}{\mathrm{d}t^2} z_n(t) - \frac{\mathrm{d}^2}{\mathrm{d}t^2} z_m(t) \right| \leq h(t)|y_n(t) - y_m(t)|, \quad n, m \in \mathbb{N}, \ t \in [0,a]. \qquad [4.89]$$

From the Arzelà–Ascoli theorem, $(y_n)_n$ has a subsequence that converges in $C([0,a])$ and this, with [4.89], implies that $(\frac{\mathrm{d}^2}{\mathrm{d}t^2} z_n)_n$ has a subsequence that is a Cauchy sequence in $C([0,a])$. This completes the proof of compactness of $G$ in $C([0,a])$.

Now, we will determine $\delta$ and prove that $T$ maps the ball $B_r = \{y \in C^2[0,\delta] : \|y\| \leq r\}$ into itself, where $r > \max\{|p|,|q|\}$. Let $y \in B_r$. Since $Ty$ and $\frac{\mathrm{d}}{\mathrm{d}t}(Ty)$ can be estimated in the same way as in the proof of theorem 4.6 (with appropriate $\delta$), we will only consider $\frac{\mathrm{d}^2}{\mathrm{d}t^2}(Ty)$

$$\begin{aligned} \left| \frac{\mathrm{d}^2}{\mathrm{d}t^2}(Ty(t)) \right| &\leq |\lambda| \int_0^2 \frac{|\phi(\gamma)|}{\Gamma(2-\gamma)} \int_0^t \frac{|y^{(2)}(x)|}{(t-x)^{\gamma-1}} \mathrm{d}x \mathrm{d}\gamma + |f(t, y(t))| + |p| \\ &\leq |\lambda| M_t \sup_{t\in[0,\delta]} |y^{(2)}(t)| + C_t \sup_{t\in[0,\delta]} |y(t)|^\alpha + |p|, \quad t \in [0,\delta], \end{aligned}$$

where

$$\begin{aligned} M_t &= \int_0^2 \frac{|\phi(\gamma)|}{\Gamma(2-\gamma)} \int_0^t (t-x)^{1-\gamma} \mathrm{d}x \mathrm{d}\gamma = \int_0^2 \frac{|\phi(\gamma)|}{\Gamma(3-\gamma)} t^{2-\gamma} \mathrm{d}\gamma, \\ C_t &= \max |h(t)|, \quad t \in [0,\delta]. \end{aligned}$$

Now, shrinking $\delta$, if it is necessary, we prove that $T$ maps the ball $Br$ into itself, and thus the proof of theorem 4.7 is finished since the fixed point $y$ solves [4.85]. ■

### 4.4.2. *Uniqueness of solutions*

In this section, we continue to assume that the constitutive function $\phi$ is continuous and that $\operatorname{supp}\phi \subset [0, d_1]$, $d_1 < 2$. We state another condition.

$(\mathrm{B})_l$ Assume

$$|f(t,u) - f(t,v)| \leq \frac{L}{t^\sigma} |u - v|, \quad t \in [0,1], \ u, v \in \mathbb{R},$$

where $0 < \sigma < d_1 - 1$ and $L > 0$ is independent of $u$ and $v$.

THEOREM 4.8.– Assume $(\mathrm{B})_l$. Then, [4.87] has a unique solution $u \in C^1([0,1])$.

Note that the assumption $0 < \sigma < d_1 - 1$ is a technical assumption and it is used for the estimate $t^{2-\sigma}, t^{1-\sigma} \leq t^{2-d_1}$, $t \in [0,1]$, and can be relaxed but we will omit this in order to simplify the proof.

PROOF.– The solution is found as a fixed point of operator $T : C^1([0,1]) \to C^1([0,1])$

$$Tu(t) = \lambda \int_0^2 \phi(\gamma) \, {}_0\mathrm{I}_t^{\gamma-2} u(t) \, \mathrm{d}\gamma + {}_0\mathrm{I}_t^2 f(t, u(t)) + pt + q.$$

Let $y$ in $C^1([0,1])$. We denote by $\frac{\mathrm{d}}{\mathrm{d}t}(Ty)$ the first derivative of $Ty$. Put $T^n = T(T^{n-1})$, $n \in \mathbb{N}$. We will show that $T^n$ maps $C^1([0,a])$ into itself and that it is the contraction operator for enough large $n$. So, in the calculations, we will have to estimate $T^n(u-v)$ and $\frac{\mathrm{d}}{\mathrm{d}t}(T^n(u-v))$ over $[0,1]$, $u, v \in C^1([0,a])$.

Let $u, v \in C^1([0,1])$, $n \in \mathbb{N}$. We will prove, for $i = 0, 1$,

$$\left| \frac{\mathrm{d}^i}{\mathrm{d}t^i}(T^n u(t)) - \frac{\mathrm{d}^i}{\mathrm{d}t^i}(T^n v(t)) \right| \leq \left( \prod_{j=1}^n K_j^i \right) t^{(2-d_1)n} \|u - v\|_1, \qquad [4.90]$$

where, for $m \in \mathbb{N}$,

$$K_m^i = |\lambda| \int_0^{d_1} \frac{\Gamma((m-1)(2-d_1)+1)}{\Gamma((m-1)(2-d_1)+3-\gamma)} |\phi(\gamma)| \mathrm{d}\gamma + L_m^i, \quad i = 0, 1, \quad [4.91]$$

$$L_m^0 = \frac{L}{((m-1)(2-d_1)+1-\sigma)((m-1)(2-d_1)+2-\sigma)},$$

$$L_m^1 = \frac{L}{((m-1)(2-d_1)+1-\sigma)}.$$

Let $m = 1$ and $i = 0, 1$. Then

$$\left| \frac{\mathrm{d}^i}{\mathrm{d}t^i} (Tu(t)) - \frac{\mathrm{d}^i}{\mathrm{d}t^i} (Tv(t)) \right|$$

$$\leq |\lambda| \int_0^2 \frac{|\phi(\gamma)|}{\Gamma(2-\gamma)} \int_0^t (t-x)^{1-\gamma} \left| u^{(i)}(x) - v^{(i)}(x) \right| \mathrm{d}x \mathrm{d}\gamma$$

$$+ \left| {}_0\mathrm{I}_t^{2-i} f(t, u(t)) - {}_0\mathrm{I}_t^{2-i} f(t, v(t)) \right|$$

$$\leq \left( |\lambda| \int_0^{d_1} \frac{1}{\Gamma(3-\gamma)} t^{2-\gamma} |\phi(\gamma)| \mathrm{d}\gamma + L_1^i t^{2-i-\sigma} \right) \|u - v\|_1$$

$$\leq K_1^i t^{2-d_1} \|u - v\|_1,$$

where $K_1^i$ is of the form [4.91] with $m = 1$. This gives the proof for $m = 1$.

We will use in the sequel that ${}_0\mathrm{I}_t^s x^\mu = \frac{\Gamma(\mu+1)}{\Gamma(\mu+s+1)} x^{\mu+s}$, $s, \mu > 0$. To prove [4.90] by induction, we assume that it is valid for $n$ and obtain, for $i = 0, 1$,

$$\left| \frac{\mathrm{d}^i}{\mathrm{d}t^i} (T^{n+1}u(t)) - \frac{\mathrm{d}^i}{\mathrm{d}t^i} (T^{n+1}v(t)) \right|$$

$$\leq |\lambda| \int_0^{d_1} |\phi(\gamma)| \frac{1}{\Gamma(2-\gamma)} \int_0^t (t-y)^{1-\gamma}$$

$$\times \left| \frac{\mathrm{d}^i}{\mathrm{d}t^i} (T^n u(y)) - \frac{\mathrm{d}^i}{\mathrm{d}t^i} (T^n v(y)) \right| \mathrm{d}y \mathrm{d}\gamma,$$

as well as

$$|{}_0\mathrm{I}_t^{2-i} f(t, T^n u(t)) - {}_0\mathrm{I}_t^{2-i} f(t, T^n v(t))|$$

$$\leq K \left( |\lambda| \int_0^2 \frac{|\phi(\gamma)|}{\Gamma(2-\gamma)} \int_0^t (t-y)^{1-\gamma} y^{(2-d_1)n} \mathrm{d}y \mathrm{d}\gamma + {}_0\mathrm{I}_t^{2-i} (L t^{(2-d_1)n-\sigma}) \right)$$

$$\times \|u - v\|_1$$

$$= K\left(|\lambda|\int_0^2 \frac{\Gamma((2-d_1)n+1)\,t^{(2-d_1)n+2-\gamma}}{\Gamma((2-d_1)n+1+2-\gamma)}|\phi(\gamma)|\,\mathrm{d}\gamma + L^i_{n+1}t^{(2-d_1)n+2-d_1}\right)$$
$$\times\|u-v\|_1$$
$$\leq KK^i_{n+1}t^{(2-d_1)(n+1)}\|u-v\|_1 \leq Kt^{(n+1)(2-d_1)}\|u-v\|_1,$$

where $K=\prod_{j=1}^n K^i_j$. Thus, we proved [4.90].

The second part in expression for $K^i_m$, $i=0,1$, given with [4.91], clearly tends to zero as $m\to 0$. To show that the first part of $K^i_m$, $i=0,1$ tends to zero as $m\to 0$, note that $\lim\limits_{\alpha\to\infty}\frac{\Gamma(\alpha)}{\Gamma(\alpha+s)}=0$, for $s>0$. Indeed, using Stirling's formula for $\Gamma(\alpha)$, we have, for $\alpha$ large enough,

$$\frac{\Gamma(\alpha)}{\Gamma(\alpha+s)}\sim\frac{\mathrm{e}^{-\alpha}\alpha^{\alpha}\sqrt{2\alpha\pi}}{\mathrm{e}^{-\alpha-s}(\alpha+s)^{\alpha+s}\sqrt{2(\alpha+s)\pi}}\sim\frac{\mathrm{e}^s}{\alpha+s}\left(\frac{\alpha}{\alpha+s}\right)^{\alpha}\sim\frac{1}{\alpha+s}.$$

Therefore, for $n$ large enough, $K^i_n$ could be made sufficiently small so that $\prod_{j=1}^n K_j<1$, $i=0,1$. This gives that for large enough $n$, $T^n$ is the contraction and there exists a fixed point $y\in C^1([0,a])$ for $T$ and it is the unique solution for [4.87]. ■

REMARK 4.4.– We do not have an appropriate condition, stronger than $(\mathrm{B})_l$, which will imply the existence and the uniqueness theorem for the classical solution to [4.85] similar to theorem 4.8. Actually, we assume much stronger conditions that imply such a theorem but with this condition the proof is simple.

We introduce a new condition.

$(\mathrm{B})_s$ Let $F=\int_0^{d_1}|\phi(\gamma)|\,\mathrm{d}\gamma$. Function $f$ is continuous in $[0,a]\times R$,

$$|f(t,u)-f(t,v)|\leq L|u-v|,\quad t\in[0,1],\ u,v\in\mathbb{R},$$

where $L\in[0,1)$ is independent of $u,v$, and $\frac{F|\lambda|}{\Gamma(3-d_1)}+L<1$.

PROPOSITION 4.2.– With the assumption $(\mathrm{B})_s$, it holds that [4.85] has a unique solution $u\in C^2([0,1])$.

PROOF.– Let $i = 0, 1, 2$, $t \in [0, 1]$. We have

$$\begin{aligned}
&\left| \frac{\mathrm{d}^i}{\mathrm{d}t^i} (Tu(t)) - \frac{\mathrm{d}^i}{\mathrm{d}t^i} (Tv(t)) \right| \\
&\quad \leq |\lambda| \int_0^{d_1} \frac{|\phi(\gamma)|}{\Gamma(2-\gamma)} \int_0^t (t-x)^{1-\gamma} \left| u^{(i)}(x) - v^{(i)}(x) \right| \mathrm{d}x \mathrm{d}\gamma \\
&\qquad + \left| {}_0\mathrm{I}_t^{2-i} f(t, u(t)) - f(t, v(t)) \right| \\
&\quad \leq \left( \frac{F|\lambda|}{\Gamma(3-d_1)} t^{2-d_1} + L \right) ||u - v||_2.
\end{aligned}$$

This implies the assertion. ■

Now, we consider the solvability of [4.80] for the special case when $f(y,t) = f(t)$, where $f$ is given. As before, we set $Y = AC^1([0,b])$ and $Z = L^r(0,b)$, where $1 \leq r < \frac{1}{\alpha}$ with $0 < \alpha < 1$. Let $\mathcal{A} : Y \to Z$ be a linear operator defined by

$$\mathcal{A}y(t) = \frac{\mathrm{d}}{\mathrm{d}t}\left( g(t) \frac{\mathrm{d}}{\mathrm{d}t} y(t) \right) + \lambda \int_0^2 \phi(\gamma)\, {}_0\mathrm{D}_t^{\gamma} y(t)\, \mathrm{d}\gamma, \quad t \in [0,b].$$

Then [4.80], with $f(t, y(t)) = f(t)$, becomes

$$\mathcal{A}y(t) = f(t). \qquad [4.92]$$

If we define $\langle \mathcal{A}y, z \rangle = \int_0^b (\mathcal{A}y(t))\, z(t)\, \mathrm{d}t$, then

$$\begin{aligned}
\langle \mathcal{A}y, z \rangle &= \int_0^b \left( \frac{\mathrm{d}}{\mathrm{d}t}\left( g(t) \frac{\mathrm{d}}{\mathrm{d}t} y(t) \right) + \lambda \int_0^2 \phi(\gamma)\, {}_0\mathrm{D}_t^{\gamma} y(t)\, \mathrm{d}\gamma \right) z(t)\, \mathrm{d}t \\
&= \int_0^b \left( \frac{\mathrm{d}}{\mathrm{d}t}\left( g(t) \frac{\mathrm{d}}{\mathrm{d}t} z(t) \right) + \lambda \int_0^2 \phi(\gamma)\, {}_t\mathrm{D}_b^{\gamma} z(t)\, \mathrm{d}\gamma \right) y(t)\, \mathrm{d}t,
\end{aligned}$$

where integration by parts is used. We write

$$\langle \mathcal{A}y, z \rangle = \langle y, \mathcal{A}^* z \rangle, \qquad [4.93]$$

where $\mathcal{A}^*$ is the adjoint operator to $\mathcal{A}$ defined by

$$\mathcal{A}^* z(t) = \frac{\mathrm{d}}{\mathrm{d}t}\left( g(t) \frac{\mathrm{d}}{\mathrm{d}t} z(t) \right) + \lambda \int_0^2 \phi(\gamma)\, {}_t\mathrm{D}_b^{\gamma} z(t)\, \mathrm{d}\gamma.$$

Let $z_0$ be a solution to $\mathcal{A}^*z = 0$. Then, from [4.92] and [4.93], we formulate the following proposition.

PROPOSITION 4.3.– Equation [4.92] has a solution in $Y$ if $\langle z_0, f\rangle = 0$.

Proposition 4.3 could be stated in the form of the Fredholm alternative. Consider the eigenvalue problem for $\mathcal{A}^*$, i.e. $\mathcal{A}^*z = \mu z$, $\mu \in \mathbb{C}$. Then, proposition 4.3 is equivalent to the following statement: either $\mu = 0$ is an eigenvalue of the operator $\mathcal{A}^*$ (with a non-zero eigenvector) or [4.92] has a solution for every $f$.

Returning now to equation [4.79], we note that it becomes

$$\frac{\mathrm{d}^2}{\mathrm{d}t^2}\varphi(t) + \Lambda \int_0^2 a^\gamma \, {}_0\mathrm{D}_t^\gamma \varphi(t)\,\mathrm{d}\gamma = -G\sin\varphi(t), \qquad [4.94]$$

with $\Lambda = \frac{\lambda}{ml^2}$, $G = \frac{g}{l}$, $\phi(\gamma) = a^\gamma$. This is a generalization of [4.95].

### 4.4.3. *Nonlinear vibrations with single term of fractional dissipation*

We consider a special case of [4.79] with $\phi(\gamma) = \delta(\gamma - \alpha)$, $\alpha \in (0,1)$. Note that it is also a nonlinear version of [4.1]. Thus, we have

$$x^{(2)}(t) + b\,{}_0\mathrm{D}_t^\alpha x(t) + k\sin x(t) = 0, \quad t \in (0,T], \qquad [4.95]$$

$$x(0) = x_0, \quad x^{(1)}(0) = v_0, \qquad [4.96]$$

where $k$ is constant and $0 < \alpha < 1$. Systems [4.95] and [4.96] represent a nonlinear mathematical pendulum with fractional damping.

Since [4.95] is a special case of [4.85], from theorem 4.6, we conclude that there exists $\delta > 0$ such that [4.95] has $x \in C^2([0,\delta])$ solution. Also, if condition $(\mathrm{B})_s$, which reduces to $k + b < 1$, is satisfied, then there exists $x \in C^2([0,1])$ as a solution to [4.95].

To find solutions to [4.95] and [4.96], we will use the expansion formulas [2.31] and [2.32] in [4.95] so that, for $t \in (0,T]$,

$$x^{(2)}(t) + b\left(\mathcal{A}(N)\frac{x(t)}{t^\alpha} + \mathcal{B}(N)t^{1-\alpha}x^{(1)}(t) - \sum_{p=1}^{N} C_{p-1}\frac{V_{p-1}(x)(t)}{t^{p+\alpha}}\right)$$
$$= -k\sin x(t), \qquad [4.97]$$

where $V_{p-1}(x)(t) = \int_0^t \tau^{p-1}x(\tau)\mathrm{d}\tau, t \in (0,T]$, and

$$\begin{aligned} \mathcal{A}(N) &= \frac{\sin(\alpha\pi)}{\alpha\pi}\frac{\Gamma(N+\alpha)}{\Gamma(N)}, \quad \mathcal{B}(N) = \frac{\Gamma(N+\alpha)}{\Gamma(\alpha)\Gamma(2-\alpha)\Gamma(N+1)}, \\ \mathcal{C}_{p-1} &= \frac{\Gamma(p+\alpha)}{\Gamma(\alpha)\Gamma(1-\alpha)\Gamma(p)}. \end{aligned}$$

With $x = x_1$, $x^{(1)} = x_1$, [4.97] becomes

$$\begin{aligned} &x_1^{(1)}(t) = x_2(t), \quad t \in (0,T], \\ &x_2^{(1)}(t) = -b\left(\mathcal{A}(N)\frac{x_1(t)}{t^\alpha} + \mathcal{B}(N)t^{1-\alpha}x_2(t) - \sum_{p=1}^{N}\mathcal{C}_{p-1}\frac{V_{p-1}(x_1)(t)}{t^{p+\alpha}}\right), \\ &\qquad t \in (0,T], \\ &V_{p-1}^{(1)}(x_1)(t) = t^{p-1}x_1(t), \quad t \in (0,T], \\ &x_1(0) = x_0, \quad x_2(0) = v_0, \quad V_p(0) = 0, \quad p = 1,2,3,\ldots,N. \end{aligned} \quad [4.98]$$

The solution for $\alpha = 0.9$, $b = 0.5$, $x_0 = 0$, $v_0 = 1$, $N = 7$ is shown in Figure 4.4, where the solution of the linearized equations [4.95] and [4.96] is also shown.

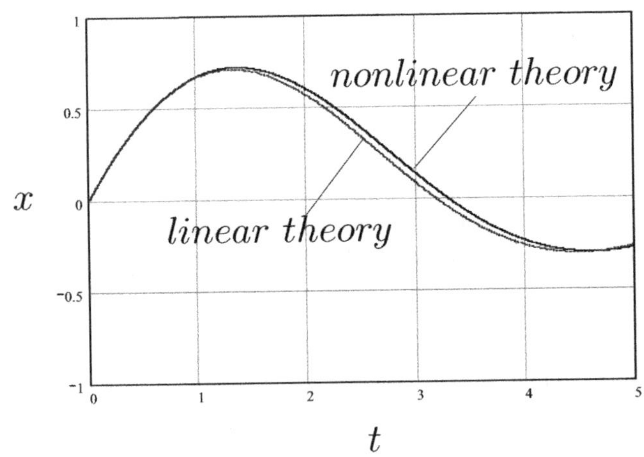


**Figure 4.4.** *Solutions to [4.95] and [4.96] and the linearization of [4.95] and [4.96]*

# Chapter 5

# Lateral Vibrations and Stability of Viscoelastic Rods

In this chapter, we will present the results from [ATA 97, ATA 02c, ATA 04c, ATA 04d, BAC 00, STA 01, STA 02, STA 04a] related to the formulation and solution of various problems from the mechanics of the deformable body, especially from the lateral vibrations of the viscoelastic rods.

Consider a rod in a deformed state, positioned in the Cartesian coordinate system $\bar{x} - O - \bar{y}$, so that one of its ends is placed at the origin, as shown in Figure 5.1. The axis of the rod is initially determined by two functions $x_0(S)$ and $y_0(S)$, $S \in [0, L]$, where $S$ is the arc length of the rod's axis and $L$ is the length of the rod.




**Figure 5.1.** *Coordinate system and load configuration*

We assume that the Bernoulli hypothesis holds (the plane cross-section remains planar during the deformation), the rod's axis is inextensible and the influence of shear stresses on deformation can be neglected. With these hypotheses, the strain of an element of the rod's axis on the distance $z$ from the neutral axis is

$$\varepsilon_z(S,t) = \frac{z}{1+\frac{z}{R(S)}}\frac{\mathrm{d}}{\mathrm{d}S}\varphi(S,t), \quad \varphi(S,t) = \theta(S,t) - \Theta(S),$$
$$S \in [0,L], \ t > 0, \tag{5.1}$$

where $\varphi$ is the rotation angle of the cross-section, and $\theta$ and $\Theta(S) = \theta(S,0)$ are the angles between the $\bar{x}$-axis and rod's axis in the deformed and in the initial state, respectively. Note that we consider a rod of constant cross-section, so that $z$ does not depend on $S$. The rotation angle of the cross-section is determined by

$$\frac{\mathrm{d}}{\mathrm{d}S}\varphi(S,t) = \frac{1}{r(S,t)} - \frac{1}{R(S)}, \quad S \in [0,L], \ t > 0, \tag{5.2}$$

where $r$ and $R(S) = r(S,0)$ denote the radius of curvature of the rod's axis in the current and in the initial state, respectively. The bending moment of the internal forces is

$$M(S) = \int_{\mathcal{A}} \sigma(S)\, z \,\mathrm{d}A, \quad S \in [0,L], \tag{5.3}$$

where $\sigma$ is the stress, at the point on the distance $z$ from the neutral axis, at the cross-section $\mathcal{A}$, which is on the distance $S$ from point $O$, and $A$ is the cross-section area.

Consider the motion of an infinitesimal element of the rod, shown in Figure 5.1, in $\bar{x} - O - \bar{y}$ plane. Applying the D'Alembert principle, the equations of motion of the rod take the form ($S \in [0,L], t > 0$)

$$\frac{\partial}{\partial S}H(S,t) + q_x(S,t) = \rho\frac{\partial^2}{\partial t^2}x(S,t), \quad \frac{\partial}{\partial S}V(S,t) + q_y(S,t) = \rho\frac{\partial^2}{\partial t^2}y(S,t), \tag{5.4}$$

$$\frac{\partial}{\partial S}M(S,t) + V(S,t)\frac{\partial}{\partial S}x(S,t) - H(S,t)\frac{\partial}{\partial S}y(S,t) + m(S,t) = J\frac{\partial^2}{\partial t^2}\theta(S,t). \tag{5.5}$$

In [5.4] and [5.5], the coordinates of an arbitrary point on the rod's axis in the deformed state, along the $\bar{x}$- and $\bar{y}$-axes, are denoted by $x$ and $y$, respectively. The angle between the tangent to the rod's axis and the $\bar{x}$-axis is denoted by $\theta$. Components of the contact forces (representing the influence of the part of the rod $[0,S)$ on the part $[S,L]$) along the $\bar{x}$- and $\bar{y}$-axes are denoted by $H$ and $V$, respectively, while $M$ is the bending moment. The projections along the $\bar{x}$- and $\bar{y}$-axes of the distributed forces per unit length acting on the rod's axis are denoted by $q_x$ and $q_y$, respectively, while $m$ represents the intensity of the distributed couples. We also use $\rho$ and $J$ to denote the line density of the rod and its moment of inertia, respectively.

Equations of motion [5.4] and [5.5] are complemented by the geometrical equations

$$\frac{\partial}{\partial S}x\left(S,t\right)=\cos\left(\theta\left(S,t\right)\right),\quad \frac{\partial}{\partial S}y\left(S,t\right)=\sin\left(\theta\left(S,t\right)\right),\quad S\in\left[0,L\right],\ t>0. \qquad [5.6]$$

We have five equations with six unknowns in systems [5.4]–[5.6]. Therefore, an additional equation must be adjoined to systems [5.4]–[5.6]. This is the constitutive equation connecting the bending moment $M$ and the change of curvature $\frac{\mathrm{d}\varphi}{\mathrm{d}S}$, given by [5.2]. For example, assuming that the material of the rod is elastic, i.e. described by Hooke's law $\sigma=E\varepsilon$, from [5.3], we obtain

$$M(S,t)=EI\frac{\partial}{\partial S}\varphi(S,t)=EI\left(\frac{1}{r\left(S,t\right)}-\frac{1}{R\left(S\right)}\right),\quad S\in\left[0,L\right],\ t>0, \qquad [5.7]$$

where $E$ is the Young modulus of elasticity and $I=\int_A \frac{z^2}{1+\frac{z}{R}}\,\mathrm{d}A$ is the second moment of inertia of the cross-section. Constitutive assumptions different from Hooke's law lead to the constitutive equation for a bending moment different from [5.7].

Systems [5.4]–[5.6], along with the constitutive moment–curvature equation, are subject to initial conditions

$$\begin{aligned}&x\left(S,0\right)=x_0\left(S\right),\quad y\left(S,0\right)=y_0\left(S\right),\quad \frac{\partial}{\partial t}x\left(S,0\right)=\dot{x}_0\left(S\right),\\&\frac{\partial}{\partial t}y\left(S,0\right)=\dot{y}_0\left(S\right),\quad S\in\left[0,L\right].\end{aligned} \qquad [5.8]$$

Boundary conditions can take various forms, as will be seen in the following. We refer to [ATA 97] for a detailed account on the theory of stability and lateral vibration of rods.

We will consider solutions that are of exponential growth (see sections 1.2 and 1.3). For this purpose, we will assume that our framework is $\mathcal{K}'_+$, the space of distributions supported by $[0,\infty)$ with exponential growth (see section 1.3). This implies that we have the Laplace transform of solutions defined for $\operatorname{Re}s>s_0$, with the suitable $s_0$ related to the growth order of the solution. Note that for a locally integrable function of the exponential growth, the Laplace transform is the classical one.

## 5.1. Lateral vibrations and creep of a fractional-type viscoelastic rod

Consider the rod of length $L$, simply supported at both ends $O$ and $B$. The end of the rod $O$ is fixed, while the end $B$ can freely move along the $\bar{x}$-axis of a rectangular Cartesian coordinate system $\bar{x}-O-\bar{y}$. The rod is loaded at $B$ by a concentrated force $F = F(t), t > 0$, as shown in Figure 5.2. The rod's axis is initially determined by the functions $x(S,0) = x_0(S)$ and $y(S,0) = y_0(S), S \in [0, L]$.

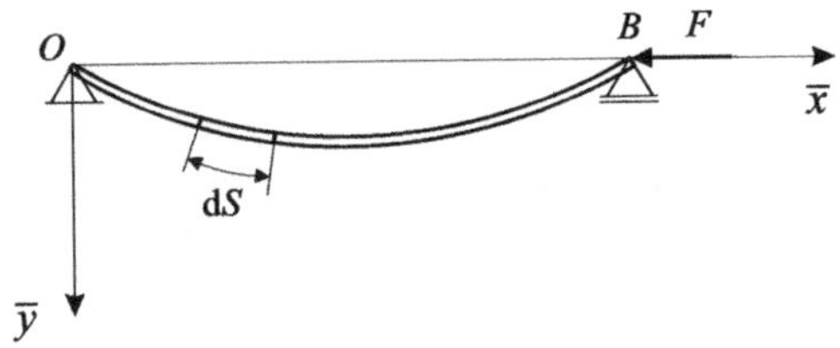


**Figure 5.2.** *Coordinate system and load configuration*

The problem of determining the critical force, i.e. the force $F$ for which the rod loses stability, is the classical one. In the case when the rod is viscoelastic, the problem has been discussed in many publications (see [DRO 87, NOW 63, VIN 87] and the references therein).

According to the setup of the problem, we have that (other than inertial) there are neither distributed forces ($q_x = q_y = 0$), nor distributed couples ($m = 0$), since we neglect the moment of inertia $J$. Therefore, the equations of motion [5.4] and [5.5], complemented by the geometrical equation [5.6], read ($S \in [0, L], t > 0$)

$$\frac{\partial}{\partial S} H(S,t) = \rho \frac{\partial^2}{\partial t^2} x(S,t), \quad \frac{\partial}{\partial S} V(S,t) = \rho \frac{\partial^2}{\partial t^2} y(S,t), \tag{5.9}$$

$$\frac{\partial}{\partial S} M(S,t) + V(S,t) \frac{\partial}{\partial S} x(S,t) - H(S,t) \frac{\partial}{\partial S} y(S,t) = 0, \tag{5.10}$$

$$\frac{\partial}{\partial S} x(S,t) = \cos(\theta(S,t)), \quad \frac{\partial}{\partial S} y(S,t) = \sin(\theta(S,t)). \tag{5.11}$$

The boundary conditions read

$$x(0,t) = 0, \ \ y(0,t) = 0, \ \ M(0,t) = 0, \ \ y(L,t) = 0, \ \ H(L,t) = -F(t), \ \ M(L,t) = 0, \ \ t > 0. \tag{5.12}$$

Consider the initially straight rod. Then, $R(S) \to \infty$ and $\Theta(S) = 0$, $S \in [0, L]$. We have that $[5.1]_2$ and [5.2] yield

$$\frac{1}{r(S,t)} = \frac{\partial}{\partial S}\theta(S,t), \quad S \in [0,L], \; t > 0. \qquad [5.13]$$

In addition, let $\dot{x}_0 = \dot{y}_0 = 0$. Then, initial conditions [5.8] become

$$x(S,0) = S, \;\; y(S,0) = 0, \;\; \frac{\partial}{\partial t}x(S,0) = 0, \;\; \frac{\partial}{\partial t}y(S,0) = 0, \;\; S \in [0,L]. \qquad [5.14]$$

We will consider three cases of the constitutive moment–curvature relation. For the corresponding constitutive stress–strain relations, we refer to [PAP 97, PRI 96, RAB 80, TOR 84].

1) The material of the rod is modeled by the fractional Kelvin–Voigt constitutive equation

$$\sigma(t) = (G_0 + G_1\, {}_0\mathrm{D}_t^{\alpha})\,\varepsilon(t), \quad t > 0, \qquad [5.15]$$

where $G_0$, $G_1$ and $\alpha \in (0,1)$ are constants. Then, by using [5.1] in [5.15] and the resulting expression in [5.3], along with $R(S) \to \infty$ and $\Theta(S) = 0$, $S \in [0, L]$, we obtain

$$M(S,t) = (G_0 I + G_1 I\, {}_0\mathrm{D}_t^{\alpha})\,\frac{1}{r(S,t)}, \quad S \in [0,L], \; t > 0. \qquad [5.16]$$

2) The stress–strain relation for the material of the rod is given by the fractional Zener model

$$(1 + a\, {}_0\mathrm{D}_t^{\alpha})\,\sigma(t) = E\,(1 + b\, {}_0\mathrm{D}_t^{\alpha})\,\varepsilon(t), \quad t > 0, \qquad [5.17]$$

where $\alpha \in (0,1)$. Recall the thermodynamical restrictions on the constants: $E > 0$, $0 < a \leq b$. By the similar calculations as in the previous case, the constitutive moment–curvature relation takes the form

$$(1 + a\, {}_0\mathrm{D}_t^{\alpha})\,M(S,t) = EI\,(1 + b\, {}_0\mathrm{D}_t^{\alpha})\,\frac{1}{r(S,t)}, \quad S \in [0,L], \; t > 0. \qquad [5.18]$$

3) The constitutive equation is a generalization of the fractional Zener model [5.17] having fractional derivatives of strain of two different orders

$$(1 + a\, {}_0\mathrm{D}_t^{\alpha})\,\sigma(t) = E\left(1 + b\, {}_0\mathrm{D}_t^{\alpha} + c\, {}_0\mathrm{D}_t^{\beta}\right)\varepsilon(t), \quad t > 0. \qquad [5.19]$$

Thermodynamical restrictions are: $E, c > 0$, $0 < a < b$, $0 < \alpha < \beta < 1$. The same procedure as before leads to the constitutive moment–curvature relation in the form

$$(1 + a\, {}_0\mathrm{D}_t^{\alpha})\,M(S,t) = EI\left(1 + b\, {}_0\mathrm{D}_t^{\alpha} + c\, {}_0\mathrm{D}_t^{\beta}\right)\frac{1}{r(S,t)}, \quad S \in [0,L], t > 0. \qquad [5.20]$$

Trivial solution to systems [5.9]–[5.12] is ($S \in [0, L]$, $t > 0$)

$$
\begin{aligned}
&x^0 (S,t) = S, \;\; y^0 (S,t) = 0, \;\; \theta^0 (S,t) = 0, \;\; H^0 (S,t) = -F (t), \\
&V^0 (S,t) = 0, \;\; M^0 (S,t) = 0.
\end{aligned} \qquad [5.21]
$$

Let $x = x^0 + \Delta x, \ldots, M = M^0 + \Delta M$, where $\Delta x, \ldots, \Delta M$ denote the perturbations. Substituting such $x, \ldots, M$ in [5.9]–[5.12] and neglecting the higher order terms in perturbations $\Delta x, \ldots, \Delta M$, we obtain $\Delta H = 0$ and $\Delta x = 0$, as well as ($S \in [0, L]$, $t > 0$)

$$
\begin{aligned}
&\frac{\partial}{\partial S} \Delta V (S,t) = \rho \frac{\partial^2}{\partial t^2} \Delta y (S,t), \\
&\frac{\partial}{\partial S} \Delta M (S,t) + \Delta V (S,t) + F (t) \frac{\partial}{\partial S} \Delta y (S,t) = 0, && [5.22] \\
&\frac{\partial}{\partial S} \Delta y (S,t) = \Delta \theta (S,t), && [5.23] \\
&\Delta y(0,t) = 0, \;\; \Delta M(0,t) = 0, \;\; \Delta y(L,t) = 0, \;\; \Delta M(L,t) = 0. && [5.24]
\end{aligned}
$$

We adjoin initial conditions to systems [5.22]–[5.24]

$$
\Delta y (S,0) = \Delta y_0 (S), \;\; \frac{\partial}{\partial t} \Delta y (S,0) = \Delta y_1 (S), \;\; S \in [0, L]. \qquad [5.25]
$$

Constitutive moment–curvature relations [5.16], [5.18] and [5.20], with [5.13] and [5.21] become ($S \in [0, L]$, $t > 0$)

$$
\Delta M (S,t) = (G_0 I + G_1 I \, {}_0\mathrm{D}_t^{\alpha}) \left( \frac{\partial}{\partial S} \Delta \theta (S,t) \right), \qquad [5.26]
$$

$$
(1 + a \, {}_0\mathrm{D}_t^{\alpha}) \Delta M (S,t) = EI (1 + b \, {}_0\mathrm{D}_t^{\alpha}) \left( \frac{\partial}{\partial S} \Delta \theta (S,t) \right), \qquad [5.27]
$$

$$
(1 + a \, {}_0\mathrm{D}_t^{\alpha}) \Delta M (S,t) = EI \left( 1 + b \, {}_0\mathrm{D}_t^{\alpha} + c \, {}_0\mathrm{D}_t^{\beta} \right) \left( \frac{\partial}{\partial S} \Delta \theta (S,t) \right), \qquad [5.28]
$$

respectively.

### 5.1.1. *Rod made of fractional Kelvin–Voigt-type material*

Following [ATA 02c, BAC 00, STA 02], we consider the lateral vibrations of a viscoelastic rod described by the fractional Kelvin–Voigt moment–curvature relation

[5.26]. In the sequel, we distinguish two separate cases: initially straight rod and rod with initial lateral deflection. Also, we will assume that the concentrated force $F$ is either constant or it has impulsive change at a specified time-instant.

5.1.1.1. *Initially straight rod*

If the rod is initially straight, systems [5.22]–[5.25], along with the moment–curvature constitutive relation [5.26], reduce to ($S \in [0, L]$, $t > 0$)

$$G_0 I \frac{\partial^4}{\partial S^4} \Delta y(S,t) + G_1 I \, {}_0\mathrm{D}_t^\alpha \left( \frac{\partial^4}{\partial S^4} \Delta y(S,t) \right) + F(t) \frac{\partial^2}{\partial S^2} \Delta y(S,t) + \rho \frac{\partial^2}{\partial t^2} \Delta y(S,t) = 0, \quad [5.29]$$

$$\Delta y(0,t) = 0, \quad \frac{\partial^2}{\partial S^2} \Delta y(0,t) = 0, \quad \Delta y(L,t) = 0, \quad \frac{\partial^2}{\partial S^2} \Delta y(L,t) = 0, \quad [5.30]$$

$$\Delta y(S,0) = \Delta y_0, \quad \frac{\partial}{\partial t} \Delta y(S,0) = \Delta y_1. \quad [5.31]$$

We rewrite systems [5.29]–[5.31] in the dimensionless form

$$\frac{\partial^4}{\partial \xi^4} u(\xi,t) + \gamma \, {}_0\mathrm{D}_t^\alpha \left( \frac{\partial^4}{\partial \xi^4} u(\xi,t) \right) + \lambda(t) \frac{\partial^2}{\partial \xi^2} u(\xi,t) + \frac{\partial^2}{\partial t^2} u(\xi,t) = 0, \quad \xi \in [0,1], \; t > 0, \quad [5.32]$$

$$u(0,t) = 0, \quad \frac{\partial^2}{\partial \xi^2} u(0,t) = 0, \quad u(1,t) = 0, \quad \frac{\partial^2}{\partial \xi^2} u(1,t) = 0, \quad t > 0, \quad [5.33]$$

$$u(\xi,0) = u_0(\xi), \quad \frac{\partial}{\partial t} u(\xi,0) = u_1(\xi), \quad \xi \in [0,1], \quad [5.34]$$

by introducing

$$\xi = \frac{S}{L}, \quad \bar{t} = t \sqrt{\frac{G_0 I}{\rho L^4}}, \quad u = \frac{\Delta y}{L}, \quad \lambda = \frac{F L^2}{G_0 I}, \quad \gamma = \frac{G_1}{G_0} \left( \frac{G_0 I}{\rho L^4} \right)^{\frac{\alpha}{2}},$$

and by omitting the bar ($\bar{t} \to t$).

If function $u$ is assumed in the form

$$u(\xi,t) = \sum_{k=1}^{\infty} T_k(t) \sin(k\pi\xi), \quad \xi \in [0,1], \; t > 0, \quad [5.35]$$

then the boundary conditions [5.33] are satisfied, while equation [5.32] reduces to

$$\frac{\mathrm{d}^2}{\mathrm{d}t^2}T_k(t) + \mu_k\, {}_0\mathrm{D}_t^{\alpha} T_k(t) + \omega_k^2 T_k(t) = 0, \quad t > 0, \; k \in \mathbb{N}, \tag{5.36}$$

$$T_k(0) = T_{k0}, \quad \frac{\mathrm{d}}{\mathrm{d}t}T_k(0) = T_{k1}, \quad k \in \mathbb{N}, \tag{5.37}$$

where we set

$$\mu_k = \gamma (k\pi)^4 \quad \text{and} \quad \omega_k^2 = (k\pi)^4 \left(1 - \frac{\lambda}{(k\pi)^2}\right), \quad k \in \mathbb{N}. \tag{5.38}$$

The initial conditions [5.37], adjoined to [5.36], will represent unknown constants, which can be determined from the expansion of the initial conditions [5.34] into the Fourier sine series.

*Dimensionless force $\lambda$ is a constant*: assume $\lambda < \pi^2$, then $\omega_k^2 > 0$, $k \in \mathbb{N}$, in [5.38]. Then, the solution to [5.36] and [5.37], for $t > 0$, $k \in \mathbb{N}$, is given by [4.5]

$$\begin{aligned} T_k(t) &= T_{k0} \sum_{i=0}^{\infty} \sum_{j=0}^{i} (-1)^i \binom{i}{j} \frac{\mu_k^j \omega_k^{2(i-j)} t^{2i-\alpha j}}{\Gamma(2i+1-\alpha j)} \\ &\quad + T_{k1} \sum_{i=0}^{\infty} \sum_{j=0}^{i} (-1)^i \binom{i}{j} \frac{\mu_k^j \omega_k^{2(i-j)} t^{2i+1-\alpha j}}{\Gamma(2i+2-\alpha j)}. \end{aligned} \tag{5.39}$$

Suppose that there exists $k^*$ such that $\lambda = (k^*\pi)^2$. Then, by [5.38], $\omega_{k^*} = 0$, so that equation [5.36] becomes

$$\frac{\mathrm{d}^2}{\mathrm{d}t^2}T_{k^*}(t) + \mu_{k^*}\, {}_0\mathrm{D}_t^{\alpha} T_{k^*}(t) = 0, \quad t > 0, \; k^* \in \mathbb{N}. \tag{5.40}$$

The solution to [5.40] and [5.37] is given by [4.11]

$$T_{k^*}(t) = T_{k0} \sum_{i=0}^{\infty} \frac{(-\mu_{k^*})^i t^{2i-\alpha i}}{\Gamma(2i+1-\alpha i)} + T_{k1} \sum_{i=0}^{\infty} \frac{(-\mu_{k^*})^i t^{2i+1-\alpha i}}{\Gamma(2i+2-\alpha i)}, \quad t > 0, \; k^* \in \mathbb{N}. \tag{5.41}$$

Note that for $k > k^*$, we have $\omega_k^2 > 0$, so that $T_k$ is determined by [5.39]. For $k = k^*$, $T_k$ is determined by [5.41], while for $k < k^*$, we have $\omega_k^2 = -\left|\omega_k^2\right|$, so that, by [5.39],

$$
\begin{aligned}
T_k(t) &= T_{k0}\sum_{i=0}^{\infty}\sum_{j=0}^{i}(-1)^j\binom{i}{j}\frac{\mu_k^j\left|\omega_k^2\right|^{i-j}t^{2i-\alpha j}}{\Gamma(2i+1-\alpha j)}\\
&\quad+T_{k1}\sum_{i=0}^{\infty}\sum_{j=0}^{i}(-1)^j\binom{i}{j}\frac{\mu_k^j\left|\omega_k^2\right|^{i-j}t^{2i+1-\alpha j}}{\Gamma(2i+2-\alpha j)}.
\end{aligned}
$$

REMARK 5.1.– Equation [5.36] is considered in [NAB 10] with the Caputo derivative, instead of the Riemann–Liouville fractional derivative. The solution is obtained by the Laplace transform method and a comparison is made with the integer-order damped oscillator equation [5.36] with $\alpha = 1$. It is found in [NAB 10] that there are nine distinct cases contrary to the usual three for the integer-order dissipation. Recall that the classical cases are: damped, overdamped and critically damped oscillations. It is shown in [NAB 10] that in three of the nine cases, the frequency of oscillations increases with the increase of the damping order and eventually falls to the limiting value, given by the ordinary damped oscillator equation. For the other six cases, the behavior is as expected, the frequency of oscillations decreases with the increase of the order of the derivative in the damping term.

*Dimensionless force $\lambda$ is a distribution*: we consider a case when the force acting on the rod consists of a constant (dimensionless) part $B$, subject to an impulsive change of intensity $A$ at the time instant $t_0$, described by the Dirac $\delta$-distribution

$$\lambda(t) = A\delta(t-t_0)+B,\quad t,t_0>0.$$

Let us rewrite equation [5.36], subject to [5.37], as

$$\frac{\mathrm{d}^2}{\mathrm{d}t^2}T_k(t)+\mu_k\,{}_0\mathrm{D}_t^{\alpha}T_k(t)+\omega_kT_k(t)+A_k\delta(t-t_0)T_k(t)=0,\quad t,t_0>0,\ k\in\mathbb{N}, \tag{5.42}$$

$$T_k(0)=T_{k0},\quad \frac{\mathrm{d}}{\mathrm{d}t}T_k(0)=T_{k1},\quad k\in\mathbb{N}, \tag{5.43}$$

where we set $\mu_k = \gamma(k\pi)^4$, $\omega_k = (k\pi)^4\left(1-\frac{B}{(k\pi)^2}\right)$ and $A_k = A(k\pi)^2$, $k\in\mathbb{N}$. We use

$$T_k(t)\,\delta(t-t_0)=T_k(t_0)\,\delta(t-t_0)=T_{k0}^{*}\delta(t-t_0),\quad t,t_0>0,\ k\in\mathbb{N},$$

(see [SCH 51, Volume 1]), so that [5.42] and [5.43] become

$$\frac{d^2}{dt^2}T_k(t) + \mu_k\, {}_0D_t^\alpha T_k(t) + \omega_k T_k(t) = -A_k T_{k0}^* \delta(t - t_0), \quad t, t_0 > 0, \ k \in \mathbb{N}, \tag{5.44}$$

$$T_k(0) = T_{k0}, \ \frac{d}{dt}T_k(0) = T_{k1}, \quad k \in \mathbb{N}. \tag{5.45}$$

By the Green formula, we have

$$\begin{aligned} &{}_0D_t^\alpha (H(t)\, T_k(t)) = H(t)\, {}_0D_t^\alpha T_k(t) \quad \text{and} \\ &\frac{d^2}{dt^2}(H(t)\, T_k(t)) = H(t)\frac{d^2}{dt^2}T_k(t) + T_{k1}\delta(t) + T_{k0}\delta^{(1)}(t), \end{aligned} \tag{5.46}$$

where $H$ is the Heaviside function. Note that [5.46] holds in the space of the exponential distributions $\mathcal{K}'_+$.

Applying the Laplace transform to [5.42], with the initial conditions [5.43] and the previous formulas taken into account, we obtain

$$\tilde{T}_k(s) = \frac{1}{s^2 + \mu_k s^\alpha + \omega_k}\left(sT_{k0} + T_{k1} - A_k T_{k0}^* e^{-st_0}\right), \quad \operatorname{Re} s > s_0, \ k \in \mathbb{N},$$

where $s_0$ will be discussed below. The inversion of the Laplace transform in the previous expression yields

$$T_k(t) = T_{k0}F_i^{(1)}(t) + T_{k1}F_i(t) - A_k T_{k0}^* H(t - t_0)F_i(t - t_0), t \geq 0, k \in \mathbb{N}, \tag{5.47}$$

where

$$F_i(t) = \mathcal{L}^{-1}\left[\tilde{F}_i(s)\right](t) = \mathcal{L}^{-1}\left[\frac{1}{s^2 + \mu_k s^\alpha + \omega_k}\right](t), \quad t \geq 0, \ i = +, 0, -,$$

so that we used $F_+$, $F_0$ and $F_-$ to denote functions corresponding to the following cases: $\omega_k > 0$, $\omega_k = 0$ and $\omega_k < 0$, respectively.

Further, we will analyze solutions in cases $\omega_k > 0$, $\omega_k = 0$ and $\omega_k < 0$, $k \in \mathbb{N}$. Analysis that follows will prove that in all cases of $\omega_k$, $k \in \mathbb{N}$, we have that $T_k \in C([0, \infty)) \cap C^1([0, t_0))$. Also, if $T_{k0} = 0$ and $A = 0$, then $T \in C^2([0, \infty)) \cap$

$L^{exp}([0,\infty))$ and $T_k^{(1)}, T_k^{(2)} \in C([0,\infty)) \cap L^{exp}([0,\infty))$, as well as if $T_{k0} \neq 0$ and $A = 0$, then $T_k \in C^1([0,\infty)) \cap C^2((0,\infty))$.

*Case* $\omega_k > 0$: Assuming $\left|\frac{\mu_k s^\alpha}{s^2+\omega_k}\right| < 1$, i.e. $\mathrm{Re}\, s > s_0$ for $s_0$ large enough, we have

$$\begin{aligned}\tilde{F}_+(s) &= \frac{1}{s^2+\mu_k s^\alpha+\omega_k} = \frac{1}{s^2+\omega_k}\frac{1}{1+\frac{\mu_k s^\alpha}{s^2+\omega_k}} \\ &= \sum_{\nu=0}^{\infty}(-\mu_k)^\nu \frac{1}{s^{(1-\alpha)\nu+1}}\left(\frac{s}{s^2+\omega_k}\right)^{\nu+1}.\end{aligned} \quad [5.48]$$

This implies that

$$\begin{aligned}F_+(t) &= \sum_{\nu=0}^{\infty}\Phi_\nu(t), \text{ with } \Phi_\nu(t) = (-\mu_k)^\nu \frac{t^{\nu(1-\alpha)}}{\Gamma(\nu(1-\alpha)+1)} \\ &\quad * \left(\cos\left(\sqrt{\omega_k}t\right)\right)^{*(\nu+1)}, \quad t \geq 0,\end{aligned} \quad [5.49]$$

where $G^{*\nu}$ stands for $\nu$-fold convolution (see [DOE 55, Part I, Satz 2, p. 305]). Series in [5.49] converges uniformly on every compact set $[0,a]$, $a > 0$. Let us examine derivatives of the function $F_+$ in $(0,a)$, $a > 0$. Fix $p \in \mathbb{N}$. Then, there exists $\nu_p$ such that $\nu_p(1-\alpha) - p > 0$. Then, $F_+$ can be written as

$$F_+(t) = \Phi_0(t) + \ldots + \Phi_{\nu_p-1}(t) + \sum_{\nu=\nu_p}^{\infty}\Phi_\nu(t), \quad t \geq 0.$$

It is easily seen that $\Phi_\nu \in C^\infty((0,\infty))$, $\nu = 0, ..., \nu_{p-1}$. Therefore, we restrict our attention to $\Phi_\nu$, $\nu \geq \nu_p$. Then

$$\Phi_\nu^{(p)}(t) = (-\mu_k)^\nu \frac{t^{\nu(1-\alpha)-p}}{\Gamma(\nu(1-\alpha)-p+1)} * \left(\cos\left(\sqrt{\omega_k}t\right)\right)^{*(\nu+1)}, \quad t \geq 0,$$

and

$$\left|\Phi_\nu^{(p)}(t)\right| \leq \mu_k^\nu \frac{t^{\nu(1-\alpha)-p+\nu+1}}{\Gamma(\nu(1-\alpha)-p+\nu+2)}, \quad t \geq 0, \quad [5.50]$$

because of

$$\left|\left(\cos\left(\sqrt{\omega_k}t\right)\right)^{*(\nu+1)}\right| \leq \frac{t^\nu}{\Gamma(\nu+1)}, \quad t \geq 0.$$

Consequently, the series $\sum_{\nu=\nu_p}^{\infty} \Phi_\nu^{(p)}(t)$ converges uniformly on any interval $[0,a]$, $a>0$. Thus, $F_+ \in C^\infty((0,\infty))$. We will prove that $F_+ \in C^2([0,\infty))$. Note that $\Phi_0 \in C^\infty([0,\infty))$. However, for $\Phi_1$, we have

$$\begin{aligned}\Phi_1^{(2)}(t) &= -\mu_k \frac{t^{-\alpha}}{\Gamma(1-\alpha)} * \left(\cos\left(\sqrt{\omega_k}t\right) - \sqrt{\omega_k}\sin\left(\sqrt{\omega_k}t\right) * \cos\left(\sqrt{\omega_k}t\right)\right), \\ \Phi_1^{(3)}(t) &= -\mu_k \frac{t^{-\alpha}}{\Gamma(1-\alpha)} - \mu_k \frac{t^{-\alpha}}{\Gamma(1-\alpha)} * \frac{\mathrm{d}}{\mathrm{d}t}\left(\cos\left(\sqrt{\omega_k}t\right)\right. \\ &\quad \left. - \sqrt{\omega_k}\sin\left(\sqrt{\omega_k}t\right) * \cos\left(\sqrt{\omega_k}t\right)\right).\end{aligned}$$

This shows that $\Phi_1 \in C^2([0,\infty))$, but not to $C^3([0,\infty))$. Since $\Phi_0 \in C^\infty([0,\infty))$, $\Phi_1 \in C^2([0,\infty))$, as well as $\Phi_\nu \in C^2([0,\infty))$, $\nu \geq 2$, it follows that $F_+ \in C^2([0,\infty))$ and $F_+^{(1)} \in C^1([0,\infty)) \cap C^\infty((0,\infty))$. By [5.50], we conclude that $F_+$, $F_+^{(1)}$ and $F_+^{(2)}$ belong to $C([0,\infty)) \cap L^{exp}([0,\infty))$, as well.

*Case $\omega_k = 0$:* In this case, we have

$$\begin{aligned}\tilde{F}_0(s) &= \frac{1}{s^2+\mu_k s^\alpha}, \quad \operatorname{Re} s > (\mu_k)^{\frac{1}{2-\alpha}}, \quad \text{so that} \\ F_0(t) &= tE_{2-\alpha,2}\left(-\mu_k t^{2-\alpha}\right), \ t \geq 0,\end{aligned}$$

where $E_{2-\alpha,2}$ is the two-parameter Mittag-Leffler function (see [ERD 53, p. 210]). Since

$$E_{2-\alpha,2}\left(-\mu_k t^{2-\alpha}\right) = \frac{1}{\mu_k t^{2-\alpha}\Gamma(\alpha)} + O\left(\left(\mu_k t^{2-\alpha}\right)^{-2}\right), \quad t \to \infty,$$

where $\alpha \in (0,2)$, we have that $F_0 \in C^{\exp}([0,\infty))$.

The first and the second derivatives of $F_0$ are

$$\begin{aligned}F_0^{(1)}(t) &= E_{2-\alpha}(-\mu_k t^{2-\alpha}) \text{ and} \\ F_0^{(2)}(t) &= -\mu_k(2-\alpha)t^{1-\alpha}\left[\frac{\mathrm{d}}{\mathrm{d}z}E_{2-\alpha}(z)\right]_{z=-\mu_k t^{2-\alpha}}, \quad t \geq 0,\end{aligned}$$

where $E_{2-\alpha} = E_{2-\alpha,1}$ is the one-parameter Mittag-Leffler function. Hence, $F_0 \in C^2([0,\infty))$ and $F_0 \notin (C^3[0,\infty))$. By the properties of the Mittag-Leffler function, it follows that $F_0, F_0^{(1)}, F_0^{(2)} \in C([0,\infty)) \cap L^{exp}([0,\infty))$.

*Case* $\omega_k < 0$: In this case

$$\tilde{F}_-(s) = \frac{1}{s^2 + \mu_k s^\alpha - (-\omega_k)}, \quad \text{so that } F_-(t) = \sum_{\nu=0}^{\infty} \Phi_\nu(t), \quad t \geq 0,$$

where $\Phi_\nu$, given by [5.49], has the form

$$\Phi_\nu(t) = (-\mu_k)^\nu \frac{t^{\nu(1-\alpha)}}{\Gamma(\nu(1-\alpha)+a)} * \left(\cosh\left(\sqrt{-\omega_k}t\right)\right)^{*(\nu+1)}, \quad t \geq 0.$$

Note that $F_-$ has the same properties as $F_+$ and $F_0$. Namely, $F_-, F_-^{(1)}, F_-^{(2)} \in C([0,\infty)) \cap L^{exp}([0,\infty))$ and $F_- \notin C^3([0,\infty))$. The proof is just the same as in the case $\omega_k > 0$.

REMARK 5.2.– From [5.44], [5.45] and [4.7], we obtain the following condition that the solution satisfies

$$\begin{aligned} T_k(t) &= T_{k0} \sum_{i=0}^{\infty} \sum_{j=0}^{i} (-1)^i \binom{i}{j} \frac{\mu_k^j \omega_k^{2(i-j)} t^{2i-\alpha j}}{\Gamma(2i+1-\alpha j)} \\ &\quad + T_{k1} \sum_{i=0}^{\infty} \sum_{j=0}^{i} (-1)^i \binom{i}{j} \frac{\mu_k^j \omega_k^{2(i-j)} t^{2i+1-\alpha j}}{\Gamma(2i+2-\alpha j)} \\ &\quad - A_k T_{k0}^* \delta(t-t_0) * \left( \sum_{i=0}^{\infty} \sum_{j=0}^{i} (-1)^i \binom{i}{j} \frac{\mu_k^j \omega_k^{2(i-j)} t^{2i+1-\alpha j}}{\Gamma(2i+2-\alpha j)} \right), \quad t > 0. \end{aligned}$$

This condition is equivalent to [5.47].

#### 5.1.1.2. *Rod with small initial deflection*

Either by supposing that the rod is light ($\rho = 0$) or by supposing that $\frac{\partial^2}{\partial t^2} x(S,t)$ and $\frac{\partial^2}{\partial t^2} y(S,t)$, $S \in [0,L]$, $t > 0$, are small (creeping motion), the equations of motion [5.9] and [5.10] with the boundary condition [5.12] simplify, since

$H = H(t) = -F(t)$ and $V = V(t) = 0$ (the latter is due to the global equilibrium conditions for the rod). Thus, we have ($S \in [0, L], t > 0$)

$$\frac{\partial}{\partial S} M(S,t) + F(t) \frac{\partial}{\partial S} y(S,t) = 0, \quad \text{or,}$$
$$\frac{\partial}{\partial S}(M(S,t) + F(t)\, y(S,t)) = 0.$$

This implies

$$M(S,t) + F(t)\, y(S,t) = 0, \quad S \in [0, L], \; t > 0, \qquad [5.51]$$

due to the boundary conditions [5.12]. The small initial deflection of the rod, along with the assumption that deflection remains small during subsequent motion, i.e. $\theta(S,t)$ is small for $S \in [0, L], t > 0$, implies that the geometrical relations [5.11] become

$$\frac{\partial}{\partial S} x(S,t) = 1, \quad \frac{\partial}{\partial S} y(S,t) = \theta(S,t), \quad S \in [0, L], \; t > 0, \qquad [5.52]$$

as well as that the initial conditions [5.14] are replaced by

$$x(S,0) = S, \quad y(S,0) = y_0(S), \quad \frac{\partial}{\partial t} x(S,0) = 0, \quad \frac{\partial}{\partial t} y(S,0) = 0,$$
$$S \in [0, L], \qquad [5.53]$$

so that there exists finite $r(S,0) = R(S)$ and $\theta(S,0) = \Theta(S), S \in [0, L]$. Therefore, the constitutive moment–curvature relation differs from [5.16]. Namely, using [5.1] and [5.2] in [5.15] and the resulting expression in [5.3], we obtain it in the form

$$M(S,t) = (G_0 I + G_1 I \, {}_0\mathrm{D}_t^{\alpha}) \left( \frac{\partial}{\partial S} \theta(S,t) - \frac{\partial}{\partial S} \Theta(S) \right),$$
$$S \in [0, L], \; t > 0. \qquad [5.54]$$

Thus, we consider systems [5.51], [5.52] and [5.54] subject to boundary [5.12] and initial conditions [5.53]. By supposing that the force does not depend on time, i.e. $F = const.$, and that deflection $y$ is decomposed into the initial deflection $y_0$ and creep deflection $y_c$ as

$$y(S,t) = y_0(S) + y_c(S,t), \quad S \in [0, L], \; t > 0,$$

systems [5.51], [5.52], [5.54], [5.12] and [5.53] describing the creep deflection become

$$G_0 I \frac{\partial^2}{\partial S^2} y_c(S,t) + G_1 I \, {}_0\mathrm{D}_t^\alpha \left( \frac{\partial^2}{\partial S^2} y_c(S,t) \right) + F y_c(S,t) = -F y_0(S), \quad S \in [0,L], \ t > 0, \qquad [5.55]$$

$$y_0(0) = y_c(0,t) = 0, \quad y_0(L) = y_c(L,t) = 0, \ t > 0. \qquad [5.56]$$

Introducing the dimensionless quantities

$$\xi = \frac{S}{L}, \ \bar{t} = \left( \frac{G_0}{G_1} \right)^{\frac{1}{\alpha}} t, \ u_0 = \frac{y_0}{L}, \ u_c = \frac{y_c}{L}, \ \lambda = \frac{FL^2}{G_0 I}$$

in [5.55] and [5.56], and after omitting the bar ($\bar{t} \to t$), we obtain

$$\frac{\partial^2}{\partial \xi^2} u_c(\xi,t) + {}_0\mathrm{D}_t^\alpha \left( \frac{\partial^2}{\partial \xi^2} u_c(\xi,t) \right) + \lambda u_c(\xi,t) = -\lambda u_0(\xi), \ \xi \in [0,1], \ t > 0, \qquad [5.57]$$

$$u_0(0) = u_c(0,t) = 0, \quad u_0(1) = u_c(1,t) = 0, \ t > 0. \qquad [5.58]$$

Systems [5.57] and [5.58] describe the creep of a viscoelastic rod with a fractional-order derivative constitutive equation.

We assume that both $u_0$ and $u_c$ are

$$u_0(\xi) = \sum_{k=1}^{\infty} u_k \sin(k\pi\xi) \ \text{ and } \ u_c(\xi,t) = \sum_{k=1}^{\infty} T_k(t) \sin(k\pi\xi), \ \xi \in [0,1], \ t > 0.$$

Then, boundary conditions [5.58] are satisfied, while equation [5.57] reduces to

$${}_0\mathrm{D}_t^\alpha T_k(t) + \left( 1 - \frac{\lambda}{(k\pi)^2} \right) T_k(t) = \frac{\lambda}{(k\pi)^2} u_k, \ t > 0, \ k \in \mathbb{N}. \qquad [5.59]$$

Note that equation [5.59] is of the form [4.12]. Therefore, by [4.15], the solution to [5.59] takes the form

$$T_k(t) = u_k \frac{\lambda}{(k\pi)^2} \sum_{n=0}^{\infty} \left( \frac{-1}{1 - \frac{\lambda}{(n\pi)^2}} \right)^n \frac{t^{\alpha(n+1)}}{\Gamma(\alpha(n+1)+1)}, \ t > 0.$$

### 5.1.2. *Rod made of fractional Zener-type material*

Following the presentation of [STA 01], we consider the lateral vibrations of the initially straight viscoelastic rod of fractional Zener-type, described by systems [5.22]–[5.25] along with the constitutive moment–curvature relation given by [5.27]. Lateral vibrations of such a system are described by

$$\frac{\partial^2}{\partial S^2}\Delta M\left(S,t\right)+F\left(t\right)\frac{\partial^2}{\partial S^2}\Delta y\left(S,t\right)+\rho\frac{\partial^2}{\partial t^2}\Delta y\left(S,t\right)=0,\quad S\in\left[0,L\right],\ t>0, \tag{5.60}$$

$$\left(1+a\,{}_0\mathrm{D}_t^{\alpha}\right)\Delta M\left(S,t\right)=EI\left(1+b\,{}_0\mathrm{D}_t^{\alpha}\right)\left(\frac{\partial^2}{\partial S^2}\Delta y\left(S,t\right)\right),\quad S\in\left[0,L\right],\ t>0, \tag{5.61}$$

$$\Delta y(0,t)=0,\ \ \Delta M(0,t)=0,\ \ \Delta y(L,t)=0,\ \ \Delta M(L,t)=0,\ \ t>0, \tag{5.62}$$

$$\Delta y\left(S,0\right)=\Delta y_0,\ \ \frac{\partial}{\partial t}\Delta y\left(S,0\right)=\Delta y_1,\ \ S\in\left[0,L\right]. \tag{5.63}$$

We will consider the case when the force $F$ consists of a constant part $F_0$ that is subject to an impulsive change of intensity $F_1$ at the time instant $t_0$

$$F\left(t\right)=F_0+F_1\delta\left(t-t_0\right),\ \ t,t_0>0,$$

where $\delta$ is the Dirac distribution. Introducing the dimensionless quantities

$$\xi=\frac{S}{L},\ \ \bar{t}=t\sqrt{\frac{EI}{\rho L^4}},\ \ u=\frac{\Delta y}{L},\ \ m=\frac{\Delta ML}{EI},\ \ \lambda=\frac{FL^2}{EI},$$

$$\mu=a\left(\frac{EI}{\rho L^4}\right)^{\frac{\alpha}{2}},\ \ \mu_1=b\left(\frac{EI}{\rho L^4}\right)^{\frac{\alpha}{2}}$$

in systems [5.60]–[5.63], after omitting bar ($\bar{t}\to t$), we obtain

$$\frac{\partial^2}{\partial \xi^2}m\left(\xi,t\right)+\lambda\left(t\right)\frac{\partial^2}{\partial \xi^2}u\left(\xi,t\right)+\frac{\partial^2}{\partial t^2}u\left(\xi,t\right)=0,\ \ \xi\in\left[0,1\right],\ t>0, \tag{5.64}$$

$$\left(1+\mu\,{}_0\mathrm{D}_t^{\alpha}\right)m\left(\xi,t\right)=\left(1+\mu_1\,{}_0\mathrm{D}_t^{\alpha}\right)\frac{\partial^2}{\partial \xi^2}u\left(\xi,t\right),\ \ \xi\in\left[0,1\right],\ t>0, \tag{5.65}$$

$$u(0,t)=0,\ \ m(0,t)=0,\ \ u(1,t)=0,\ \ m(1,t)=0,\ \ t>0, \tag{5.66}$$

$$u(\xi,0)=u_0(\xi),\ \ \frac{\partial}{\partial t}u(\xi,0)=u_1(\xi),\ \ \xi\in\left[0,1\right], \tag{5.67}$$

with $\lambda(t) = A\delta(t - t_0) + B, t, t_0 > 0$, where $A = \frac{F_1 L^2}{EI}$ and $B = \frac{F_0 L^2}{EI}$.

Assuming $u$ and $m$ in the form ($\xi \in [0,1], t > 0$)

$$u(\xi, t) = \sum_{k=1}^{\infty} T_k(t) \sin(k\pi\xi) \quad \text{and} \quad m(\xi, t) = \sum_{k=1}^{\infty} V_k(t) \sin(k\pi\xi), \qquad [5.68]$$

we have that the boundary conditions [5.66] are satisfied, so that [5.64] and [5.65], with initial conditions adjoined, become ($t, t_0 > 0$, $k \in \mathbb{N}$)

$$\frac{\mathrm{d}^2}{\mathrm{d}t^2} T_k(t) - (k\pi)^2 (A\delta(t - t_0) + B) T_k(t) - (k\pi)^2 V_k(t) = 0, \qquad [5.69]$$

$$\mu_1 (k\pi)^2 \, {}_0\mathrm{D}_t^\alpha T_k(t) + (k\pi)^2 T_k(t) + \mu \, {}_0\mathrm{D}_t^\alpha V_k(t) + V_k(t) = 0, \qquad [5.70]$$

$$T_k(0) = T_{k0}, \quad \frac{\mathrm{d}}{\mathrm{d}t} T_k(0) = T_{k1}. \qquad [5.71]$$

Note that initial conditions [5.71] will represent unknown constants in [5.68], which are to be determined from [5.67].

Since $\delta$-distribution can be treated as a measure, systems [5.69]–[5.71] have a meaning if $T_k \in C([0, \infty))$. Then, also ($t, t_0 > 0$, $k \in \mathbb{N}$)

$$\delta(t - t_0) T_k(t) = T_k(t_0) \delta(t - t_0) = T_{k0}^* \delta(t - t_0),$$

see [SCH 51, Volume 1]. By the Green formula, we have

$$ {}_0\mathrm{D}_t^\alpha (H(t) T_k(t)) = H(t) \, {}_0\mathrm{D}_t^\alpha T_k(t) \quad \text{and}$$

$$\frac{\mathrm{d}^2}{\mathrm{d}t^2} (H(t) T_k(t)) = H(t) \frac{\mathrm{d}^2}{\mathrm{d}t^2} T_k(t) + T_{k1} \delta(t) + T_{k0} \delta^{(1)}(t),$$

where $H$ is the Heaviside function.

Note that in [KEM 98b, POD 99], a single fractional differential equation is treated. Systems [5.69]–[5.71] can be reduced to a single higher order equation by differentiation (if we assume that the functions involved are differentiable a sufficient number of times) and elimination of one function. Systems [5.69]–[5.71] were treated in [ATA 02c, STA 01] by the use of the generalized Laplace transform in the space of hyperfunctions (see [KOM 87, KOM 96]).

5.1.2.1. *Solutions to systems [5.69]–[5.71]*

Applying the Laplace transform to [5.69] and [5.70], with the initial conditions [5.71] taken into account, we obtain ($\operatorname{Re} s > s_0,\ k \in \mathbb{N}$)

$$-\left(k\pi\right)^2 \tilde{V}_k\left(s\right) - \left(B\left(k\pi\right)^2 - s^2\right)\tilde{T}_k\left(s\right) = sT_{k0} + T_{k1} + \left(k\pi\right)^2 AT_{k0}^{*}\mathrm{e}^{-st_0},$$

$$\left(1+\mu s^{\alpha}\right)\tilde{V}_k\left(s\right) + \left(k\pi\right)^2\left(1+\mu_1 s^{\alpha}\right)\tilde{T}_k\left(s\right) = 0.$$

The solution to previous system is ($\operatorname{Re} s > s_0,\ k \in \mathbb{N}$)

$$\tilde{T}_k\left(s\right) = \frac{s^{\alpha}+\frac{1}{\mu}}{D_k\left(s\right)}\left(\mu sT_{k0} + \mu T_{k1} + \left(k\pi\right)^2\mu AT_{k0}^{*}\mathrm{e}^{-st_0}\right), \qquad [5.72]$$

$$\tilde{V}_k\left(s\right) = -\left(k\pi\right)^2\frac{s^{\alpha}+\frac{1}{\mu_1}}{D_k\left(s\right)}\left(\mu_1 sT_{k0} + \mu_1 T_{k1} + \left(k\pi\right)^2\mu_1 AT_{k0}^{*}\mathrm{e}^{-st_0}\right), \qquad [5.73]$$

where

$$D_k\left(s\right) = \mu s^{2+\alpha} + s^2 + a_k s^{\alpha} + d_k,\quad \operatorname{Re} s > s_0,\ k \in \mathbb{N}, \qquad [5.74]$$

with

$$a_k = \left(k\pi\right)^2\left(\left(k\pi\right)^2\mu_1 - \mu B\right),\quad d_k = \left(k\pi\right)^2\left(\left(k\pi\right)^2 - B\right),\quad k \in \mathbb{N}. \qquad [5.75]$$

We have ($\operatorname{Re} s > s_0,\ k \in \mathbb{N}$)

$$\frac{1}{D_k\left(s\right)} = \frac{1}{\mu}\frac{1}{\left(s^2+\frac{a_k}{\mu}\right)\left(s^{\alpha}+\frac{1}{\mu}\right)+c},\quad \text{with } c = \frac{1}{\mu}\left(d_k - \frac{a_k}{\mu}\right)$$

$$= \frac{1}{\mu}\frac{1}{\left(s^2+\frac{a_k}{\mu}\right)\left(s^{\alpha}+\frac{1}{\mu}\right)}\left(1+\tilde{\phi}_k\left(s\right)\right), \qquad [5.76]$$

where we chose $s_0 > 0$ so that $\operatorname{Re} s > s_0$ implies $\left|\left(s^2+\frac{a_k}{\mu}\right)\left(s^{\alpha}+\frac{1}{\mu}\right)\right| > c$, so that

$$\tilde{\phi}_k\left(s\right) = \sum_{\nu=1}^{\infty}\left(-c\right)^{\nu}\left(\frac{1}{s^2+\frac{a_k}{\mu}}\right)^{\nu}\left(\frac{1}{s^{\alpha}+\frac{1}{\mu}}\right)^{\nu},\quad \operatorname{Re} s > s_0,\ k \in \mathbb{N}. \qquad [5.77]$$

Considering [5.77], let

$$\tilde{w}(s) = \frac{1}{s^{\alpha} + \frac{1}{\mu}} \quad \operatorname{Re} s > 0, \quad \text{then} \quad w(t) = \alpha t^{\alpha-1} E_{\alpha}^{(1)}\left(-\frac{1}{\mu}t^{\alpha}\right), \quad t \geq 0, \quad [5.78]$$

where $E_{\alpha}$ is the one-parameter Mittag-Leffler function (see section 1.5). Recall that $E_{\alpha}$ is an entire function satisfying

$$E_{\alpha}(z) = -\frac{1}{\Gamma(1-\alpha)}\frac{1}{z} + O\left(\frac{1}{|z|^2}\right), \quad |\arg(-z)| < \left(1 - \frac{1}{2}\alpha\right)\pi, \quad z \to \infty.$$

By [BER 68, p. 36], we obtain

$$E_{\alpha}^{(1)}(z) = \frac{1}{\Gamma(1-\alpha)}\frac{1}{z^2} + O\left(\frac{1}{|z|^3}\right), \quad |\arg(-z)| < \left(1 - \frac{3}{4}\alpha\right)\pi, \quad z \to \infty.$$

Consequently, for $w$, given by [5.78], we have

$$w(t) \sim \frac{1}{\Gamma(\alpha)} t^{\alpha-1} \sim O\left(t^{\alpha-1}\right), \quad t \to 0 \quad \text{and}$$

$$w(t) \sim -\frac{\mu^2}{\Gamma(-\alpha)} t^{-(1+\alpha)} \sim o\left(t^{\alpha-1}\right), \quad t \to \infty,$$

Then, there exists a constant $C_1$ such that

$$|w(t)| \leq C_1 t^{\alpha-1}, \quad t \geq 0.$$

In the analysis of [5.77], we distinguish three cases, for $t \geq 0$, $k \in \mathbb{N}$, depending on the sign of $a_k$

$$\mathcal{L}^{-1}\left[\left(\frac{1}{s^2 + \frac{a_k}{\mu}}\right)^{\nu}\left(\frac{1}{s^{\alpha} + \frac{1}{\mu}}\right)^{\nu}\right](t)$$

$$= \begin{cases} \left(\sqrt{\frac{\mu}{a_k}}\right)^{\nu}\left(\sin\left(\sqrt{\frac{a_k}{\mu}}t\right) * w(t)\right)^{*\nu}, & \text{for } a_k > 0, \\ (t * w(t))^{*\nu}, & \text{for } a_k = 0, \\ \left(\sqrt{\frac{\mu}{-a_k}}\right)^{\nu}\left(\sinh\left(\sqrt{\frac{-a_k}{\mu}}t\right) * w(t)\right)^{*\nu}, & \text{for } a_k < 0, \end{cases}$$

where $f^{*\nu}$ means $\nu$-fold convolution of $f$ and $f^{*1}=f$.

Let $a_k>0$. Then, the formal inverse Laplace transform of [5.77] yields

$$\phi_k(t)=\sum_{\nu=1}^{\infty}(-c)^{\nu}\left(\sqrt{\frac{\mu}{a_k}}\right)^{\nu}\left(\sin\left(\sqrt{\frac{a_k}{\mu}}t\right)*w(t)\right)^{*\nu},\quad t\geq 0,\ k\in\mathbb{N}. \qquad [5.79]$$

If $\nu=1$, then

$$\left|\left(\sin\left(\sqrt{\frac{a_k}{\mu}}t\right)*w(t)\right)\right|\leq C_1\int_0^t t^{\alpha-1}\mathrm{d}t=C_1\frac{t^{\alpha}}{\alpha}=C_2\frac{t^{\alpha}}{\Gamma(\alpha+1)},\quad t\geq 0.$$

Also, for any $\nu\in\mathbb{N}$ and $t\geq 0$, we have

$$\left|\left(\sin\left(\sqrt{\frac{a_k}{\mu}}t\right)*w(t)\right)^{*\nu}\right|\leq C_2^{\nu}\left(\frac{t^{\alpha}}{\Gamma(\alpha+1)}\right)^{*\nu}\leq C_2^{\nu}\mathcal{L}^{-1}\left[\frac{1}{s^{(\alpha+1)\nu}}\right](t)$$
$$\leq C_2^{\nu}\frac{t^{\nu(\alpha+1)-1}}{\Gamma(\nu(\alpha+1))}.$$

By [5.79] and the previous inequality, for $t\geq 0$, $k\in\mathbb{N}$, it holds

$$|\phi_k(t)|\leq t^{-1}\sum_{\nu=1}^{\infty}\left(|c|\sqrt{\frac{\mu}{a_k}}t^{\alpha+1}\right)^{\nu}\frac{1}{\Gamma(\nu\alpha+1)}\leq t^{-1}\left(E_{\alpha}\left(|c|\sqrt{\frac{\mu}{a_k}}t^{\alpha+1}\right)-1\right).$$

Hence, series [5.79], which determines $\phi_k$ is absolutely convergent for $t\geq 0$ and $\phi_k$, is bounded on every compact set $K\subset[0,\infty)$.

We have the following properties of $F_{\nu}(t)=(-c)^{\nu}\left(\sqrt{\frac{\mu}{a_k}}\right)^{\nu}\left(\sin\sqrt{\frac{a_k}{\mu}}t*w(t)\right)^{*\nu}$, $t\geq 0$, $\nu\in\mathbb{N}$:

– $\mathcal{L}[F_{\nu}(t)](s)=(-c)^{\nu}\left(\frac{1}{s^2+\frac{a_k}{\mu}}\right)^{\nu}\left(\frac{1}{s^{\alpha}+\frac{1}{\mu}}\right)^{\nu}$.

– $\int_0^{\infty}|F_{\nu}(t)|\,\mathrm{e}^{-x_0t}\mathrm{d}t\leq C_2^{\nu}|c|^{\nu}\left(\sqrt{\frac{\mu}{a_k}}\right)^{\nu}\frac{1}{x_0^{\nu(\alpha+1)}}$.

– The series $\sum_{\nu=1}^{\infty}\int_0^{\infty}|F_{\nu}(t)|\,\mathrm{e}^{-x_0t}\mathrm{d}t\leq\sum_{\nu=1}^{\infty}C_2^{\nu}|c|^{\nu}\left(\sqrt{\frac{\mu}{a_k}}\right)^{\nu}\frac{1}{x_0^{\nu(\alpha+1)}}$ converges if $x_0^{\alpha+1}>C_2|c|\sqrt{\frac{a_k}{\mu}}$.

Thus, by [DOE 55, Volume 1, Theorem 2, p. 305], we have shown that the formal expression [5.79] for $\phi_k$ is the inverse Laplace transform of $\tilde{\phi}_k$, given by [5.77].

In the cases when $a_k = 0$ and $a_k < 0$, the procedure is analogous to the one presented for $a_k > 0$. Note that $\phi_k$ takes the following forms:

$$\phi_k(t) = \begin{cases} \sum_{\nu=1}^{\infty} (-c)^\nu \left(\sqrt{\frac{\mu}{a_k}}\right)^\nu \left(\sin\left(\sqrt{\frac{a_k}{\mu}}t\right) * w(t)\right)^{*\nu}, & \text{for } a_k > 0, \\ \sum_{\nu=1}^{\infty} (-c)^\nu (t * w(t))^{*\nu}, & \text{for } a_k = 0, \\ \sum_{\nu=1}^{\infty} (-c)^\nu \left(\sqrt{\frac{\mu}{-a_k}}\right)^\nu \left(\sinh\left(\sqrt{\frac{-a_k}{\mu}}t\right) * w(t)\right)^{*\nu}, & \text{for } a_k < 0, \end{cases}$$

$t \geq 0, \ k \in \mathbb{N}.$

Now, we examine the function

$$\tilde{f}_k(s) = \frac{s^\alpha + \frac{1}{\mu}}{D_k(s)} = \frac{1}{\mu}\frac{1}{s^2 + \frac{a_k}{\mu}}\left(1 + \tilde{\phi}_k(s)\right), \quad \operatorname{Re} s > s_0, \ k \in \mathbb{N}, \qquad [5.80]$$

where we used [5.76]. We apply the inverse Laplace transform

$$f_k(t) = \frac{1}{2\pi \mathrm{i}} \int_{\sigma - \mathrm{i}\infty}^{\sigma + \mathrm{i}\infty} \tilde{f}_k(s) \mathrm{e}^{st} \mathrm{d}s, \quad \sigma > s_0, \ t \geq 0, \ k \in \mathbb{N}, \qquad [5.81]$$

to [5.80]. Since the integral in [5.81] converges uniformly for $0 \leq t_0 \leq t \leq t_1 < \infty$, $f_k$ is a continuous function in $[0, \infty)$. Consequently, $f_k$ is bounded in the interval $[0, \varepsilon], \varepsilon \in (0, \infty)$. We refer to [GOR 97b, GOR 95] for the methods of calculation of the inverse Laplace transform. However, we will obtain an analytic form of $f_k$, more suitable than the integral representation [5.81].

By the inverse Laplace transform applied to [5.80], for $t \geq 0$, we have ($t \geq 0$, $k \in \mathbb{N}$)

$$\begin{aligned} f_k(t) &= \mathcal{L}^{-1}\left[\tilde{f}_k(s)\right](t) = \frac{1}{\mu}\left(\mathcal{L}^{-1}\left[\frac{1}{s^2 + \frac{a_k}{\mu}}\right](t) + \right. \\ &\quad \left. \mathcal{L}^{-1}\left[\frac{1}{s^2 + \frac{a_k}{\mu}}\right](t) * \phi_k(t)\right), \end{aligned} \qquad [5.82]$$

$$\begin{aligned} f_k^{(1)}(t) &= \mathcal{L}^{-1}\left[s\tilde{f}_k(s)\right](t) = \frac{1}{\mu}\left(\mathcal{L}^{-1}\left[\frac{s}{s^2 + \frac{a_k}{\mu}}\right](t) \right. \\ &\quad \left. + \mathcal{L}^{-1}\left[\frac{s}{s^2 + \frac{a_k}{\mu}}\right](t) * \phi_k(t)\right), \end{aligned} \qquad [5.83]$$

because in all three cases ($a_k > 0$, $a_k = 0$, $a_k < 0$), we have $f_k(0) = 0$ and therefore $s\tilde{f}_k(s) = \mathcal{L}\left[f_k^{(1)}(t)\right](s)$.

By [5.80], for $\operatorname{Re} s > s_0$, $k \in \mathbb{N}$, we also have

$$\begin{aligned} \frac{1}{D_k(s)} &= \frac{1}{s^\alpha + \frac{1}{\mu}}\tilde{f}_k(s) = \mathcal{L}\left[w(t) * f_k(t)\right](s) \quad \text{and} \\ \frac{s}{D_k(s)} &= \mathcal{L}\left[w(t) * f_k^{(1)}(t)\right](s). \end{aligned}$$

Applying the inverse Laplace transform to [5.72] and [5.73], for $t, t_0 \geq 0$, $k \in \mathbb{N}$, we obtain the solution to systems [5.69]–[5.71] in the form

$$\begin{aligned} T_k(t) &= \mu T_{k0} f_k^{(1)}(t) + \mu T_{k1} f_k(t) + (k\pi)^2 \mu A T_{k0}^* f_k(t - t_0) H(t - t_0), \\ V_k(t) &= -(k\pi)^2 \mu_1 T_{k0}\left(f_k^{(1)}(t) + \left(\frac{1}{\mu_1} - \frac{1}{\mu}\right)\left(w(t) * f_k^{(1)}(t)\right)\right) \\ &\quad -(k\pi)^2 \mu_1 T_{k1}\left(f_k(t) + \left(\frac{1}{\mu_1} - \frac{1}{\mu}\right)(w(t) * f_k(t))\right) \\ &\quad -(k\pi)^4 \mu_1 A T_{k0}^*\left[f_k(\tau) + \left(\frac{1}{\mu_1} - \frac{1}{\mu}\right)(w(\tau) * f_k(\tau))\right]_{\tau = t - t_0} H(t - t_0), \end{aligned} \qquad [5.84]$$

where $f_k$ and $f_k^{(1)}$ are given by [5.82] and [5.83]. Note that by [5.83], $f_k^{(1)}$, $f_k^{(2)}$ and $f_k^{(3)}$ belong to $C([0,\infty))$.

Collecting the results, we state the following theorem for systems [5.69]—[5.71] (see [STA 01]).

THEOREM 5.1.– Solution to systems [5.69]–[5.71] is given by [5.84]. It is continuous on $[0, \infty)$. If $A = 0$, then the solution belongs to $C^2([0,\infty))$. In the general case ($A \neq 0$), solution [5.84] is a solution in the sense of distributions.

Regarding systems [5.64]–[5.67], we have the following theorem (see [STA 01]).

THEOREM 5.2.– Solutions to systems [5.64]–[5.67] are given by [5.68]. The series given by [5.68] are convergent for any $\xi \in [0,1]$, $t > 0$, and they are solutions to systems [5.64]–[5.67] if

$$|C_k| \leq \begin{cases} \frac{1}{k^{5+\varepsilon}}, & a_k > 0, \\ \frac{1}{k^{3+\varepsilon}}, & a_k = 0, \\ \frac{1}{k^{5+\varepsilon}} \mathrm{e}^{-\sqrt{\frac{\mu_1}{\mu_2}}k^2}, & a_k < 0, \end{cases} \quad \varepsilon > 0,$$

where $a_k$ is given by [5.75].

5.1.2.2. *Two special cases*

In some special cases, the analytic form of the solutions to systems [5.64]–[5.67] can be remarkably simple. For example if $\mu = \mu_1 \neq 0$, then the constitutive moment–curvature relation [5.65] reduces to ($\xi \in [0,1]\,, t > 0$)

$$(1 + \mu\, {}_0\mathrm{D}_t^{\alpha}) \left( m(\xi,t) - \frac{\partial^2}{\partial \xi^2} u(\xi,t) \right) = 0, \quad \text{i.e.,} \quad m(\xi,t) = \frac{\partial^2}{\partial \xi^2} u(\xi,t),$$

which represents the elastic moment–curvature relation. Thus, with $\mu = \mu_1 \neq 0$, systems [5.64]–[5.67] describe the lateral vibration of an elastic rod and the stability criterion (boundness of solution) is already known (see [ATA 97]).

Let $k = k^*$ such that $B = (k^*\pi)^2$. Then, by [5.75], $d_{k^*} = 0$, so the function $\tilde{f}_{k^*}$, given by [5.80], takes the form

$$\tilde{f}_{k^*}(s) = \frac{1}{\mu s^2} \frac{1}{1 + \frac{1}{\mu s^{2-\alpha}} \frac{a_{k^*}}{s^\alpha + \frac{1}{\mu}}} = \frac{1}{\mu s^2} \sum_{\nu=0}^{\infty} \left( -\frac{a_{k^*}}{\mu} \right)^{\nu} \frac{1}{s^{(2-\alpha)\nu}} \left( \frac{1}{s^\alpha + \frac{1}{\mu}} \right)^{\nu},$$

$$\operatorname{Re} s > s_0, \; k \in \mathbb{N},$$

with suitable $s_0$. The inverse Laplace transform of $\tilde{f}_k$ is

$$f_{k^*}(t) = \frac{1}{\mu} t + \frac{1}{\mu} t \sum_{\nu=1}^{\infty} \left( -\frac{a_{k^*}}{\mu} \right)^{\nu} \left( \frac{t^{(2-\alpha)\nu+1}}{\Gamma((2-\alpha)\nu+2)} * w^{*\nu}(t) \right),$$

$$t \geq 0, \; k \in \mathbb{N},$$

with $w$ given by [5.78].

5.1.2.3. *Stability of $T_k$ and $V_k$*

The stability of $T_k$ and $V_k$, given by [5.84], that are the solutions to [5.64]–[5.67], depends on the real part of the zeros of $D_k$ given by [5.74]. Namely, if the zeros of $D_k$ are located in the left complex half-plane, then the amplitudes of $T_k$ and $V_k$ will decrease with time. In this case, we say that the solutions are stable. If there is a zero of $D_k$ located in the right complex half-plane, then the amplitudes of $T_k$ and $V_k$ will increase with time. In this case, we say that the solutions are unstable. For the theory related to the analysis of stability, we refer to [DOE 55, Volume II, Chapter 7]. We divide the analysis of the real part of zeros of $D_k$ in a sequence of cases, according to the parameters $a_k$, $d_k$ and $\alpha$ (see [5.75]).

We note that in studying the solution of a particular fractional differential equation by the Fourier transform method (thus, obtaining $L^2$ solutions), Beyer and Kempfle in [BEY 95] studied zeros of a complex function obtained if $\mu = 0$ in $D_k$. Kempfle and Beyer in [KEM 98a, KEM 98b] localized the zeros of an expression of the same type as $D_k$, in the context of causality, by using the Rouché theorem. Our analysis of zeros is different.

PROPOSITION 5.1.– If complex numbers $w_j$, $j = 1, ..., p$, satisfy $w_j \neq 0$ and $\gamma \leq \arg w_j < \gamma + \pi$, $j = 1, ..., p$, where $\gamma$ is a real constant, then $\sum_{j=1}^{p} w_j$ cannot vanish.

*Case $d_k < 0$*: Suppose that $s = \rho$ is a real positive zero of $D_k$, given by [5.74]. Then, equation $D_k(\rho) = 0$ is

$$\mu\rho^{2+\alpha} + \rho^2 = -a_k\rho^\alpha + (-d_k), \quad \rho > 0.$$

It is easily seen that the graphs of two functions

$$y_1(\rho) = \mu\rho^{2+\alpha} + \rho^2 \text{ and } y_2(\rho) = -a_k\rho^\alpha + (-d_k), \quad \rho > 0, \qquad [5.85]$$

always have at least one common point $\rho > 0$ for all real $a_k$ and $\alpha \in (0, 1)$ (see Figure 5.3). Thus, in case $d_k < 0$, function $D_k$ has a positive real zero, implying the instability of $T_k$ and $V_k$.

*Case $d_k = 0$*: Then $D_k$, given by [5.74], reduces to

$$D_k(s) = s^\alpha\left(\mu s^2 + s^{2-\alpha} + a_k\right) = 0, \quad s \in \mathbb{C}.$$

If we apply proposition 5.1 to $D_k$, then it follows that $D_k$, for $d_k = 0$ and $a_k \geq 0$, has no zeros with positive real part. If there would exist such a zero $s_0 = \rho e^{i\varphi}$, $\rho > 0$, $\varphi \in \left(-\frac{\pi}{2}, \frac{\pi}{2}\right)$, we would have by proposition 5.1 with $\gamma = 0$ or $\gamma = -\pi$ that $D_k(s) \neq 0$, which is a contradiction.

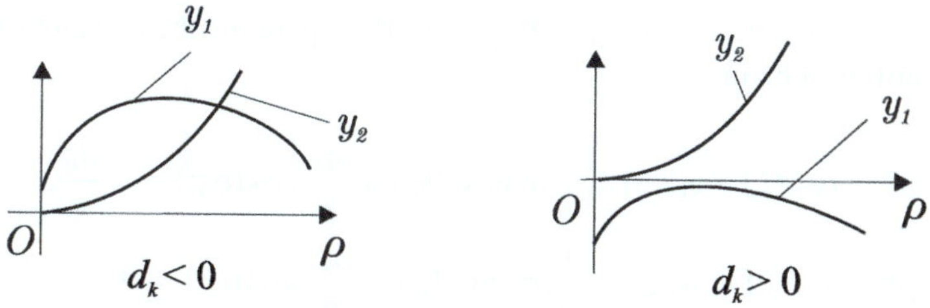


**Figure 5.3.** *The graphs of the functions [5.85]*

If $a_k < 0$, then the graphs of the two functions [5.85] always have a common point. Consequently, for $d_k = 0$ and $a_k < 0$, there always exists a real positive zero.

Thus, if $d_k = 0$ and $a_k \geq 0$, function $D_k$ has no zeros with positive real part, which implies that $T_k$ and $V_k$ are stable, while if $d_k = 0$ and $a_k < 0$, then $D_k$ has a positive real zero implying the instability of $T_k$ and $V_k$.

*Case $d_k > 0$:* Let $a_k \geq 0$ and suppose that $s = \rho$ is a real positive zero of $D_k$, given by [5.74]. Then, equation $D_k(\rho) = y_1(\rho) - y_2(\rho) = 0$, $\rho > 0$, where $y_1$ and $y_2$ are given by [5.85] and plotted in Figure 5.3, has no zero, which is real and positive. If $a_k < 0$ and if it is near to zero, then both possibilities can appear.

To see whether there are complex zeros $s_0$ with real positive part, we examine the equation $D_k(s) = 0$, $s \in \mathbb{C}$. Then, we have that $D_k(s_0) = 0$ is equivalent to

$$\left(s_0^2 + \frac{a_k}{\mu}\right)\left(s_0^\alpha + \frac{1}{\mu}\right) = \frac{1}{\mu}\left(\frac{a_k}{\mu} - d_k\right), \quad s_0 \in \mathbb{C}. \tag{5.86}$$

Let $s_0 = \rho e^{i\varphi}$, $\rho > 0$, $\varphi \in \left(-\frac{\pi}{2}, \frac{\pi}{2}\right)$, $\varphi \neq 0$, be a solution to equation [5.86]. Suppose that $\frac{a_k}{\mu} - d_k$ is positive. Then, we have

$$\rho^2 e^{2i\varphi} + \frac{a_k}{\mu} = r_1 e^{i\theta} \quad \text{and} \quad \rho^\alpha e^{i\alpha\varphi} + \frac{1}{\mu} = r_2 e^{-i\theta}, \tag{5.87}$$

where $\theta \neq 0$, because $\varphi \neq 0$. We analyze [5.87]: if $\varphi \in \left(0, \frac{\pi}{2}\right)$, then $\arg\left(\rho^2 e^{2i\varphi} + \frac{a_k}{\mu}\right) \in (0, \pi)$, and $\theta$ has to satisfy $\theta \in (0, \pi)$, but according to [5.87], we also have $\arg\left(\rho^\alpha e^{i\alpha\varphi} + \frac{1}{\mu}\right) \in \left(0, \frac{\pi}{2}\right)$ and $-\theta$ has to satisfy $-\theta \in \left(0, \frac{\pi}{2}\right)$. Such $\theta$ does not exist. The same conclusion holds if $\varphi \in \left(-\frac{\pi}{2}, 0\right)$.

Thus, we have that if $d_k > 0$, $a_k \geq 0$ and $d_k < \frac{a_k}{\mu}$, then function $D_k$ has no zeros with positive real part. The same conclusion holds if $d_k = \frac{a_k}{\mu}$.

Let $s = \rho e^{i\varphi}, \rho > 0, \varphi \in \left(-\frac{\pi}{2}, \frac{\pi}{2}\right), \varphi \neq 0$. By separating the real and imaginary parts in [5.86], we obtain

$$\rho^{2+\alpha} \cos\left((2+\alpha)\varphi\right) + \frac{1}{\mu}\rho^2 \cos(2\varphi) + \frac{a_k}{\mu} \cos(\alpha\varphi) = -\frac{d_k}{\mu}, \qquad [5.88]$$

$$\rho^{2+\alpha} \sin\left((2+\alpha)\varphi\right) + \frac{1}{\mu}\rho^2 \sin(2\varphi) + \frac{a_k}{\mu} \sin(\alpha\varphi) = 0. \qquad [5.89]$$

Systems [5.88] and [5.89], for a given set of parameters $\alpha$, $\mu$, $a_k$ and $d_k$, decide on the existence of $\rho$ and $\varphi$. If such a solution exists, the system is unstable.

5.1.2.3.1. Summary and discussion

In Table 5.1, we summarize the results on the existence of real positive zeros of $D_k$, given by [5.74], as well as of the complex zeros of $D_k$ with positive real part.

| | | | Real positive zero of $D_k$ | Zero of $D_k$ with positive real part |
|---|---|---|---|---|
| $d_k < 0$ | | | Exists | |
| $d_k = 0$ | $a_k \geq 0$ | | Does not exist | Does not exist |
| $d_k = 0$ | $a_k < 0$ | | Exists | |
| $d_k > 0$ | $a_k \geq 0$ | $d_k \leq \frac{a_k}{\mu}$ | Does not exist | Does not exist |
| $d_k > 0$ | $a_k \geq 0$ | $d_k > \frac{a_k}{\mu}$ | | Determined by [5.88] and [5.89] |
| $d_k > 0$ | $a_k < 0$ | | | Determined by [5.88] and [5.89] |

**Table 5.1.** *Dependence of the position of zeros of $D_k$ on parameters $d_k$, $a_k$ and $\frac{a_k}{\mu} - d_k$*

We comment on the physical meaning of these results. Note that $d_k < 0$ corresponds to the case when $\pi^2 < B$ (see [5.75]). In the case when the rod is loaded by the concentrated force $F_0$, we have $B = \lambda = \frac{F_0 L^2}{EI}$. Thus, $\pi^2 < B$ is equivalent to $F_0 > \pi^2 \frac{EI}{L^2}$. From Table 5.1, we conclude that in this case, there is always a real root of $D_k$ with positive real part, implying instability. Note that $F_{cr} = \pi^2 \frac{EI}{L^2}$ represents the Euler critical force for the elastic rod corresponding to [5.18]. Thus, the viscoelastic rod is unstable if the compressive force $F_0$ is larger than the Euler critical force defined for the "elastic" part of the constitutive equation (obtained from [5.18] by setting $a = b = 0$).

In the case when $d_k = 0$, i.e. $F = F_{cr}$, and $a_k = \pi^2 \left(\mu_1 \pi^2 - B\mu\right) < 0$, we have instability, indicating that viscous terms, in this case, are not sufficient to damp the vibration.

Finally, in the remaining cases, the question of stability (instability) can be answered if we solve the system of equations [5.88] and [5.89].

### 5.1.3. *Viscoelastic rod with two different fractional derivatives of strain*

Lateral vibrations of the initially straight viscoelastic rod, made of the material that follows the constitutive moment–curvature relation given by [5.28], are described by systems [5.22]–[5.25]

$$\frac{\partial^2}{\partial S^2}\Delta M(S,t) + F(t)\frac{\partial^2}{\partial S^2}\Delta y(S,t) + \rho\frac{\partial^2}{\partial t^2}\Delta y(S,t) = 0, \quad S \in [0,L], \ t > 0, \tag{5.90}$$

$$(1 + a\,{}_0\mathrm{D}_t^{\alpha})\,\Delta M(S,t) = EI\left(1 + b\,{}_0\mathrm{D}_t^{\alpha} + c\,{}_0\mathrm{D}_t^{\beta}\right)\left(\frac{\partial^2}{\partial S^2}\Delta y(S,t)\right), \quad S \in [0,L], \ t > 0, \tag{5.91}$$

$$\Delta y(0,t) = 0, \quad \Delta M(0,t) = 0, \quad \Delta y(L,t) = 0, \quad \Delta M(L,t) = 0, \quad t > 0, \tag{5.92}$$

$$\Delta y(S,0) = \Delta y_0, \quad \frac{\partial}{\partial t}\Delta y(S,0) = \Delta y_1, \quad S \in [0,L]. \tag{5.93}$$

Assume, as in the previous section, that the force $F$ consists of a constant part $F_0$ and of an impulsive change of intensity $F_1$ at the time instant $t_0$

$$F(t) = F_0 + F_1\delta(t - t_0), \quad t, t_0 > 0.$$

Following [STA 04a], we consider the system

$$\frac{\partial^2}{\partial \xi^2}m(\xi,t) + \lambda(t)\frac{\partial^2}{\partial \xi^2}u(\xi,t) + \frac{\partial^2}{\partial t^2}u(\xi,t) = 0, \quad \xi \in [0,1], \ t > 0, \tag{5.94}$$

$$(1 + \mu\,{}_0\mathrm{D}_t^{\alpha})\,m(\xi,t) = \left(1 + \mu_1\,{}_0\mathrm{D}_t^{\alpha} + \mu_2\,{}_0\mathrm{D}_t^{\beta}\right)\frac{\partial^2}{\partial \xi^2}u(\xi,t), \quad \xi \in [0,1], \ t > 0, \tag{5.95}$$

$$u(0,t) = 0, \quad m(0,t) = 0, \quad u(1,t) = 0, \quad m(1,t) = 0, \quad t > 0, \tag{5.96}$$

$$u(\xi,0) = u_0, \quad \frac{\partial}{\partial t}u(\xi,0) = u_1, \quad \xi \in [0,1], \tag{5.97}$$

with $\lambda(t) = A\delta(t - t_0) + B$. Systems [5.94]–[5.97] are obtained by introducing the dimensionless quantities

$$\xi = \frac{S}{L}, \quad \bar{t} = t\sqrt{\frac{EI}{\rho L^4}}, \quad u = \frac{\Delta y}{L}, \quad m = \frac{\Delta M L}{EI}, \quad \lambda = \frac{FL^2}{EI},$$

$$\mu = a\left(\frac{EI}{\rho L^4}\right)^{\frac{\alpha}{2}}, \quad \mu_1 = b\left(\frac{EI}{\rho L^4}\right)^{\frac{\alpha}{2}}, \quad \mu_2 = c\left(\frac{EI}{\rho L^4}\right)^{\frac{\beta}{2}}$$

in systems [5.90]–[5.93] and omitting the bar ($\bar{t} \to t$).

Assuming $u$ and $m$ in the form ($\xi \in [0,1]$, $t > 0$)

$$u(\xi,t) = \sum_{k=1}^{\infty} T_k(t) \sin(k\pi\xi) \quad \text{and} \quad m(\xi,t) = \sum_{k=1}^{\infty} V_k(t) \sin(k\pi\xi), \qquad [5.98]$$

we have that the boundary conditions [5.96] are satisfied, so that [5.94] and [5.95], with initial conditions adjoined, become ($t, t_0 > 0$, $k \in \mathbb{N}$)

$$\frac{\mathrm{d}^2}{\mathrm{d}t^2} T_k(t) - (k\pi)^2 (A\delta(t-t_0) + B) T_k(t) - (k\pi)^2 V_k(t) = 0, \qquad [5.99]$$

$$\mu_1 (k\pi)^2 \, {}_0\mathrm{D}_t^{\alpha} T_k(t) + \mu_2 (k\pi)^2 \, {}_0\mathrm{D}_t^{\beta} T_k(t) + (k\pi)^2 T_k(t) + \mu \, {}_0\mathrm{D}_t^{\alpha} V_k(t) + V_k(t) = 0, \qquad [5.100]$$

$$T_k(0) = T_{k0}, \quad \frac{\mathrm{d}}{\mathrm{d}t} T_k(0) = T_{k1}. \qquad [5.101]$$

Note that initial conditions [5.101] will represent unknown constants in [5.98], which may be determined from initial conditions [5.97].

We will assume that $T_k$ and $V_k$ are elements of $L^{exp}([0,\infty))$ (see section 1.2), so that their Laplace transforms are analytic functions for $\mathrm{Re}\, s > s_0$. Applying the Laplace transform to [5.99] and [5.100], with the initial conditions [5.101] taken into account, for $\mathrm{Re}\, s > s_0$, $k \in \mathbb{N}$, with suitable $s_0$, we get

$$\left(s^2 - B(k\pi)^2\right) \tilde{T}_k(s) - (k\pi)^2 \tilde{V}_k(s) = sT_{k0} + T_{k1} + (k\pi)^2 A T_{k0}^{*} \mathrm{e}^{-st_0}, \qquad [5.102]$$

$$(k\pi)^2 \left(1 + \mu_1 s^{\alpha} + \mu_2 s^{\beta}\right) \tilde{T}_k(s) + (1 + \mu s^{\alpha}) \tilde{V}_k(s) = 0, \qquad [5.103]$$

because $\delta(t-t_0) T_k(t) = T_k(t_0)\delta(t-t_0) = T_{k0}^{*}\delta(t-t_0)$. The solutions to systems [5.102] and [5.103] are ($\mathrm{Re}\, s > s_0$, $k \in \mathbb{N}$)

$$\tilde{T}_k(s) = \frac{1+\mu s^{\alpha}}{D_k(s)} \left(sT_{k0} + T_{k1} + (k\pi)^2 A T_{k0}^{*} \mathrm{e}^{-st_0}\right), \qquad [5.104]$$

$$\tilde{V}_k(s) = -(k\pi)^2 \frac{1 + \mu_1 s^{\alpha} + \mu_2 s^{\beta}}{D_k(s)} \left(sT_{k0} + T_{k1} + (k\pi)^2 A T_{k0}^{*} \mathrm{e}^{-st_0}\right), \qquad [5.105]$$

where

$$D_k(s) = \mu s^{2+\alpha} + s^2 + b_k s^{\beta} + a_k s^{\alpha} + d_k, \quad \mathrm{Re}\, s > s_0, \; k \in \mathbb{N}, \qquad [5.106]$$

with

$$a_k = (k\pi)^2 \left(\mu_1 (k\pi)^2 - B\mu\right), \quad b_k = \mu_2 (k\pi)^4, \quad d_k = (k\pi)^2 \left((k\pi)^2 - B\right),$$
$$k \in \mathbb{N}. \qquad [5.107]$$

Define the functions in [5.104] and [5.105] by ($\mathrm{Re}\, s > s_0$)

$$f_0 (s) = \mathcal{L}[F_0 (t)] (s) = \frac{1}{D_k (s)}, \quad f_1 (s) = \mathcal{L}[F_1 (t)] (s) = \frac{s}{D_k (s)},$$
$$f_\alpha (s) = \mathcal{L}[F_\alpha (t)] (s) = \frac{s^\alpha}{D_k (s)}, \quad f_{\alpha+1} (s) = \mathcal{L}[F_{\alpha+1} (t)] (s) = \frac{s^{\alpha+1}}{D_k (s)},$$
$$f_\beta (s) = \mathcal{L}[F_\beta (t)] (s) = \frac{s^\beta}{D_k (s)}, \quad f_{\beta+1} (s) = \mathcal{L}[F_{\beta+1} (t)] (s) = \frac{s^{\beta+1}}{D_k (s)}.$$

Then, by inverting the Laplace transform in [5.104] and [5.105], according to previously define functions, we obtain ($t, t_0 > 0$, $k \in \mathbb{N}$)

$$\begin{aligned} T_k (t) &= T_{k0} (F_1 (t) + \mu F_{\alpha+1} (t)) + T_{k1} (F_0 (t) + \mu F_\alpha (t)) \\ &\quad + (k\pi)^2 A T_{k0}^* (F_0 (t - t_0) + \mu F_\alpha (t - t_0)) H (t - t_0), \\ V_k (t) &= -(k\pi)^2 T_{k0} (F_1 (t) + \mu_1 F_{1+\alpha} (t) + \mu_2 F_{1+\beta} (t)) \\ &\quad - (k\pi)^2 T_{k1} (F_0 (t) + \mu_1 F_\alpha (t) + \mu_2 F_\beta (t)) \\ &\quad - (k\pi)^4 A T_{k0}^* (F_0 (t - t_0) + \mu_1 F_\alpha (t - t_0) + \mu_2 F_\beta (t - t_0)) H (t - t_0), \end{aligned}$$
$$[5.108]$$

provided that functions $F_0$, $F_1$, $F_\alpha$, $F_{\alpha+1}$, $F_\beta$ and $F_{\beta+1}$ exist. Proving the existence of these functions will be the subject of the analysis that follows.

#### 5.1.3.1. *Analysis of the inverse Laplace transform*

We will prove that the solutions to systems [5.99] and [5.100] are continuous functions for $t > 0$, with derivatives that exist in the sense of distributions. For this purpose, we prove that for $0 \leq p \leq 2$, there exists a continuous function $F_p (t)$, $t > 0$ such that

$$f_p (s) = \mathcal{L}[F_p (t)] (s) = \frac{s^p}{D_k (s)}, \quad \mathrm{Re}\, s > s_0, \qquad [5.109]$$

where $D_k$ is given by [5.106].

We will also analyze the behavior of the function $F_p(t)$, as $t \to \infty$. This analysis will show that we do not have a classical solution in the case $A \neq 0$. For this purpose, we use the following proposition.

PROPOSITION 5.2.– [DOE 55, Part I, Chapter 15, Paragraph 5, theorem 2] Let $f(s)$ be a holomorphic function for $\operatorname{Re} s \geq s_0$ and $f(s) = O\left(\frac{1}{s^\mu}\right)$, $\mu > 1$, when $s \to \infty$. Then, $\frac{1}{2\pi \mathrm{i}} \int_{s_0 - \mathrm{i}\infty}^{s_0 + \mathrm{i}\infty} f(s) \, \mathrm{e}^{st} \mathrm{d}s$ converges for $t \geq 0$, and defines the locally integrable function $F(t)$, $t \geq 0$ such that $\mathcal{L}[F(t)](s) = f(s)$ and $F(t) = O\left(t^{\mu-1}\right)$, $t \to \infty$.

*Case $0 \leq p < 1 + \alpha$*: To prove that continuous function $F_p$ exists, such that [5.109] holds, we use [DOE 55, Part I, Kapitel 7, Satz 5, Paragraph 2] and show that the conditions of the theorem hold.

First, suppose that $f_p(s)$ is holomorphic for $\operatorname{Re} s > s_0 \geq 0$. Let $s_1 > s_0$ and $t > 0$. Let us prove that the integral

$$I(t) = \int_{s_1 - \mathrm{i}\omega}^{s_1 + \mathrm{i}\omega} f_p(s) \, \mathrm{e}^{st} \mathrm{d}s$$

converges when $\omega \to \infty$ and that this convergence is uniform if $t \in [T_0, T_1]$, $0 < T_0 < T_1 < \infty$. This follows from the estimates

$$\begin{aligned} |I(t)| &= \left| \mathrm{e}^{s_1 t} \int_{-\omega}^{\omega} f_p(s_1 + \mathrm{i}\tau) \, \mathrm{e}^{\mathrm{i}\tau t} \mathrm{d}\tau \right| \leq \mathrm{e}^{s_1 T_1} \int_{-\omega}^{\omega} \frac{|s_1 + \mathrm{i}\tau|^p}{|D_k(s_1 + \mathrm{i}\tau)|} \mathrm{d}\tau \\ &\leq \mathrm{e}^{s_1 T_1} \left( \int_{-\omega}^{0} \frac{|s_1 + \mathrm{i}\tau|^p}{|D_k(s_1 + \mathrm{i}\tau)|} |\mathrm{d}\tau| + \int_{0}^{\omega} \frac{|s_1 + \mathrm{i}\tau|^p}{|D_k(s_1 + \mathrm{i}\tau)|} \mathrm{d}\tau \right) \\ &= \mathrm{e}^{s_1 T_1} \left( I_1(t) + I_2(t) \right), \end{aligned}$$

where $D_k$ is given by [5.106]. Let us estimate $I_2$. It is easily seen that $\lim_{|s| \to \infty} \frac{s^{2+\alpha}}{D_k(s)} = \mu$. Then, there exists $\rho_0 \in \mathbb{R}_+$ such that

$$\left| \frac{s^{2+\alpha}}{D_k(s)} \right| \leq 2\mu, \quad |s| \geq \rho_0. \qquad [5.110]$$

Inserting [5.110] in $I_2$, we have

$$\int_{\omega_1}^{\omega_2} \frac{|s_1 + \mathrm{i}\tau|^p}{|D_k(s_1 + \mathrm{i}\tau)|} \mathrm{d}\tau \leq 2\mu \int_{\omega_1}^{\omega_2} \frac{|s_1 + \mathrm{i}\tau|^p}{|s_1 + \mathrm{i}\tau|^{2+\alpha}} \mathrm{d}\tau \to 0, \quad \omega_1, \omega_2 \to \infty,$$

because of $0 \leq p < 1+\alpha$. Thus, $I_2$ converges. The same proof holds for $I_1$. Moreover, this estimate shows the uniform convergence of $I$ in any interval $[T_0, T_1]$. Similarly, we prove that $\int_{s_1-\mathrm{i}\omega}^{s_1+\mathrm{i}\omega} \frac{|f_p(s)|}{1+|s|}\mathrm{d}s$ converges when $\omega \to \infty$. Because of [5.110], there exists $M \in \mathbb{R}_+$ such that

$$|f_p(s)| = \frac{1}{|s|^{2+\alpha-p}} \frac{|s|^{2+\alpha}}{|D_k(s)|} \leq M, \quad \mathrm{Re}\, s > s_0 > s_1.$$

We also have

$$f_p(x+\mathrm{i}y) = \frac{|x+\mathrm{i}y|^p}{|D_k(x+\mathrm{i}y)|} = \frac{1}{|x+\mathrm{i}y|^{2+\alpha-p}} \frac{|x+\mathrm{i}y|^{2+\alpha}}{|D_k(x+\mathrm{i}y)|} \to 0,$$

when $x \to \infty$ for every $y \in \mathbb{R}$.

Thus, by the cited theorem, there exists a continuous function $F_p(t)$, $t > 0$, such that $\mathcal{L}[F_p(t)](s) = f_p(s)$, given by

$$F_p(t) = \mathrm{VP}\left(\frac{1}{2\pi\mathrm{i}} \int_{s_1-\mathrm{i}\infty}^{s_1+\mathrm{i}\infty} f_p(s)\, \mathrm{e}^{st} \mathrm{d}s\right), \quad t > 0, \qquad [5.111]$$

where VP denotes the principal value.

By [5.109], we have

$$f_p(s) = \frac{s^p}{D_k(s)} = \frac{s^p}{s^{2+\alpha}} \frac{s^{2+\alpha}}{D_k(s)} = O\left(\frac{1}{s^{2+\alpha-p}}\right), \quad |s| \to \infty,$$

so that, by proposition 5.2, we obtain

$$F_p(t) = O\left(t^{1+\alpha-p}\right), \quad t \to \infty.$$

*Case* $p = 1+\alpha$: For this case, we use [DOE 55, Part I, Kapitel 7, Paragraph 2, Satz 5]. So, let us show that the conditions of this theorem hold. We write $f_p$ in the form

$$f_p(s) = f_p(s) - \frac{1}{\mu s} + \frac{1}{\mu s} = \tilde{\phi}(s) + \frac{1}{\mu s}, \quad \mathrm{Re}\, s > s_0.$$

Thus, there exists a continuous function $\phi(t)$, $t > 0$, such that $\mathcal{L}[\phi(t)](s) = f_p(s) - \frac{1}{\mu s}$, $\operatorname{Re} s > s_0$. This implies that $F_p(t) = \phi(t) + \frac{1}{\mu}$, $t > 0$, is the continuous function and that $\mathcal{L}[F_p] = f$, since

$$\begin{aligned}\tilde{\phi}(s) &= f_p(s) - \frac{1}{\mu s} = \frac{s^{1+\alpha}}{\mu s^{2+\alpha} + s^2 + b_k s^\beta + a_k s^\alpha + d_k} - \frac{1}{\mu s} \\ &= \frac{-s^2 - b_k s^\beta - a_k s^\alpha - d_k}{\mu s\left(\mu s^{2+\alpha} + s^2 + b_k s^\beta + a_k s^\alpha + d_k\right)} \sim \frac{1}{\mu s^{1+\alpha}}, \quad |s| \to \infty. \qquad [5.112]\end{aligned}$$

According to [5.112], by proposition 5.2, it follows that $\phi(t) = O(t^\alpha)$, $t \to \infty$, so that

$$F_p(t) \to \frac{1}{\mu}, \quad t \to \infty.$$

*Case* $p = 1 + \beta$: Let $k \in \mathbb{N}$ be such that $k\alpha > \beta$. We can always find $a_1, \ldots, a_k$, so that

$$f_p(s) + \frac{1}{a_1 s^{1-(\beta-\alpha)}} + \ldots + \frac{1}{a_k s^{1-(\beta-k\alpha)}}, \quad \operatorname{Re} s > s_0,$$

has sufficient properties that [DOE 55, Part I, Kapitel 7, Paragraph 2, Satz 5] can be applied to find continuous function $\phi(t)$, $t > 0$ given by

$$\mathcal{L}[\phi(t)](s) = f_p(s) + \frac{1}{a_1 s^{1-(\beta-\alpha)}} + \ldots + \frac{1}{a_k s^{1-(\beta-k\alpha)}}, \quad \operatorname{Re} s > s_0.$$

Then

$$\begin{aligned}F_p(t) &= \phi(t) - \frac{1}{a_1 \Gamma(1-(\beta-\alpha))} t^{-(\beta-\alpha)} - \ldots \\ &\quad - \frac{1}{a_k \Gamma(1-(\beta-k\alpha))} t^{-(\beta-k\alpha)}, \quad t > 0.\end{aligned}$$

We apply the same procedure as in case $p = 1 + \alpha$ and obtain

$$F_p(t) = O\left(t^{-(\beta-\alpha)}\right), \quad t \to \infty.$$

Consequently, $\lim_{t\to 0^+} T_k(t) = T_{k0}$, $k \in \mathbb{N}$ and $V_k(t) \sim t^{-(\beta-\alpha)}$, as $t \to 0$. If $T_{k0} = 0$, then $\lim_{t\to 0^+} T_k(t) = 0$ and $\lim_{t\to 0^+} V_k(t) = 0$, as well.

*Case $p = 2$*: Consider, for $\operatorname{Re} s > s_0$,

$$\begin{aligned}\tilde{\phi}(s) &= f_p(s) - \frac{1}{\mu s^\alpha + 1} = \frac{s^2}{\mu s^{2+\alpha} + s^2 + b_k s^\beta + a_k s^\alpha + d_k} - \frac{1}{\mu s^\alpha + 1} \\ &= \frac{-b_k s^\beta - a_k s^\alpha - d_k}{(\mu s^\alpha + 1)(\mu s^{2+\alpha} + s^2 + b_k s^\beta + a_k s^\alpha + d_k)}.\end{aligned}$$

We see that $\tilde{\phi}$ has sufficient properties to apply [DOE 55, Part I, Kapitel 7, Satz 5, Paragraph 2], as in the previous cases. Hence

$$F_p(t) = \phi(t) + \frac{1}{\mu} w(t), \quad t > 0,$$

where $w(t) = \alpha t^{\alpha-1} E_\alpha^{(1)}\left(-\frac{1}{\mu} t^\alpha\right)$, with $E_\alpha$ being the one-parameter Mittag-Leffler function.

5.1.3.2. *Stability of $T_k$ and $V_k$*

To analyze the asymptotic behavior of solutions to systems [5.99]–[5.100], we discuss the set $Z$ on which $D_k$, given by [5.106], vanishes. If any element of $Z$ is located in the right complex half-plane, i.e. $\operatorname{Re} s_0 > 0$, $s_0 \in Z$, we have instability of solutions. If all elements of $Z$ are located in the left complex half-plane, i.e. $\operatorname{Re} s_0 < 0$, $s_0 \in Z$, we have stability of solutions. We present analysis of the elements of $Z$ in three cases: $d_k < 0$, $d_k = 0$ and $d_k > 0$, where $d_k$ is given by [5.107].

*Case $d_k < 0$*: We will show that there exists a $\rho_0 \geq 0$ such that $D_k(\rho_0) = 0$, where $D_k$ is given by [5.106]. Let us write

$$\begin{aligned}&D_k(\rho) = y_1(\rho) - y_2(\rho), \quad \rho \geq 0, \quad \text{where} \\ &y_1(\rho) = \rho^2(\mu_1 \rho^\alpha + 1) \quad \text{and} \quad y_2(\rho) = -b_k \rho^\beta - a_k \rho^\alpha - d_k.\end{aligned} \qquad [5.113]$$

From Figure 5.4, where we plot $y_1$ and $y_2$, we conclude that there always exists a $\rho_0$ such that $y_1(\rho_0) = y_2(\rho_0)$.

*Case $d_k > 0$*: It is easily seen that for $\rho_0 \geq 0$, we have $D_k(\rho_0) > 0$, where $D_k$ is given by [5.106]. Suppose that $D_k(\rho e^{i\theta}) = 0$, $\rho > 0$, $|\theta| \leq \frac{\pi}{2}$ and consider its

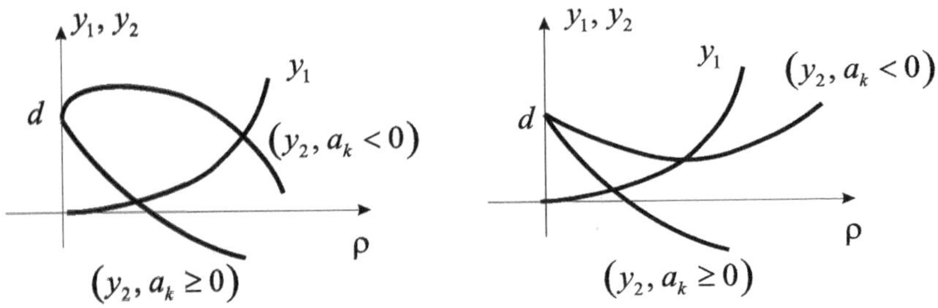


**Figure 5.4.** *Functions $y_1$ and $y_2$ given by [5.113]*

imaginary part

$$\begin{aligned}\operatorname{Im}\left(D_k\left(\rho\,\mathrm{e}^{\mathrm{i}\theta}\right)\right) &= \mu\rho^{2+\alpha}\sin\left((2+\alpha)\,\theta\right)+\rho^2\sin\left(2\theta\right)+b_k\rho^\beta\sin\left(\beta\theta\right)\\ &\quad +a_k\rho^\alpha\sin\left(\alpha\theta\right).\end{aligned}$$

If $\theta \in \left(-\frac{\pi}{2+\alpha}, \frac{\pi}{2+\alpha}\right)$, then all addends in $\operatorname{Im}\left(D_k\left(\rho\,\mathrm{e}^{\mathrm{i}\theta}\right)\right)$ are either positive or negative. To have $\operatorname{Im}\left(D_k\left(\rho\,\mathrm{e}^{\mathrm{i}\theta}\right)\right) = 0$, $\theta$ has to satisfy either $\theta \in \left(\frac{\pi}{2+\alpha}, \frac{\pi}{2}\right]$, or $\theta \in \left[-\frac{\pi}{2}, -\frac{\pi}{2+\alpha}\right)$ (see Figure 5.5).

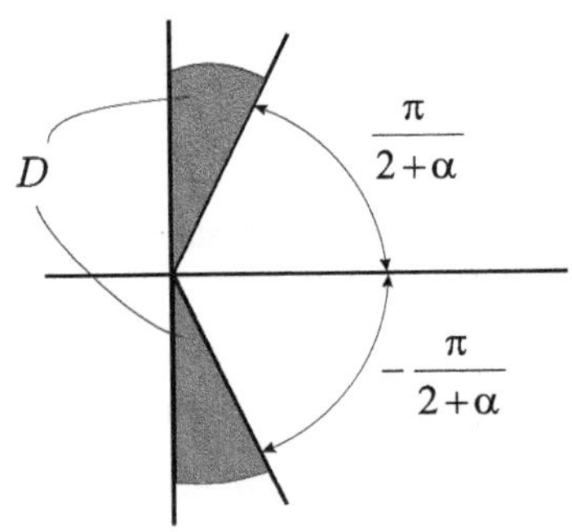


**Figure 5.5.** *The possible instability region corresponding to the case $d > 0$, $\mu_1 > \mu_2$*

If there exist $\rho_0$ and $\theta_0$ such that $D_k\left(\rho_0\,\mathrm{e}^{\mathrm{i}\theta_0}\right) = 0$, then $\rho_0$ and $\theta_0 \in D \equiv \left\{\theta \mid \theta \in \left(\frac{\pi}{2+\alpha}, \frac{\pi}{2}\right] \text{ or } \theta \in \left[-\frac{\pi}{2}, -\frac{\pi}{2+\alpha}\right)\right\}$ have to satisfy

$$\mu\rho_0^{2+\alpha} \sin\left((2+\alpha)\,\theta_0\right) + \rho_0^2 \sin\left(2\theta_0\right) + b_k\rho_0^\beta \sin\left(\beta\theta_0\right) + a_k\rho_0^\alpha \sin\left(\alpha\theta_0\right) = 0,$$
$$\mu\rho_0^{2+\alpha} \cos\left((2+\alpha)\,\theta_0\right) + \rho_0^2 \cos\left(2\theta_0\right) + b_k\rho^\beta \cos\left(\beta\theta_0\right) + a_k\rho^\alpha \cos\left(\alpha\theta_0\right)$$
$$+d_k = 0.$$

By using the Rouché theorem, we can make the domain to which $\rho_0$ and $\theta_0$ belong narrower, if they exist. Let $r$ denote a positive number such that

$$d_k > \left|\mu\rho^{2+\alpha}\mathrm{e}^{\mathrm{i}(2+\alpha)\theta} + \rho^2\mathrm{e}^{2\mathrm{i}\theta} + b_k\rho^\beta\mathrm{e}^{\mathrm{i}\beta\theta} + a_k\rho^\alpha\mathrm{e}^{\mathrm{i}\alpha\theta}\right|, \quad 0 \leq \rho < r.$$

If $0 \leq \rho < r$, $\theta \in D$, then $\rho_0$ and $\theta_0$ cannot belong to $D_1 = \operatorname{int} D \cap \{z \mid |z| < r\}$, shown in Figure 5.6(a). To prove this, it is enough to take any simple closed Jordan curve $C \subset D_1$ and apply the Rouché theorem. Let $R$ denote a positive number such that

$$\left|\mu\rho^{2+\alpha}\mathrm{e}^{\mathrm{i}(2+\alpha)\theta} + \rho^2\mathrm{e}^{2\mathrm{i}\theta}\right| > \left|b_k\rho^\beta\mathrm{e}^{\mathrm{i}\beta\theta} + a_k\rho^\alpha\mathrm{e}^{\mathrm{i}\alpha\theta} + d_k\right|. \quad [5.114]$$

If $\rho > R$, $\theta \in D$, then $\rho_0, \theta_0$ cannot belong to $D_2 = \operatorname{int} D \cap \{z \mid |z| > R\}$, shown in Figure 5.6(b). In proving this property, we once again use the Rouché theorem. Let $C_1$ be any simple closed Jordan curve belonging to $D_2$. On $C_1$, inequality [5.114] is satisfied. Consider

$$\mu\rho^{2+\alpha}\mathrm{e}^{\mathrm{i}(2+\alpha)\theta} + \rho^2\mathrm{e}^{2\mathrm{i}\theta} = \rho^2\mathrm{e}^{2\mathrm{i}\theta}\left(\mu\rho^\alpha\mathrm{e}^{\mathrm{i}\alpha\theta} + 1\right) = 0. \quad [5.115]$$

Since $\mu\rho^\alpha\mathrm{e}^{\mathrm{i}\alpha\theta} + 1 \neq 0$, $\theta \in \left(\frac{\pi}{2+\alpha}, \frac{\pi}{2}\right)$, equation [5.115] has no zeros in interior of $C_1$. Consequently, $\rho_0$ and $\theta_0$ cannot belong to $D_2$.

*Case $d_k = 0$*: By [5.106], we have

$$D_k\left(s\right) = s^\alpha\left(\mu_1 s^2 + s^{2-\alpha} + b_k s^{\beta-\alpha} + a_k\right), \quad s \in \mathbb{C},$$

so that $s = 0$ is a critical point of $D_k$ for all $k \in \mathbb{N}$, since $D_k\left(0\right) = 0$.

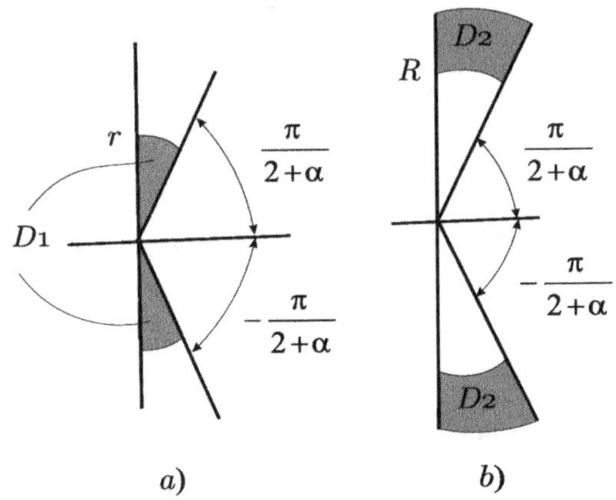


**Figure 5.6.** *Regions $D_1$ and $D_2$*

If $a_k \geq 0$, then we apply proposition 5.1, which yields

$$D_k(s) \neq 0, \quad \operatorname{Re} s > 0.$$

We also show that $D_k(s) \neq 0$ if $s = \rho\, e^{\pm i\frac{\pi}{2}}$, $\rho \geq 0$, i.e. $s$ belongs to the imaginary axis and $s \neq 0$. Suppose that

$$\left.\frac{D_k(s)}{s^{\alpha}}\right|_{s=\rho\, e^{\pm i\frac{\pi}{2}}} = -\mu_1\rho^2 - \rho^{2-\alpha}e^{\pm i\alpha\frac{\pi}{2}} + b_k\rho^{\beta-\alpha}e^{\pm i(\beta-\alpha)\frac{\pi}{2}} + a_k = 0,$$

then the imaginary part of the previous equation is

$$\rho^{2-\alpha}\sin\frac{\alpha\pi}{2} + b_k\rho^{\beta-\alpha}\sin\frac{(\beta-\alpha)\pi}{2} = 0,$$

which is the sum of two positive numbers. Thus, the assumption that $D_k\left(\rho\, e^{\pm i\frac{\pi}{2}}\right) = 0$ is wrong.

If $a_k < 0$, then there exists $s = \rho \geq 0$ such that $D_k(\rho) = 0$. Let

$$D_k(\rho) = y_1(\rho) - y_2(\rho), \quad \rho \geq 0, \quad \text{where} \qquad [5.116]$$
$$y_1(\rho) = \rho^{2-\alpha}(\mu_1\rho^{\alpha} + 1) \quad \text{and} \quad y_2(\rho) = -b_k\rho^{\beta-\alpha} - a_k.$$

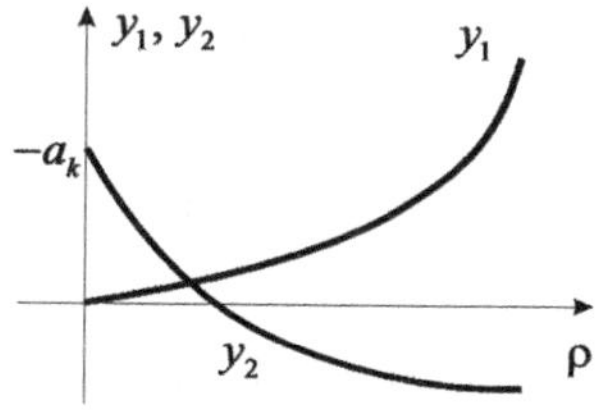


**Figure 5.7.** *Functions $y_1$ and $y_2$ given by [5.116]*

From Figure 5.7, where we schematically plot $y_1$ and $y_2$, it follows that there exists $\rho > 0$, such that $D_k(\rho) = 0$.

We discuss the behavior (the stability) of the functions $F_p(t) = \mathcal{L}^{-1}[f_p(s)](t)$ at infinity if $0 \le p \le 1 + \beta$.

We start with the case $0 \le p < 1 + \alpha$. Consider the integration path $C$ shown in Figure 5.8.

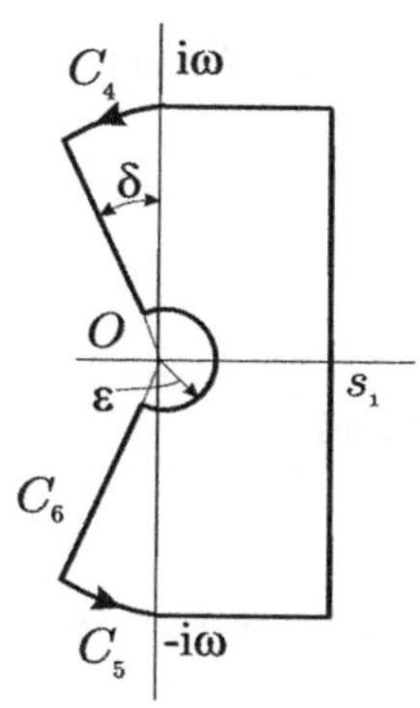


**Figure 5.8.** *Integration path used in determining $F_p$*

We can always find $\delta > 0$, such that the function $f$ remains holomorphic on the contour $C$ and inside of it. If we prove that

$$I_4 = \int_{C_4} f(s)\,\mathrm{e}^{st}\mathrm{d}s \to 0 \text{ and } I_5 = \int_{C_5} f(s)\,\mathrm{e}^{st}\mathrm{d}s \to 0 \text{ as } \omega \to \infty, \quad [5.117]$$

then

$$F_p(t) = \frac{1}{2\pi\mathrm{i}} \int_{C_6} f(s)\,\mathrm{e}^{st}\mathrm{d}s, \quad t > 0. \quad [5.118]$$

We prove only $[5.117]_1$, since the proof for $[5.117]_2$ is analogous

$$
\begin{aligned}
|I_4| &= \left| \int_{\frac{\pi}{2}}^{\frac{\pi}{2}+\delta} \frac{\omega^{p_1} e^{ip\gamma}}{D_k(\omega e^{i\gamma})} e^{\omega t e^{i\gamma}} i\omega e^{i\gamma} d\gamma \right| \\
&\leq \int_{\frac{\pi}{2}}^{\frac{\pi}{2}+\delta} \frac{\omega^{p+1}}{\omega^{2+\alpha}} \left| \frac{\omega e^{i\gamma}}{D_k(\omega e^{i\gamma})} \right| e^{\omega t \cos\gamma} d\gamma \leq M_2 \frac{1}{\omega^{1+\alpha-p}}.
\end{aligned}
$$

Hence, $I_4 \to 0$ as $\omega \to \infty$.

To obtain the asymptotic behavior of the function $F_p(t)$ as $t \to \infty$, we use the following result [DOE 55, Part II, Kapitel 7, Paragraph 4, Satz 1] that in our case leads to the estimate

$$
F_p(t) \sim \frac{1}{a} t^{-p+\alpha-1} \frac{1}{\Gamma(\alpha - p)} \to 0, \quad \text{as} \quad t \to \infty.
$$

Next, we analyze the case $p = 1 + \alpha$. We get

$$
\tilde{\phi}(s) = f(s) - \frac{1}{\mu s} \sim -\frac{1}{\mu s}, \quad s \to 0.
$$

In this case, $A = -\frac{1}{\mu}$, $\lambda = -1$ and

$$
\phi(t) \sim -\frac{1}{\mu} \quad \text{so that} \quad F_p(t) \to 0 \quad \text{as} \quad t \to \infty.
$$

Finally, suppose that $p = 1 + \beta$. Then, we have

$$
\begin{aligned}
\tilde{\phi}(s) &= f_p(s) - \frac{1}{\mu s^{1-(\beta-\alpha)} + s^{1-\beta}} = -\frac{b_k s^{\beta} + a_k s^{\alpha}}{\left(\mu s^{1-(\beta-\alpha)} + s^{1-\beta}\right) D_k(s)} \\
&\sim -\frac{b}{\mu} \frac{1}{s^{1+2(1-(\beta-\alpha))}}, \quad \text{as} \quad |s| \to \infty.
\end{aligned}
$$

Then, there exists $\phi = \mathcal{L}^{-1}\left[\tilde{\phi}\right]$. We have

$$
\tilde{\phi}(s) \sim -\frac{1}{s^{1-\beta}}, \quad |\arg s| \leq \frac{\pi}{2} + \delta, \quad \text{as} \quad s \to 0.
$$

By [DOE 55, Part II, Kapitel 7, Paragraph 4, Satz 1], we obtain

$$\phi(t) \sim -\frac{1}{\Gamma(1-\beta)} t^{-\beta} \quad \text{as} \ t \to \infty.$$

Knowing that

$$\begin{aligned} \frac{1}{\mu s^{1-(\beta-\alpha)} + s^{1-\beta}} &= \frac{1}{\mu s^{1-\beta}\left(s^{\alpha} + \frac{1}{\mu}\right)} \\ &= \frac{1}{\mu}\frac{1}{\Gamma(1-\beta)} \mathcal{L}\left[t^{-\beta} * w(t)\right](s), \quad \operatorname{Re} s > s_0, \end{aligned}$$

where $w(t) = \alpha t^{\alpha-1} E_{\alpha}^{(1)}\left(-\frac{1}{\mu}t^{\alpha}\right)$, $t > 0$, with $E_{\alpha}$ being the one-parameter Mittag-Leffler function, by the property of $w$ that $|w(t)| \leq Ct^{\alpha-1}$, $t > 0$, we get

$$\left|t^{-\beta} * w(t)\right| \leq \int_0^t \frac{\tau^{\alpha-1}}{(t-\tau)^{\beta}} \mathrm{d}\tau \leq \frac{1}{t^{\beta-\alpha}} \int_0^1 \frac{1}{u^{1-\alpha}(1-u)^{\beta}} \mathrm{d}u, \quad t > 0.$$

Thus, we have

$$F_p(t) = \phi_p(t) + \frac{1}{\mu}\frac{1}{\Gamma(1-\beta)} t^{-\beta} * w(t) \to 0, \quad \text{as} \ t \to \infty.$$

Consequently, for $0 \leq p \leq 1+\beta$, we proved that

$$F_p(t) \to 0, \quad \text{as} \ t \to \infty.$$

According to [5.108] and the asymptotic behavior of $F_p$, $0 \leq p \leq 1+\beta$, that we proved, it follows that $T_k(t)$ and $V_k(t)$ tend to zero when $t \to \infty$.

5.1.3.3. *Summary and discussion*

With regard to the boundness (stability) of $T_k$ and $V_k$, given by [5.108], we obtained the following results:

– If $d_k < 0$, i.e. $B > (k\pi)^2$, $0 < \alpha < \beta < 1$ and $\mu, \mu_1 \in \mathbb{R}_+$, then $T_k$ and $V_k$ are not bounded when $t \to \infty$. We note that the case $B = \pi^2$ corresponds to Euler buckling load obtained on the basis of the "elastic" part of the constitutive equation, obtained from [5.19] when $a = b = c = 0$. The results presented correspond to this case when we use $k = 1$.

– If $d_k = 0$, i.e. $B = (k\pi)^2$, $0 < \alpha < \beta < 1$, $\mu_1 \geq \mu$, then the functions $T_k$ and $V_k$ tend to zero when $t \to \infty$. Thus, when the axial force is equal to the Euler buckling force and the thermodynamical restrictions hold, vibrations of the rod are stable.

– If $d_k > 0$, i.e. $B < (k\pi)^2$, then $T_k$ and $V_k$ can be stable or unstable depending on the parameters $B$, $\alpha$, $\beta$, $\mu$ and $\mu_1$. It is important to note that for the case when $\alpha$, $\beta$, $\mu$ and $\mu_1$ satisfy the thermodynamical restrictions, numerical analysis shows that the condition $B < \pi^2$ implies stability. Analytical results show only that the values of $s$ for which $D_k(s) = 0$, which govern stability, cannot have very small or very large norms in the complex plane.

– If the thermodynamical restrictions are violated, the rod becomes unstable.

## 5.2. Stability of Beck's column on viscoelastic foundation

We consider an elastic column loaded by the follower-type concentrated force at the free end $B$, whose motion is restricted by viscoelastic foundation, as shown in Figure 5.9. For the case when the motion of the column is not restricted by the (elastic or viscoelastic) foundation, the column becomes the well-known Beck's column.

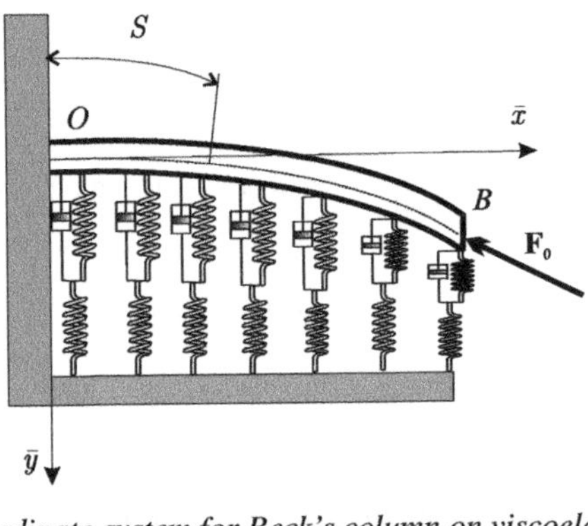


**Figure 5.9.** *Coordinate system for Beck's column on viscoelastic foundation*

The choice of the type of foundation is motivated by the Herrmann–Smith paradox. Namely, in [SMI 72], Smith and Herrmann analyzed the stability of a column loaded by a follower force. It was assumed that the column is positioned on the Winkler-type elastic foundation. They obtained, intuitively unexpected, behavior of the critical load for flatter: the critical load was independent of the foundation modulus. The frequency of vibration of the beam increases with increasing foundation modulus, but the magnitude of the critical load is not affected. In [VOL 77], the problem was reconsidered, where it was argued that, like in the situation of the Ziegler paradox (see [ZIE 52]) (small internal viscosity decreases the stability boundary), in the case of a non-conservative load, the paradox is explained as: "a consequence of uncritical application of the small-oscillation method". The authors in [VOL 77] assumed that the column is made of linearly viscoelastic material and showed that the paradox disappears, i.e. the critical load depends on the rigidity of the base. This leads the authors of [VOL 77] to state: "the elastic idealization of a deformable system in non-conservative stability problems becomes physically meaningless". A summary of results on the stability of columns subject to follower loads is given in [LAN 00], while the complete solution of the Herrmann–Smith problem, allowing the variable stiffness of the foundation, for the case of the elastic column is presented in [KIR 02].

Our analysis will show that the Herrmann–Smith paradox remains if we keep the column elastic but assume that the foundation is viscoelastic. We will show this for a particular type of viscoelastic foundation that is described by the fractional derivative type constitutive equation. The model of viscoelastic foundation is used, for example in [FEN 98], to model rail pads in a study of the stability of railway tracks. We follow the analysis presented in [ATA 04c].

According to the setup of the problem, we have that the projection along the $\bar{x}$-axis of distributed forces, other than inertial, is zero ($q_x = 0$). Except for the inertial forces, along the $\bar{y}$-axis, there are distributed forces $q_y = -q_F$ that are the consequence of the foundation–rod interaction, expressed through the fractional Zener constitutive equation

$$\left(1 + a\,{}_0\mathrm{D}_t^{\alpha}\right) q_F\left(S,t\right) = E_0\left(1 + b\,{}_0\mathrm{D}_t^{\alpha}\right) y\left(S,t\right), \quad S \in \left[0,L\right], \; t > 0, \qquad [5.119]$$

where $\alpha \in (0,1)$ and coefficients satisfy thermodynamical restrictions $E_0 > 0$ and $0 < a \leq b$. Note that for the case $a = b$, the constitutive equation [5.119] leads to $q_F = E_0 y$, so that the foundation becomes elastic as in [SMI 72]. There are no distributed couples ($m = 0$) and we neglect the moment of inertia $J$. Therefore, the equations of motion [5.4] and [5.5], and the geometrical equations [5.6] read ($S \in \left[0,L\right], t > 0$)

$$\frac{\partial}{\partial S}H\left(S,t\right) = \rho\frac{\partial^2}{\partial t^2}x\left(S,t\right), \quad \frac{\partial}{\partial S}V\left(S,t\right) - q_F\left(S,t\right) = \rho\frac{\partial^2}{\partial t^2}y\left(S,t\right), \qquad [5.120]$$

$$\frac{\partial}{\partial S}M\left(S,t\right) + V\left(S,t\right)\frac{\partial}{\partial S}x\left(S,t\right) - H\left(S,t\right)\frac{\partial}{\partial S}y\left(S,t\right) = 0, \qquad [5.121]$$

$$\frac{\partial}{\partial S}x\left(S,t\right) = \cos\left(\theta\left(S,t\right)\right), \quad \frac{\partial}{\partial S}y\left(S,t\right) = \sin\left(\theta\left(S,t\right)\right). \qquad [5.122]$$

Since the rod is initially straight, we have that $R(S) \to \infty$ and $\Theta(S) = 0$, implying $\frac{1}{r(S,t)} = \frac{\partial}{\partial S}\theta(S,t)$, by [5.1] and [5.2]. This, along with the assumption that the rod is elastic, leads to the constitutive moment–curvature relation

$$M(S,t) = EI\frac{\partial}{\partial S}\theta(S,t), \quad S \in [0,L], \ t > 0, \qquad [5.123]$$

where $EI$ is the bending stiffness of the rod.

The boundary conditions read

$$x(0,t) = 0, \quad y(0,t) = 0, \quad \theta(0,t) = 0,$$

$$H(L,t) = -F_0 \cos(\theta(L,t)), \quad V(L,t) = -F_0 \sin(\theta(L,t)), \quad M(L,t) = 0, \quad t > 0. \qquad [5.124]$$

In addition to the assumption that the rod is initially straight, suppose that $\dot{x}_0 = \dot{y}_0 = 0$. Then, the initial conditions [5.8] become

$$x(S,0) = S, \quad y(S,0) = 0, \quad \frac{\partial}{\partial t}x(S,0) = 0, \quad \frac{\partial}{\partial t}y(S,0) = 0, \quad S \in [0,L]. \qquad [5.125]$$

The trivial solution to systems [5.119]–[5.125] is

$$x^0(S,t) = S, \quad y^0(S,t) = 0, \quad \theta^0(S,t) = 0, q_F^0(S,t) = 0,$$

$$H^0(S,t) = -F_0, \quad V^0(S,t) = 0, \quad M^0(S,t) = 0, \quad S \in [0,L], \ t > 0.$$

Let $x = x^0 + \Delta x, \ldots, M = M^0 + \Delta M$, where $\Delta x, \ldots, \Delta M$ denote the perturbations. Substituting such $x, \ldots, M$ in [5.119]–[5.125] and neglecting the higher order terms in perturbations $\Delta x, \ldots, \Delta M$, we obtain $\Delta x = 0$ and $\Delta H = 0$, as well as

$$EI\frac{\partial^4}{\partial S^4}\Delta y(S,t) + F_0\frac{\partial^2}{\partial S^2}\Delta y(S,t) + \rho\frac{\partial^2}{\partial t^2}\Delta y(S,t) + \Delta q_F(S,t) = 0, \quad S \in [0,L], \quad t > 0, \qquad [5.126]$$

$$\left(1 + a\,{}_0\mathrm{D}_t^\alpha\right)\Delta q_F(S,t) = E_0\left(1 + b\,{}_0\mathrm{D}_t^\alpha\right)\Delta y(S,t), \quad S \in [0,L], \ t > 0, \qquad [5.127]$$

$$\Delta y(0,t) = 0, \quad \frac{\partial}{\partial S}\Delta y(0,t) = 0, \quad \frac{\partial^2}{\partial S^2}\Delta y(L,t) = 0, \quad \frac{\partial^3}{\partial S^3}\Delta y(L,t) = 0, \quad t > 0, \qquad [5.128]$$

$$\Delta y(S,0) = \Delta y_0(S), \quad \frac{\partial}{\partial t}\Delta y(S,0) = \Delta y_1(S), \quad S \in [0,L]. \qquad [5.129]$$

Introducing the dimensionless quantities

$$\xi = \frac{S}{L}, \quad \bar{t} = \sqrt{\frac{EI}{\rho L^4}} t, \quad u = \frac{\Delta y}{L}, \quad \lambda = \frac{F_0 L^2}{EI}, \quad q = \frac{\Delta q_F}{E_0 L},$$

$$\beta = \frac{E_p L^4}{EI}, \quad \bar{a} = a \left( \frac{EI}{\rho L^4} \right)^{\frac{\alpha}{2}}, \quad \bar{b} = b \left( \frac{EI}{\rho L^4} \right)^{\frac{\alpha}{2}}$$

in systems [5.126]–[5.129] and omitting the bar ($\bar{t} \to t$, $\bar{a} \to a$, $\bar{b} \to b$), we obtain

$$\frac{\partial^4}{\partial \xi^4} u(\xi, t) + \lambda \frac{\partial^2}{\partial \xi^2} u(\xi, t) + \frac{\partial^2}{\partial \tau^2} u(\xi, t) + \beta q(\xi, t) = 0, \quad \xi \in [0,1], \ t > 0, \tag{5.130}$$

$$(1 + a\, {}_0\mathrm{D}_t^{\alpha}) \, q(\xi, t) = (1 + b\, {}_0\mathrm{D}_t^{\alpha}) \, u(\xi, t), \quad \xi \in [0,1], \ t > 0, \tag{5.131}$$

$$u(0,t) = 0, \quad \frac{\partial}{\partial \xi} u(0,t) = 0, \quad \frac{\partial^2}{\partial \xi^2} u(1,t) = 0, \quad \frac{\partial^3}{\partial \xi^3} u(1,t) = 0, \quad t > 0, \tag{5.132}$$

$$u(\xi, 0) = u_0(\xi), \quad \frac{\partial}{\partial t} u(\xi, 0) = u_1(\xi), \quad \xi \in [0,1]. \tag{5.133}$$

### 5.2.1. *Solution to systems [5.130]–[5.133]*

We consider solutions in the framework of $\mathcal{K}'_+$ (see section 1.3), which means that we assume that functions under consideration are of the exponential growth, as well as that their Laplace transforms are analytic functions for $\operatorname{Re} s > s_0$, for suitable $s_0$.

Assume that functions $u$ and $q$ are of the form

$$u(\xi, t) = U(\xi) T(t) \ \text{ and } \ q(\xi, t) = F(\xi) V(t), \quad \xi \in [0,1], \ t > 0. \tag{5.134}$$

Then, [5.130] and [5.131] reduce to ($\xi \in [0,1], t > 0$)

$$T(t) \frac{\mathrm{d}^4}{\mathrm{d}\xi^4} U(\xi) + \lambda T(t) \frac{\mathrm{d}^2}{\mathrm{d}\xi^2} U(\xi) + U(\xi) \frac{\mathrm{d}^2}{\mathrm{d}t^2} T(t) + \beta F(\xi) V(t) = 0, \tag{5.135}$$

$$(V(t) + a\, {}_0\mathrm{D}_t^{\alpha} V(t)) \, F(\xi) = (T(t) + b\, {}_0\mathrm{D}_t^{\alpha} T(t)) \, U(\xi). \tag{5.136}$$

On the basis of [5.136], we set

$$F(\xi) = AU(\xi), \quad \xi \in [0,1],$$

where $A \in \mathbb{R}\setminus\{0\}$ is a constant. Equation [5.135] becomes ($\xi \in [0,1], t > 0$)

$$\frac{\frac{d^4}{d\xi^4}U(\xi)}{U(\xi)} + \lambda\frac{\frac{d^2}{d\xi^2}U(\xi)}{U(\xi)} = -\left(\frac{\frac{d^2}{dt^2}T(t)}{T(t)} + A\beta\frac{V(t)}{T(t)}\right) = \varkappa,$$

where $\varkappa$ is another constant.

Thus, the system equivalent to [5.135] and [5.136] reads

$$\frac{d^4}{d\xi^4}U(\xi) + \lambda\frac{d^2}{d\xi^2}U(\xi) - \varkappa U(\xi) = 0, \quad \xi \in [0,1], \qquad [5.137]$$

$$\frac{d^2}{dt^2}T(t) + \varkappa T(t) + A\beta V(t) = 0, \quad t > 0, \qquad [5.138]$$

$$A\left(1 + a\,{}_0D_t^\alpha\right)V(t) = \left(1 + b\,{}_0D_t^\alpha\right)T(t), \quad t > 0. \qquad [5.139]$$

Boundary conditions [5.132], with decomposition [5.134] taken into account, become

$$U(0) = 0, \quad \frac{d}{d\xi}U(0) = 0, \quad \frac{d^2}{d\xi^2}U(1) = 0, \quad \frac{d^3}{d\xi^3}U(1) = 0, \qquad [5.140]$$

while we assume initial conditions

$$T(0) = T_0, \quad \frac{d}{dt}T(0) = T_1. \qquad [5.141]$$

5.2.1.1. *Solution to [5.137] and [5.140]*

Equation [5.137], with boundary conditions [5.140], has the same analytical form as the corresponding equation and boundary conditions for Beck's column without foundation (see [ATA 97]). The solution to [5.137] is

$$U(\xi) = C_1\cosh(r_1\xi) + C_2\sinh(r_1\xi) + C_3\cos(r_2\xi) + C_4\sin(r_2\xi), \quad \xi \in [0,1], \qquad [5.142]$$

where $C_i$, $i = 1, ..., 4$, are constants and

$$r_1 = \sqrt{\sqrt{\left(\frac{\lambda}{2}\right)^2 + \varkappa} - \frac{\lambda}{2}}, \quad r_2 = \sqrt{\sqrt{\left(\frac{\lambda}{2}\right)^2 + \varkappa} + \frac{\lambda}{2}}. \qquad [5.143]$$

Equation [5.142], with boundary conditions [5.140] used, implies the connection between $\lambda$ and $\varkappa$ given by

$$\Delta(\lambda, \varkappa) = \lambda^2 + 2\varkappa(1 + \cosh r_1 \cos r_2) + \lambda\sqrt{\varkappa}\sinh r_1 \sin r_2 = 0. \qquad [5.144]$$

When $\lambda = 0$, there are an infinite number of positive real values of $\varkappa$ lying in pairs between $\left(\frac{\pi}{2} + 2k\pi\right)^2$ and $\left(\frac{3\pi}{2} + 2k\pi\right)^2$, $k = 0, 1, ...,$ and satisfying $\Delta(\lambda, \varkappa) = 0$ (see [MAC 04]). In the applications, we are interested in the lowest values of $\lambda$ for which the rod loses stability. Thus, we consider the first two values of $\varkappa$. If $\lambda$ increases, then the difference between two corresponding values of $\varkappa$ decreases, until they become equal at a critical value of $\lambda = \lambda_{cr} \approx 20.05$. If $\lambda$ increases beyond $\lambda_{cr}$, $\varkappa$ becomes complex, with real part equal to $\varkappa_{cr} \approx 121.25$ (see [SEY 02, p. 240]). More precisely, if $\lambda \in (0, \lambda_{cr} + \varepsilon)$, with $\varepsilon > 0$ and small enough, then the values of $\varkappa$ determined from [5.144] are either positive numbers or complex numbers with the real parts close to $\varkappa_{cr}$.

Let $\varkappa = \varkappa_{cr} + \mathrm{i}q$, $q \in \mathbb{R}$, $q \neq 0$. According to [5.143], we have to take the square root of a complex number twice. Recall the general procedure. If $c + \mathrm{i}d$, then

$$\sqrt{c + \mathrm{i}d} = x_j + \mathrm{i}y_j, \quad j = 1, 2, \qquad [5.145]$$

where

$$x_{1,2} = \pm\frac{1}{\sqrt{2}}\sqrt{\sqrt{c^2 + d^2} + c}, \quad y_{1,2} = \pm\frac{1}{\sqrt{2}}\frac{d}{|d|}\sqrt{\sqrt{c^2 + d^2} - c}.$$

By combining [5.143] and [5.145], we conclude that there is always a root given by [5.143] with positive real part, implying instability.

5.2.1.2. *Solutions to systems [5.138], [5.139] and [5.141]*

Applying the Laplace transform to systems [5.138] and [5.139], with the initial conditions [5.141] taken into account, leads to

$$\left(s^2 + \varkappa\right)\tilde{T}(s) + A\beta\tilde{V}(s) = sT_0 + T_1, \quad \mathrm{Re}\, s > s_0, \qquad [5.146]$$

$$\left(1 + bs^\alpha\right)\tilde{T}(s) - A\left(1 + as^\alpha\right)\tilde{V}(s) = 0, \quad \mathrm{Re}\, s > s_0. \qquad [5.147]$$

The solution to systems [5.146] and [5.147] is

$$\tilde{T}(s) = \frac{s^\alpha + \frac{1}{a}}{\left(s^\alpha + \frac{1}{a}\right)\left(s^2 + p\right) + c}\left(T_0 s + T_1\right), \quad \mathrm{Re}\, s > s_0, \qquad [5.148]$$

$$\tilde{V}(s) = \frac{b}{Aa}\frac{s^\alpha + \frac{1}{a} - \frac{b-a}{ab}}{\left(s^\alpha + \frac{1}{a}\right)\left(s^2 + p\right) + c}\left(T_0 s + T_1\right), \quad \mathrm{Re}\, s > s_0, \qquad [5.149]$$

where $s_0$ is determined so that $\left|\left(s^{\alpha}+\frac{1}{a}\right)\left(s^2+p\right)\right| > c$ and where we introduced $p = \varkappa + \beta\frac{b}{a}$ and $c = -\beta\frac{b-a}{a}$.

Consider the function

$$\tilde{f}(s) = \frac{s^{\alpha}+\frac{1}{a}}{\left(s^{\alpha}+\frac{1}{a}\right)\left(s^2+p\right)+c} = \frac{1}{s^2+p}\left(1+\tilde{\phi}(s)\right), \quad \operatorname{Re} s > s_0, \quad \text{with} \qquad [5.150]$$

$$\tilde{\phi}(s) = \sum_{\nu=1}^{\infty}(-c)^{\nu}\left(\frac{1}{s^{\alpha}+\frac{1}{a}}\right)^{\nu}\left(\frac{1}{s^2+p}\right)^{\nu}, \quad \operatorname{Re} s > s_0. \qquad [5.151]$$

We have proved in [STA 01] (see section 5.1.2.1) that there is a continuous function

$$\phi(t) = \sum_{\nu=1}^{\infty}(-c)^{\nu}\left(\mathcal{L}^{-1}\left[\frac{1}{s^2+p}\right](t) * w(t)\right)^{*\nu}, \quad t \geq 0, \qquad [5.152]$$

where $*$ denotes convolution and $F^{*\nu}$ means $\nu$-fold convolution of $F$, such that its Laplace transform is $\tilde{\phi}$, given by [5.151]. In [5.152], we used

$$w(t) = \mathcal{L}^{-1}\left[\frac{1}{s^{\alpha}+\frac{1}{a}}\right](t) = \alpha t^{\alpha-1}E_{\alpha}^{(1)}\left(-\frac{t^{\alpha}}{a}\right), \quad t \geq 0, \qquad [5.153]$$

where $E_{\alpha}$ is the one-parameter Mittag-Leffler function.

Thus, the inverse Laplace transform of function $\tilde{f}$, given by [5.150], reads

$$f(t) = \frac{1}{\sqrt{p}}\sin\left(\sqrt{p}t\right) + \frac{1}{\sqrt{p}}\sin\left(\sqrt{p}t\right) * \phi(t), \quad t \geq 0. \qquad [5.154]$$

Since $f(0) = 0$, we have $\mathcal{L}\left[f^{(1)}(t)\right](s) = s\tilde{f}(s) = \frac{s}{s^2+p}$, so that

$$f^{(1)}(t) = \cos\left(\sqrt{p}t\right) + \cos\left(\sqrt{p}t\right) * \phi(t), \quad t \geq 0. \qquad [5.155]$$

Inverting the Laplace transform in [5.148] and [5.149], we obtain the solution to systems [5.138], [5.139] and [5.141] in the form ($t \geq 0$)

$$\begin{aligned} T(t) &= T_0 f^{(1)}(t) + T_1 f(t), \quad t \geq 0, \qquad [5.156] \\ V(t) &= \frac{b}{Aa}\left(T_0 f^{(1)}(t) + T_1 f(t) - \frac{b-a}{ab} w(t) * \left(T_0 f^{(1)}(t) + T_1 f(t)\right)\right), \\ & t \geq 0, \end{aligned}$$

where $f$ and $f^{(1)}$ are given by [5.154] and [5.155], respectively. In obtaining [5.157], we used [5.150] yielding

$$\begin{aligned} \frac{1}{\left(s^{\alpha}+\frac{1}{a}\right)\left(s^{2}+p\right)+c} &= \frac{\tilde{f}(s)}{s^{\alpha}+\frac{1}{a}} = \mathcal{L}\left[w(t) * f(t)\right](s), \\ \frac{s}{\left(s^{\alpha}+\frac{1}{a}\right)\left(s^{2}+p\right)+c} &= \frac{s\tilde{f}(s)}{s^{\alpha}+\frac{1}{a}} = \mathcal{L}\left[w(t) * f^{(1)}(t)\right](s), \end{aligned} \qquad [5.157]$$

where $w$ is given by [5.153].

### 5.2.2. *Properties of functions T and V*

We analyze the properties of $T$ and $V$, given by [5.156] and [5.157], respectively, obtained as solutions to systems [5.138], [5.139] and [5.141]. We have that $f \in L^{exp}([0,\infty))$, (see section 1.2), and prove that $f \in C^{3}([0,\infty))$. By [5.155], it follows that $f \in C^{1}([0,\infty))$. Let us consider $f^{(2)}$ and by differentiating [5.155], we have ($t \geq 0$)

$$\begin{aligned} f^{(2)}(t) &= \frac{\mathrm{d}}{\mathrm{d}t}\left(\cos\left(\sqrt{p}t\right)+\cos\left(\sqrt{p}t\right) * \phi(t)\right) \\ &= \frac{1}{\sqrt{p}}\sin\left(\sqrt{p}t\right)+\phi(t)+\frac{1}{\sqrt{p}}\sin\left(\sqrt{p}t\right) * \phi(t), \end{aligned} \qquad [5.158]$$

where $\phi$ is given by [5.152]. Function $\phi$ has the property that $\left|\phi^{(i)}(t)\right| = O(t^{\alpha})$, as $t \to 0$, $i \in \{0,1\}$, since $w$ is a smooth function on $(0,\infty)$ and $w(t) = O\left(t^{\alpha-1}\right)$, as $t \to 0$. Consequently, by [5.158], we have that $f \in C^{3}([0,\infty))$. Now it is easily seen that $T \in C^{2}([0,\infty))$ and that ${}_{0}\mathrm{D}_{t}^{\alpha}V \in C([0,\infty))$. Thus, functions $T$ and $V$, given by [5.156] and [5.157], respectively, are classical solutions to systems [5.138], [5.139] and [5.141].

#### 5.2.2.1. *Asymptotic behavior of functions $T$ and $V$*

We have shown that if the function $U$, given by [5.142], satisfies boundary conditions [5.140], then $\varkappa$ is either positive real number or complex number. Therefore, we discuss the asymptotic behavior of the functions $T$ and $V$ in [5.156] and [5.157] only for these values of $\varkappa$. The asymptotic behavior depends on the real parts of zeros (that have to lie in the left complex half-plane in order that the system is stable) of denominator of functions $\tilde{T}$ and $\tilde{V}$, given by [5.148] and [5.149]

$$D(s) = \left(s^{\alpha}+\frac{1}{a}\right)\left(s^{2}+p\right)+c, \qquad [5.159]$$

with $p = \varkappa + \beta \frac{b}{a}$ and $c = -\beta \frac{b-a}{a}$, it becomes

$$D(s) = \frac{1}{a}\left(as^{\alpha+2} + s^2 + (\varkappa a + \beta b)s^\alpha + \varkappa + \beta\right). \qquad [5.160]$$

However, we can also use the analytical form of $T$ and $V$, [5.156] and [5.157], for the discussion of the asymptotic behavior.

Let $\varkappa > 0$. Zeros of a complex function $D$, given by [5.160], using different approaches, were studied in, e.g., [GOR 97a, KEM 98b, MAR 49]. However, in [STA 01], the elementary analysis was used and we will follow this approach. We have that $a\varkappa + \beta b > 0$ and $\varkappa + \beta > 0$. It is proved that in this case, $D$, given by [5.160], has neither real positive, nor complex zeros with positive real part (see [MAR 49, p. 518]). Consequently, if $\varkappa > 0$, then the solutions $T$ and $V$, given by [5.156] and [5.157], are stable.

Let $\varkappa = \varkappa_{cr} + \mathrm{i}q$, $q \in \mathbb{R}$, $q \neq 0$. Then, by [5.159], we have that $D(s) = 0$ implies

$$\left(s^\alpha + \frac{1}{a}\right)\left(s^2 + \varkappa_{cr} + \beta\frac{b}{a} + \mathrm{i}q\right) = \beta\frac{b-a}{a}. \qquad [5.161]$$

Suppose that $\rho_0 \geq 0$ satisfies [5.161]. Then

$$\left(\rho_0^\alpha + \frac{1}{a}\right)\left(\rho_0^2 + \varkappa_{cr} + \beta\frac{b}{a} + \mathrm{i}q\right) = \beta\frac{b-a}{a}. \qquad [5.162]$$

Then, on the right-hand side of [5.162], we have a positive real number (since $a \leq b$) and on the left-hand side of [5.162], we have a complex number. Since this is not possible, it follows that there is no positive real ($\rho_0 \geq 0$) solution of [5.161]. We look for the complex solution to [5.159]. By [5.145], for $\varkappa = \varkappa_{cr} + \mathrm{i}q$, we have

$$\begin{aligned}\sqrt{p} &= \pm\frac{1}{\sqrt{2}}\sqrt{\sqrt{\left(\varkappa_{cr} + \beta\frac{b}{a}\right)^2 + q^2} + \left(\varkappa_{cr} + \beta\frac{b}{a}\right)} \\ &\quad \pm \mathrm{i}\frac{1}{\sqrt{2}}\frac{q}{|q|}\sqrt{\sqrt{\left(\varkappa_{cr} + \beta\frac{b}{a}\right)^2 + q^2} - \left(\varkappa_{cr} + \beta\frac{b}{a}\right)}.\end{aligned}$$

By using [5.154] and [5.155], we conclude that in this case, we have oscillations with increasing amplitude, i.e. a flutter.

5.2.2.2. *Approximation of a series in function $\phi$*

Consider the function $\phi$, given by [5.152], in the form of a series of repeated convolutions. We will estimate the error when we take the finite sum in [5.152]. For this purpose, we define

$$\phi_m(t) = \sum_{\nu=1}^{m} (-c)^\nu \left( \mathcal{L}^{-1}\left[\frac{1}{s^2+p}\right](t) * w(t) \right)^{*\nu}, \quad t \in [0, t_0],$$

and consider only the case $\varkappa > 0$, since then the solution is stable. It is easily seen (see [STA 01, p. 511]) that for any $\nu \in \mathbb{N}$ and $t \in [0, t_0]$, $p > 0$, it holds

$$\left| \left( \mathcal{L}^{-1}\left[\frac{1}{s^2+p}\right](t) * w(t) \right)^{*\nu} \right| \leq C^\nu(t_0) \frac{t^{(\alpha+1)\nu-1}}{\Gamma(\nu(\alpha+1))}.$$

Then, with $C_0 = cC(t_0)$, we have

$$\begin{aligned}
&\left| \sum_{\nu=m+1}^{\infty} (-c)^\nu \left( \mathcal{L}^{-1}\left[\frac{1}{s^2+p}\right](t) * w(t) \right)^{*\nu} \right| \leq \sum_{\nu=m+1}^{\infty} C_0^\nu \frac{t^{(\alpha+1)\nu-1}}{\Gamma(\nu(\alpha+1))} \\
&\quad \leq \sum_{k=1}^{\infty} \frac{C_0^{k+m} t^{(\alpha+1)(k+m)-1}}{\Gamma(k(\alpha+1)+m(\alpha+1))} \\
&\quad \leq C_0^m t^{m(\alpha+1)-1} \sum_{k=1}^{\infty} \frac{\left(C_0 t^{\alpha+1}\right)^k}{\Gamma(k(\alpha+1)+m(\alpha+1))} \\
&\quad \leq C_0^m t^{m(\alpha+1)-1} C_0 t^{\alpha+1} \sum_{n=0}^{\infty} \frac{\left(C_0 t^{\alpha+1}\right)^n}{\Gamma(n(\alpha+1)+(m+1)(\alpha+1))} \\
&\quad \leq C_0^{m+1} t^{(m+1)(\alpha+1)-1} \sum_{n=0}^{\infty} \frac{\left(C_0 t^{\alpha+1}\right)^n}{\Gamma(n(\alpha+1)+(m+1)(\alpha+1))} \\
&\quad \leq C_0^{m+1} t^{(m+1)(\alpha+1)-1} E_{\alpha+1,(m+1)(\alpha+1)}\left(C_0 t^{\alpha+1}\right),
\end{aligned}$$

where $E_{u,v}$ is the two-parameter Mittag-Leffler function (see [ERD 53, p. 210] and section 1.5).

5.2.2.3. *Summary on stability*

For $0 \leq \lambda \leq 20.0509$ (then $\varkappa > 0$), we have the stability of the rod, i.e. solution to systems [5.138], [5.139] and [5.141] is bounded. If $\lambda \in (20.0599, 20.0599 + \varepsilon)$ with $\varepsilon > 0$ small enough (then $\varkappa = \varkappa_{cr} + iq$, with $\varkappa_{cr} \approx 121.25$, see [SEY 02]), then the solution to systems [5.138], [5.139] and [5.141] becomes unstable with both $T$ and $V$, representing oscillations with increasing amplitude. The stability results just stated show that the viscoelastic foundation of fractional derivative type, for Beck's column, does not increase the stability bound. This result, often called the Herrmann–Smith paradox, was known to hold for the Winkler-type of foundation (see [SMI 72]). Here, we show that it also holds for viscoelastic foundations of the fractional derivative type. Therefore, the solution of the Herrmann–Smith paradox is in introducing viscoelastic effects in a rod material, not in the foundation.

## 5.3. Compressible elastic rod on a viscoelastic foundation

The stability of an elastic rod of finite length with a compressible axis, positioned on a viscoelastic foundation of fractional derivative type, will be analyzed. The results will represent a generalization of the classical problem of determining the stability boundary of an elastic rod on an elastic foundation (for example a Winkler-type foundation). It will be shown that the use of viscoelastic foundation for an axially loaded rod produces two effects. First, the foundation has a stabilizing effect, and second leads to dissipation of energy in the lateral motion, thus causing quick disappearance of vibrations. Stability bounds obtained here will be compared with the stability bounds that follow from the classical Euler (static) theory. Following [ATA 04d], the mechanical model of a system is shown in Figure 5.10.

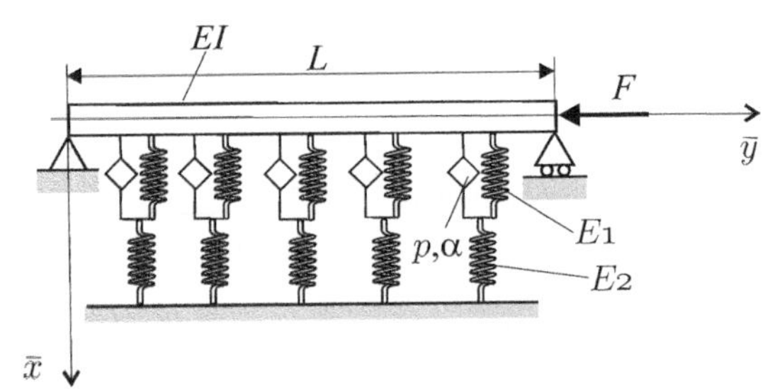


**Figure 5.10.** *Rod on a viscoelastic foundation*

The equations of motion of the rod [5.4] and [5.5] take the form ($S \in [0, L]$, $t > 0$)

$$\frac{\partial}{\partial S}H(S,t) = \rho\frac{\partial^2}{\partial t^2}x(S,t), \quad \frac{\partial}{\partial S}V(S,t) - q_F(S,t) = \rho\frac{\partial^2}{\partial t^2}y(S,t), \tag{5.163}$$

$$\frac{\partial}{\partial S}M(S,t) + V(S,t)\frac{\partial}{\partial S}x(S,t) - H(S,t)\frac{\partial}{\partial S}y(S,t) = 0, \tag{5.164}$$

since, by the setup of the problem, the projection along the $\bar{x}$-axis of distributed forces, other than inertial, is zero ($q_x = 0$), while, except for the inertial forces, along the $\bar{y}$-axis there are distributed forces $q_y = -q_F$ that are the consequence of the foundation–rod interaction. Also, there are no distributed couples ($m = 0$) and we neglect the moment of inertia $J$.

The foundation–rod interaction is expressed through the fractional Zener constitutive equation

$$(1 + a\,{}_0\mathrm{D}_t^{\alpha})\, q_F(S,t) = E_0 (1 + b\,{}_0\mathrm{D}_t^{\alpha})\, y(S,t), \quad S \in [0,L], \; t > 0, \qquad [5.165]$$

where $\alpha \in (0,1)$ and coefficients satisfy thermodynamical restrictions $E_0 > 0$ and $0 < a \leq b$.

Because of the rod's axis compressibility, the geometrical relations [5.6] are changed, so that they read ($S \in [0,L], t > 0$)

$$\begin{aligned} \frac{\partial}{\partial S} x(S,t) &= (1 + \varepsilon(S,t)) \cos(\theta(S,t)), \\ \frac{\partial}{\partial S} y(S,t) &= (1 + \varepsilon(S,t)) \sin(\theta(S,t)), \end{aligned} \qquad [5.166]$$

where we used $\varepsilon$ to denote the axial strain of the rod's axis

$$\varepsilon = \frac{\mathrm{d}s - \mathrm{d}S}{\mathrm{d}S},$$

with $\mathrm{d}s$ ($\mathrm{d}S$) denoting the length of an element of the rod's axis in the deformed (undeformed) state. Assuming that the rod is elastic, for the axial strain we have ($S \in [0,L], t > 0$)

$$\varepsilon(S,t) = \frac{1}{EA} \left( H(S,t) \cos(\theta(S,t)) + V(S,t) \sin(\theta(S,t)) \right), \qquad [5.167]$$

with $E$ being the modulus of elasticity of the rod and $A$ its cross-sectional area.

Since the rod is initially straight, we have that $R(S) \to \infty$ and $\Theta(S) = 0$, implying $\frac{1}{r(S,t)} = \frac{\partial}{\partial S}\theta(S,t)$, by [5.1] and [5.2]. This, along with the assumption that the rod is elastic, leads to the constitutive moment–curvature relation

$$M(S,t) = EI \frac{\partial}{\partial S} \theta(S,t), \quad S \in [0,L], \; t > 0. \qquad [5.168]$$

Boundary conditions, according to the setup of the problem, read

$$\begin{aligned} x(0,t) &= 0, \; y(0,t) = 0, \; M(0,t) = 0, \; y(L,t) = 0, \; H(L,t) = -F_0, \\ M(L,t) &= 0, \; t > 0. \end{aligned} \qquad [5.169]$$

In addition to the assumption that the rod is initially straight, suppose that $\dot{x}_0 = \dot{y}_0 = 0$. Then, initial conditions [5.8] become

$$x\left(S,0\right)=S,\ \ y\left(S,0\right)=0,\ \ \frac{\partial}{\partial t}x\left(S,0\right)=0,\ \ \frac{\partial}{\partial t}y\left(S,0\right)=0,\ \ S\in\left[0,L\right].\text{[5.170]}$$

Trivial solution to systems [5.163]–[5.170] is

$$x^0\left(S,t\right)=S,\ \ y^0\left(S,t\right)=0,\ \ \theta^0\left(S,t\right)=0,\ \ \varepsilon^0\left(S,t\right)=-\frac{F_0}{EA},$$
$$q_F^0\left(S,t\right)=0,\ \ H^0\left(S,t\right)=-F_0,\ \ V^0\left(S,t\right)=0,\ \ M^0\left(S,t\right)=0,\ \ S\in\left[0,L\right],\ t>0.$$

Let $x = x^0 + \Delta x, \ldots, M = M^0 + \Delta M$, where $\Delta x, \ldots, \Delta M$ denote the perturbations. Substituting such $x, \ldots, M$ in [5.163]–[5.170] and neglecting the higher order terms in perturbations $\Delta x, \ldots, \Delta M$, we obtain two systems.

The first system is for $\Delta x$ and it reads

$$\frac{\partial^2}{\partial S^2}\Delta x\left(S,t\right)=\frac{\rho}{EA}\frac{\partial^2}{\partial t^2}\Delta x\left(S,t\right),\ \ S\in\left[0,L\right],\ t>0,$$
$$\Delta x(0,t)=0,\ \ \frac{\partial}{\partial S}\Delta x(0,t)=-\frac{F_0}{EA},\ \ t>0,$$
$$x\left(S,0\right)=S,\ \ \frac{\partial}{\partial t}x\left(S,0\right)=0,\ \ S\in\left[0,L\right].$$

The previous system consists of the wave equation for the axial motion of the rod element, subject to boundary and initial conditions, and it will not be the subject of our analysis.

The second system, with the initial conditions adjoined, is for $\Delta y$ and it takes the form ($S \in [0, L], t > 0$)

$$EI\frac{\partial^4}{\partial S^4}\Delta y\left(S,t\right)+F_0\left(1-\frac{F_0}{EA}\right)\frac{\partial^2}{\partial S^2}\Delta y\left(S,t\right)$$
$$+\rho\left(1-\frac{F_0}{EA}\right)^2\frac{\partial^2}{\partial t^2}\Delta y\left(S,t\right)+\left(1-\frac{F_0}{EA}\right)^2\Delta q_F\left(S,t\right)=0, \qquad \text{[5.171]}$$

$$\left(1+a\,{}_0\mathrm{D}_t^{\alpha}\right)\Delta q_F\left(S,t\right)=E_0\left(1+b\,{}_0\mathrm{D}_t^{\alpha}\right)\Delta y\left(S,t\right), \qquad \text{[5.172]}$$

$$\Delta y(0,t)=0,\ \ \frac{\partial^2}{\partial S^2}\Delta y(0,t)=0,\ \ \Delta y(L,t)=0,\ \ \frac{\partial^2}{\partial S^2}\Delta y(L,t)=0, \qquad \text{[5.173]}$$

$$\Delta y\left(S,0\right)=\Delta y_0\left(S\right),\ \ \frac{\partial}{\partial t}\Delta y\left(S,0\right)=\Delta y_1\left(S\right). \qquad \text{[5.174]}$$

Systems [5.171]–[5.174] transform into ($\xi \in [0,1], t > 0$)

$$\frac{\partial^4}{\partial \xi^4} u(\xi,t) + \lambda \left(1 - \frac{\lambda}{\mu}\right) \frac{\partial^2}{\partial \xi^2} u(\xi,t) + \left(1 - \frac{\lambda}{\mu}\right)^2 \frac{\partial^2}{\partial t^2} u(\xi,t) + \beta \left(1 - \frac{\lambda}{\mu}\right)^2 q(\xi,t) = 0, \quad [5.175]$$

$$(1 + a\,{}_0\mathrm{D}_t^\alpha)\, q(\xi,t) = (1 + b\,{}_0\mathrm{D}_t^\alpha)\, u(\xi,t), \quad [5.176]$$

$$u(0,t) = 0, \quad \frac{\partial^2}{\partial \xi^2} u(0,t) = 0, \quad u(1,t) = 0, \quad \frac{\partial^2}{\partial \xi^2} u(1,t) = 0, \quad t > 0, \quad [5.177]$$

$$u(\xi,0) = u_0(\xi), \quad \frac{\partial}{\partial t} u(\xi,0) = u_1(\xi), \quad \xi \in [0,1], \quad [5.178]$$

by introducing the dimensionless quantities

$$\xi = \frac{S}{L}, \quad \bar{t} = \sqrt{\frac{EI}{\rho L^4}}\, t, \quad u = \frac{\Delta y}{L}, \quad \lambda = \frac{F_0 L^2}{EI}, \quad q = \frac{\Delta q_F}{E_0 L}, \quad \beta = \frac{E_0 L^4}{EI},$$

$$\mu = \frac{A}{I} L^2, \quad \bar{a} = a \left(\frac{EI}{\rho L^4}\right)^{\frac{\alpha}{2}}, \quad \bar{b} = b \left(\frac{EI}{\rho L^4}\right)^{\frac{\alpha}{2}},$$

and omitting the bar ($\bar{t} \to t, \bar{a} \to a, \bar{b} \to b$).

Assuming $u$ and $q$ in the form ($\xi \in [0,1], t > 0$)

$$u(\xi,t) = \sum_{k=1}^{\infty} T_k(t) \sin(k\pi\xi) \quad \text{and} \quad q(\xi,t) = \sum_{k=1}^{\infty} V_k(t) \sin(k\pi\xi), \quad [5.179]$$

we have that the boundary conditions [5.177] are satisfied, so that [5.175] and [5.176], with initial conditions adjoined, become ($t > 0$, $k \in \mathbb{N}$)

$$\frac{\mathrm{d}^2}{\mathrm{d}t^2} T_k(t) + p_k T_k(t) + \beta V_k(t) = 0, \quad [5.180]$$

$$V_k(t) + a\,{}_0\mathrm{D}_t^\alpha V_k(t) = T_k(t) + b\,{}_0\mathrm{D}_t^\alpha T_k(t), \quad [5.181]$$

$$T_k(0) = T_{k0}, \quad \frac{\mathrm{d}}{\mathrm{d}t} T_k(0) = T_{k1}, \quad [5.182]$$

where

$$p_k = (k\pi)^2 \frac{(k\pi)^2 - \lambda\left(1 - \frac{\lambda}{\mu}\right)}{\left(1 - \frac{\lambda}{\mu}\right)^2}.$$

Note that the initial conditions [5.182] will represent unknown constants in [5.179], which may be determined from [5.178]. Also note that the special case when the slenderness ratio $\mu$ of the rod tends to infinity ($\mu \to \infty$) corresponds to the case of the incompressible rod ($EA \to \infty$) and then

$$p_k = (k\pi)^2 \left( (k\pi)^2 - \lambda \right). \qquad [5.183]$$

To solve systems [5.180]–[5.182], we use the Laplace transform method, so that systems [5.180]–[5.182] are transformed into

$$\left(s^2 + p_k\right) \tilde{T}_k(s) + \beta \tilde{V}_k(s) = T_{k1} + sT_{k0}, \quad \operatorname{Re} s > s_0, \ k \in \mathbb{N}, \qquad [5.184]$$

$$(1 + bs^\alpha) \tilde{T}_k(s) - (1 + as^\alpha) \tilde{V}_k(s) = 0, \quad \operatorname{Re} s > s_0, \ k \in \mathbb{N}, \qquad [5.185]$$

where $s_0$ will be specified later and where we used the fact that $\left[{}_0\mathrm{I}_t^{1-\alpha} f(t)\right]_{t=0} = 0$, since $f(t)$ is bounded when $t \to 0$ (see [SAM 93]). The solutions of systems [5.184] and [5.185] are

$$\begin{aligned} \tilde{T}_k(s) &= \frac{as^\alpha + 1}{D_k(s)} \left(T_{k1} + T_{k0}s\right) \quad \text{and} \\ \tilde{V}_k(s) &= \frac{bs^\alpha + 1}{D_k(s)} \left(T_{k1} + sT_{k0}\right), \quad \operatorname{Re} s > s_0, \ k \in \mathbb{N}, \end{aligned} \qquad [5.186]$$

where

$$\begin{aligned} D_k(s) &= as^{\alpha+2} + s^2 + (p_k a + \beta b) s^\alpha + p_k + \beta, \quad \operatorname{Re} s > s_0, \ k \in \mathbb{N}, \\ &= \left(s^2 + A_k\right)(as^\alpha + 1) - B, \ \text{ with } A_k = p_k + \beta\frac{b}{a}, \quad B = \beta\frac{b-a}{a} > 0. \end{aligned} \qquad [5.187]$$

Note that due to the thermodynamical restrictions $0 < a \leq b$, we have $B \geq 0$.

### 5.3.1. *Zeros of* $D_k$

We estimate the zeros of $D_k$, given by [5.187], in the complex plane. We have that $D_k = 0$ if and only if

$$\left(s^2 + A_k\right)(as^\alpha + 1) = B > 0, \quad s \in \mathbb{C}. \qquad [5.188]$$

PROPOSITION 5.3.–

(i) Let $s = r\,\mathrm{e}^{\mathrm{i}\gamma}$, $r > 0$, $\gamma \in (-\pi, \pi]$, $\gamma \neq 0$, be a zero of $D_k$, which is given by [5.187]. Then, $|\gamma| > \frac{\pi}{2}$.

(ii) $B - A_k > 0$ $(\beta + p_k < 0)$ is a necessary and sufficient condition for the existence of $r^* > 0$ such that $D_k(r^*) = 0$.

(iii) $B - A_k = 0$ $(\beta + p_k = 0)$ is the necessary and sufficient condition that $r = 0$ is a solution to $D_k(r) = 0$. If in addition $A_k > 0$ $(p_k + \beta\frac{b}{a} > 0)$, then $r = 0$ is the unique solution to $D_k(r) = 0$, but if $A_k < 0$ $(p_k + \beta\frac{b}{a} < 0)$, then there exists additionally $r^* > 0$ such that $D_k(r^*) = 0$.

PROOF.– Let

$$s^2 + A_k = \rho\,\mathrm{e}^{\mathrm{i}\theta}. \qquad [5.189]$$

Then, from [5.188], we obtain

$$as^\alpha + 1 = \frac{B}{\rho}\mathrm{e}^{\mathrm{i}(-\theta+2n\pi)}, \quad n \in \mathbb{N}. \qquad [5.190]$$

Equations [5.189] and [5.190] lead to

$$r^2 \sin(2\gamma) = \rho \sin\theta \;\text{ and }\; ar^\alpha \sin(\alpha\gamma) = -\frac{B}{\rho}\sin\theta, \qquad [5.191]$$

where $s = r\,\mathrm{e}^{\mathrm{i}\gamma}$ with $\gamma \neq 0$. If $\gamma \neq 0$, from [5.191] it follows that

$$\frac{r^{2-\alpha}}{a}\frac{\sin(2\gamma)}{\sin(\alpha\gamma)} = -\frac{\rho^2}{B} < 0. \qquad [5.192]$$

Equation [5.192] implies $|\gamma| > \frac{\pi}{2}$.

If $\gamma = 0$, then from [5.191] we have $\theta = 0$. Suppose that there exists $s = r > 0$ such that $D_k(r) = 0$. Then, from [5.189] and [5.190] we obtain

$$r^2 + A_k = \rho \;\text{ and }\; ar^\alpha + 1 = \frac{B}{\rho}.$$

Thus, $r^2 = \rho - A_k > 0$ so that $\rho > A_k$. Also, $ar^\alpha = \frac{B}{\rho} - 1 > 0$ so that $\rho < B$. Consequently, if there exists $r > 0$ such that $D_k(r) = 0$, then $A_k < \rho < B$, or

$B - A_k > 0$. By [5.187], $B - A_k > 0$ gives $\beta + p_k < 0$. Now, we show that the condition $B - A_k > 0$ $(\beta + p_k < 0)$ is not only necessary, but also sufficient for the existence of $r > 0$, such that $D_k(r) = 0$. Thus, the condition $D_k(r) = 0$, by [5.187], is written as

$$\begin{aligned} D_k(r) &= y_1(r) - y_2(r) = 0, \quad \text{where } y_1(r) = r^2(ar^\alpha + 1) \text{ and} \\ y_2(r) &= B - A_k - A_k a r^\alpha. \end{aligned} \quad [5.193]$$

From Figure 5.11, it follows that, independently of the sign of $A_k$, the curves $y_1$ and $y_2$ intersect at a point $r = r^* > 0$ if $B - A_k > 0$.

It is easily seen that $B - A_k = 0$ is a necessary and sufficient condition that $r = 0$ is a solution to $D_k(r) = 0$. This case is analyzed more closely. Suppose that $B - A_k = 0$. Then $D_k(r) = 0$, by [5.187], implies

$$r^\alpha\left(ar^2 + r^{2-\alpha} + aA_k\right) = 0, \quad \text{i.e. } r = 0 \text{ and } ar^2 = -A_k a + r^{2-\alpha}. \quad [5.194]$$

If $A_k > 0$ and $B - A_k = 0$, then $r = 0$ is the unique solution to $D_k(r) = 0$. If $A_k < 0$ and $B - A_k = 0$, then suppose that $r > 0$ exists such that from $ar^2 = -A_k a - r^{2-\alpha}$. From Figure 5.12, we conclude that such $r$ exists. It is denoted by $r^*$. ■

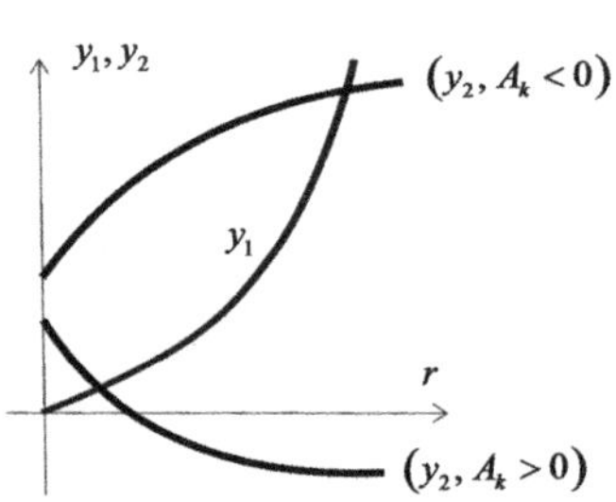


Figure 5.11. *Graphs of functions $y_1$ and $y_2$*

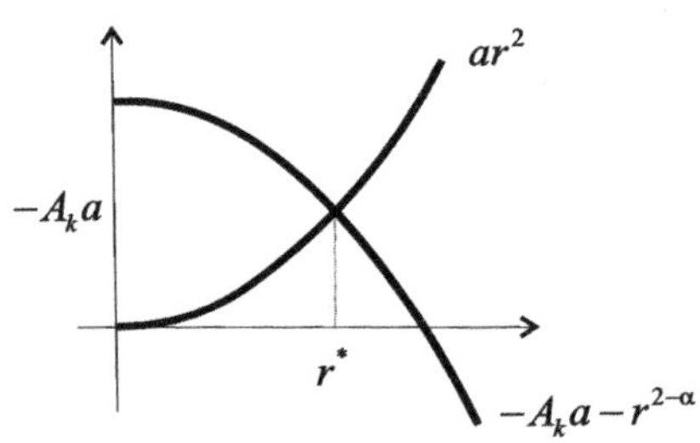


Figure 5.12. *Solution to $ar^2 = -A_k a - r^{2-\alpha}$*

### 5.3.2. *Existence of $T_k$ and $V_k$*

It will be shown that there exist functions $T_k$ and $V_k$, that are the inverse Laplace transforms of $\tilde{T}_k$ and $\tilde{V}_k$, given by [5.186]. The analysis of existence will be presented only for $T_k$, since the analysis of existence of $V_k$ is analogous.

Let us present the analytic expression for $T_k$ given by the series. Consider the function

$$\begin{aligned}\tilde{f}_k(s) &= \frac{as^\alpha+1}{D_k(s)} = \frac{as^\alpha+1}{(s^2+A_k)(as^\alpha+1)-B}\\ &= \frac{1}{s^2+A_k}\frac{1}{1-\frac{B}{(s^2+A_k)(as^\alpha+1)}}\\ &= \frac{1}{s^2+A_k}\sum_{k=0}^{\infty}\left(\frac{B}{a}\right)^k\left(\frac{1}{s^2+A_k}\right)^k\left(\frac{1}{s^\alpha+\frac{1}{a}}\right)^k, \quad \operatorname{Re}s>s_0,\end{aligned} \qquad [5.195]$$

where $s_0$ is determined so that $\left|\left(s_0^2+A_k\right)\left(as_0^\alpha+1\right)\right|>|B|$. Since

$$\begin{aligned}W(t) &= \mathcal{L}^{-1}\left[\frac{1}{s^2+A_k}\right](t) = \begin{cases}\frac{1}{\sqrt{A_k}}\sin\left(\sqrt{A_k}t\right), & \text{if } A_k>0,\\ \frac{1}{\sqrt{-A_k}}\sinh\left(\sqrt{-A_k}t\right), & \text{if } A_k<0,\end{cases} \quad t\geq 0,\\ w(t) &= \mathcal{L}^{-1}\left[\frac{1}{s^\alpha+\frac{1}{a}}\right](t) = \alpha t^{\alpha-1}E_\alpha^{(1)}\left(-\frac{t^\alpha}{a}\right), \quad t\geq 0,\end{aligned}$$

where $E_\alpha$ is the one-parameter Mittag-Leffler function (see [ERD 53] and section 1.5), we have that the inverse Laplace transform of [5.195] reads

$$f_k(t) = W(t) * \sum_{k=0}^{\infty}\left(\frac{B}{a}\right)^k (W(t)*w(t))^{*k}, \quad t\geq 0, \qquad [5.196]$$

where $F^{*k}$ denoted the $k$-fold convolution and $F^{*1}=F$. The proof that [5.196] holds can be found in [STA 01]. Since $f_k(0)=0$, it holds $s\tilde{f}_k(s)=\mathcal{L}\left[f_k^{(1)}(t)\right](s)$, so that the inverse Laplace transform of $[5.186]_1$ yields

$$T_k(t) = T_{k0}f_k^{(1)}(t) + T_{k1}f_k(t), \quad t\geq 0.$$

Inversion of the Laplace transform in [5.186] can also be done by the definition

$$T_k(t) = \lim_{\omega\to\infty}\frac{1}{2\pi \mathrm{i}}\int_{s_1-\mathrm{i}\omega}^{s_1+\mathrm{i}\omega}\tilde{T}_k(s)\,\mathrm{e}^{st}\mathrm{d}s, \quad t\geq 0, \qquad [5.197]$$

with $s_1>s_0$, so that $T_k$ is obtained in the integral representation.

In order to do so, let us transform $\tilde{T}_k$, given by [5.186]$_1$, by using [5.187], as $(\mathrm{Re}\, s > s_0)$

$$
\begin{aligned}
\tilde{T}_k(s) &= \frac{as^\alpha + 1}{D_k(s)}\left(T_{k1} + T_{k0}s\right) = \frac{T_{k0}}{s}\frac{s^2\left(as^\alpha + 1\right)}{D_k(s)} + \frac{T_{k1}}{s}\frac{s\left(as^\alpha + 1\right)}{D_k(s)} \\
&= \frac{T_{k0}}{s}\left(1 - \frac{p_k\left(as^\alpha + 1\right) + \beta\left(bs^\alpha + 1\right)}{D_k(s)}\right) + \frac{T_{k1}}{s}\frac{s\left(as^\alpha + 1\right)}{D_k(s)} \\
&= \frac{T_{k0}}{s} - \tilde{\mathcal{T}}_k(s), \quad \text{where} \\
\tilde{\mathcal{T}}_k(s) &= \frac{\left(T_{k0}p_k - T_{k1}s\right)\left(as^\alpha + 1\right) + T_{k0}\beta\left(bs^\alpha + 1\right)}{sD_k(s)}.
\end{aligned}
\qquad [5.198]
$$

We will prove that every addend in [5.198] is the Laplace transform on $\mathbb{R}_+$ of a function of exponential growth by using [DOE 55, Part I, Kapitel 7, Satz 3]. Consequently, $\tilde{T}_k$ will have the same property.

PROPOSITION 5.4.– Let $s_0 = r \geq 0$ be a singular point of $\tilde{T}_k$ such that for all other singular points $s_i$ of $\tilde{T}_k$, we have $\mathrm{Re}\, s_i < r$. Then

$$
T_k(t) = \frac{1}{2\pi\mathrm{i}}\int_\Sigma \tilde{T}_k(s)\,\mathrm{e}^{st}\mathrm{d}s, \quad t \geq 0, \qquad [5.199]
$$

holds, where $\Sigma$ is given in Figure 5.13.

PROOF.– To prove the claim, we use [DOE 55, Part II, Kapitel 7]. First, we prove that, for $s_1 > s_0$,

$$
T_k(t) = \lim_{\omega\to\infty}\frac{1}{2\pi\mathrm{i}}\int_{s_1-\mathrm{i}\omega}^{s_1+\mathrm{i}\omega}\tilde{T}_k(s)\,\mathrm{e}^{st}\mathrm{d}s = \frac{1}{2\pi\mathrm{i}}\int_\Sigma \tilde{T}_k(s)\,\mathrm{e}^{st}\mathrm{d}s, \quad t \geq 0.
$$

The curve $\Sigma$ consists of two straight lines having angles $\frac{\pi}{2} + \delta$ and $-\left(\frac{\pi}{2} + \delta\right)$ and an arc of a circle with center in $s_0$ and of arbitrary small radius, as it is shown in Figure 5.13.

The point $s_0$ ($\mathrm{Re}\, s_0 > 0$ or $s_0 = 0$) is a singular point of $\tilde{T}_k$, and $\tilde{T}_k$ is analytic between $\Sigma$ and $\{s \in \mathbb{C} \mid \mathrm{Re}\, s = s_1\}$. By proposition 5.3, all points where $\tilde{T}_k$ is singular are of the form $s = 0$, $s = r > 0$ and $s = r\,\mathrm{e}^{\mathrm{i}\gamma}$, $|\gamma| > \frac{\pi}{2}$. Then, there exists a $\delta > 0$ such that $\tilde{\mathcal{T}}_k$ and $\tilde{T}_k$ are analytic on the right-hand side of curve $\Sigma$ (see Figure 5.13). Let $\Omega$ denote the closed domain with the border $\partial\Omega$ consisting of the straight line "1", arcs "2" and "3" with arbitrary $R>0$ and of the curve $\Sigma$. It is well known (see [DOE 55, Part I, p. 226]) that

$$\frac{1}{2\pi \mathrm{i}}\int_{\Sigma} s^{\lambda}\mathrm{e}^{ts}\mathrm{d}s=\frac{t^{-\lambda-1}}{\Gamma\left(-\lambda\right)},\quad t>0,\ \lambda\in\mathbb{C}. \qquad [5.200]$$

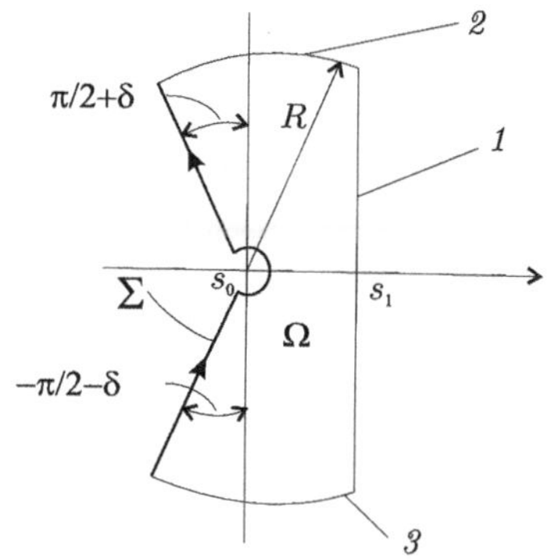


**Figure 5.13.** *Integration path for [5.199]*

By using [5.200] with $\lambda=-1$, it follows that

$$\frac{1}{2\pi \mathrm{i}}\int_{\Sigma} s^{-1}\mathrm{e}^{st}\mathrm{d}s=1. \qquad [5.201]$$

Now consider the integral

$$\frac{1}{2\pi \mathrm{i}}\int_{\partial\Omega} \tilde{T}_k\left(s\right)\mathrm{e}^{st}\mathrm{d}s=0, \qquad [5.202]$$

where $\tilde{T}_k$ is given by [5.198]. By the properties of $\tilde{T}_k$, when the radius $R$ of the arcs "2" and "3" tends to infinity, integral [5.200] becomes

$$\frac{1}{2\pi \mathrm{i}}\int_{\Sigma} \tilde{T}_k\left(s\right)\mathrm{e}^{st}\mathrm{d}s=\frac{1}{2\pi \mathrm{i}}\int_{s_1-\mathrm{i}\omega}^{s_1+\mathrm{i}\omega} \tilde{T}_k\left(s\right)\mathrm{e}^{st}\mathrm{d}s. \qquad [5.203]$$

From [5.201] and [5.203], it follows

$$\begin{aligned} T_k\left(t\right) &= \mathcal{L}^{-1}\left[\tilde{T}_k\left(s\right)\right]\left(t\right)=T_{k0}-\frac{1}{2\pi \mathrm{i}}\int_{s_1-\mathrm{i}\omega}^{s_1+\mathrm{i}\omega}\tilde{T}_k\left(s\right)\mathrm{e}^{st}\mathrm{d}s \\ &= \frac{1}{2\pi \mathrm{i}}\int_{\Sigma}\tilde{T}\left(s\right)\mathrm{e}^{st}\mathrm{d}s. \end{aligned} \qquad [5.204]$$

■

### 5.3.3. *Asymptotic behavior of* $T_k$

Following proposition 5.3 in the asymptotic analysis, consider three cases: $\beta + p_k < 0$, $\beta + p_k = 0$ with $A_k < 0$, $\beta + p_k < 0$ with $A_k > 0$ and $\beta + p_k > 0$. The analysis is based on the following theorem.

THEOREM 5.3.– [DOE 55, Part II, p. 157] Let the function $f(s)$ be analytic in the sector $|\arg(s - s_0)| \leq \Psi$, $\frac{\pi}{2} < \Psi < \pi$, of a neighborhood of $s_0$, except in $s_0$. Let $f(s)$ be locally integrable on the rays $\arg(s - s_0) = \pm\psi$. The function $F(t)$ is given by $\mathcal{L}^{-1}[f(s)](t)$, for $t \geq T$, where $T$ is a positive number. Suppose that for a $\lambda \in C$, we have

$$f(s) \sim M(s - s_0)^{\lambda}, \quad \text{as } s \to s_0, \quad \text{uniformly in } |\arg(s - s_0)| \leq \Psi, \quad [5.205]$$

then

$$F(t) \sim M\frac{t^{-\lambda-1}}{\Gamma(-\lambda)}e^{s_0 t}, \quad t \to \infty. \quad [5.206]$$

If $\lambda = 0, 1, 2, \ldots$, we take $\frac{1}{\Gamma(-\lambda)} = 0$ and

$$F(t) = o\left(t^{-\lambda-1}e^{s_0 t}\right), \quad t \to \infty. \quad [5.207]$$

To find the asymptotic behavior of $T_k$, the cited theorem and proposition 5.4 are used. The following cases will be distinguished:

*Case I*: Suppose that $\beta + p_k < 0$ or $\beta + p_k = 0$ and $p_k + \beta\frac{b}{a} < 0$. By proposition 5.3, there exists $r^* > 0$, such that $\Delta(r^*) = 0$. Let $M$ denote $M = \lim_{s \to r^*}\left((s - r^*)\tilde{T}(s)\right)$, then, by [5.206], we have $T_k(t) \sim Me^{r^* t}$, as $t \to \infty$.

*Case II*: Suppose that $\beta + p_k > 0$. Since $s = 0$ is a point of ramification and $\lim_{s \to 0}\tilde{T}_k(s) = \frac{T_{k1}}{p_k + \beta}$, then, by [5.207], we have $T_k(t) = o\left(t^{-1}\right)$, as $t \to \infty$.

*Case III*: Suppose that $\beta + p_k = 0$ with $A_k > 0$. Now, $\tilde{T}_k(s) \sim T_{k1}s^{-\alpha}$, as $s \to 0$. Hence, $T_k(t) \sim T_{k1}\frac{t^{\alpha-1}}{\Gamma(\alpha)}$, as $t \to \infty$.

The previous results are summarized in the following proposition.

PROPOSITION 5.5.– The asymptotic behavior of $T_k(t)$, obtained as a solution to systems [5.180]–[5.182], when $t \to \infty$, is

$$T_k(t) \sim Me^{r^* t}, \text{ if } \beta + p_k < 0, \text{ or } \beta + p_k = 0 \text{ with } p_k + \beta\frac{b}{a} < 0, \quad [5.208]$$

$$T_k(t) \sim o\left(t^{-1}\right), \text{ if } \beta + p_k > 0, \quad [5.209]$$

$$T_k(t) \sim T_{k1}\frac{t^{\alpha-1}}{\Gamma(\alpha)}, \text{ if } \beta + p_k = 0 \text{ with } p_k + \beta\frac{b}{a} > 0. \quad [5.210]$$

### 5.3.4. *Summary of the stability analysis*

The stability analysis of an elastic rod with a compressible axis positioned on a viscoelastic foundation of fractional derivative type is presented. To examine the influence of the viscoelastic foundation on stability, note that the incompressible rod without foundation, i.e. $B = 0$ in [5.187], is stable, according to Liapunov stability criteria, when the dimensionless force $\lambda < \pi^2$, while in the case when $\lambda > \pi^2$, the instability of a rod without foundation is obtained (see [ATA 97, p. 363]).

For the incompressible rod with a viscoelastic foundation, the following cases are distinguished:

1) Suppose $\lambda < \pi^2$, so that $p_k = (k\pi)^2\left((k\pi)^2 - \lambda\right) > 0$ (see [5.183]). Then, proposition 5.5 applies and since $\beta + p_k > 0$, by [5.209], we have that the rod is asymptotically stable, i.e. the vibrations of the rod diminish.

2) Suppose $\lambda > \pi^2$, i.e. $p_k < 0$, and $\beta + p_k > 0$. Then, according to proposition 5.5, [5.209], the rod is stable, although without foundation ($\beta = 0$), the rod would be unstable. Thus, the foundation stabilizes the rod.

3) Suppose $\lambda > \pi^2$, i.e. $p_k < 0$, $\beta + p_k = 0$ and that the thermodynamic restriction is satisfied ($b > a$), then $p_k + \beta\frac{b}{a} > 0$. According to proposition 5.5, [5.210], we have that the rod is stable.

4) Suppose $\lambda > \pi^2$, i.e. $p_k < 0$, $\beta + p_k = 0$ and that the thermodynamic restriction is violated ($b < a$), then $p_k + \beta\frac{b}{a} < 0$. Applying proposition 5.5, [5.208], we have that the rod is unstable. The same holds true if $\beta + p_k < 0$ independently of the thermodynamic restriction.

On the basis of the analysis presented, it follows that the value of the sum of parameters $p_k$ and $\beta$ determines the stability of the lateral vibration of the rod: if $p_k + \beta > 0$, the rod is stable and if $p_k + \beta < 0$, the rod is unstable. The case $p_k + \beta = 0$ corresponds to stable vibrations if the thermodynamic restriction $b > a$ is satisfied, otherwise the vibrations are unstable.

We compare obtained results with the results of the static stability analysis of an elastic rod on an elastic foundation. Note that the case $a = b$ in [5.175] corresponds to an elastic foundation. For such a foundation, the Euler stability criterion leads to the critical force $F_{cr} = \frac{\pi^2 EI}{L^2} + \frac{\mu L^2}{\pi^2}$ (see [CHE 76, p. 109]), where $\mu$ is, in our notation, $\mu = bE_0$. Thus

$$F < \frac{\pi^2 EI}{L^2} + \frac{bE_0 L^2}{\pi^2} \qquad [5.211]$$

guarantees (static) stability of a rod on elastic foundation. The presented dynamic analysis implies the stability conditions $p_1 + \beta > 0$, or $\pi^2 \left(\pi^2 - \frac{FL^2}{EI}\right) + b\frac{E_0 L^4}{EI} > 0$. This condition is satisfied if [5.211] holds. Thus, the condition for the stability of a rod on an elastic foundation can be obtained, as a special case of our results.

The results, summarized in proposition 5.5, could be applied to determine the stability boundary for the rod without foundation, i.e. for $B = 0$. The stability condition requires $p_k > 0$ (see [5.208]), which is equivalent to $\lambda < \pi^2$.

# Chapter 6

# Fractional Diffusion-Wave Equations

In this Chapter, we present the equations that will be treated as well as presenting their backgrounds. For the analysis, we can use the framework of the generalized functions, which gives flexibility in calculations. Solutions of equations will always be interpreted as the functions of polynomial or exponential growth.

The nonlinear classical heat conduction equation in one spatial dimension reads

$$\frac{\partial}{\partial t} T(x,t) = \frac{\partial}{\partial x}\left(\mathcal{D}(T,x,t)\frac{\partial}{\partial x}T(x,t)\right), \tag{6.1}$$

where $T$ is the temperature at the point $x$ of an infinite, semi-infinite or finite spatial domain at time instant $t > 0$ and $\mathcal{D}$ is the temperature-dependent thermal diffusivity. In this case, when the thermal diffusivity is constant, equation [6.1] becomes the classical heat conduction (diffusion) equation

$$\frac{\partial}{\partial t} T(x,t) = \mathcal{D}\frac{\partial^2}{\partial x^2}T(x,t). \tag{6.2}$$

The classical wave equation is of the form

$$\frac{\partial^2}{\partial t^2} u(x,t) = c^2\frac{\partial^2}{\partial x^2}u(x,t), \tag{6.3}$$

where $u$ is displacement of the point $x$ of an infinite, semi-infinite or finite spatial domain at time instant $t > 0$ and $c$ is the speed of wave propagation.

The generalization of the heat conduction (diffusion) equation [6.2] and wave equation [6.3] to a fractional-order diffusion-wave equation was proposed by Mainardi and his colleagues in [GOR 00, MAI 96, MAI 97a, MAI 08, MAI 97b] (see also [ALK 05]), as

$$
{}_0\mathrm{D}_t^{\alpha} T(x,t) = \mathcal{D}\frac{\partial^2}{\partial x^2}T(x,t), \qquad [6.4]
$$

where either $\alpha \in (0,1)$ or $\alpha \in (1,2)$. Letting $\alpha \in (0,1)$, we obtain subdiffusion, while letting $\alpha \in (1,2)$, we obtain superdiffusion. In the latter case, the equation of fractional order generalizes both the diffusion [6.2] and wave equation [6.3].

We refer to [MAI 97b] for the background of the diffusion-wave equation [6.4]. Extensive overview of the diffusion-wave equation with single time- and space-fractional derivative can be found in [MAI 96, MAI 97a, MAI 01], where solutions are obtained by the use of the Laplace and Fourier transforms. New function of the Mittag-Leffler type is introduced in [YU 06] in order to solve the space–time-fractional diffusion-wave equation. The question of the speed of propagation of impulses of the space-fractional diffusion equation was considered in [ENG 10]. We refer to [PET 08, PET 05] for application of the diffusion-wave equation [6.4] to the problem of stresses due to the polymerization of dental materials. We also refer to [MET 04] for an extensive review on the diffusion-wave phenomena.

The nonlinear fractional diffusion-wave equation, obtained by the fractionalization (replacement of the classical time derivative by the fractional time derivative) of [6.1], reads

$$
{}_0\mathrm{D}_t^{\alpha} T(x,t) = \frac{\partial}{\partial x}\left(\mathcal{D}(T,x,t)\frac{\partial}{\partial x}T(x,t)\right). \qquad [6.5]
$$

Equations of this type have been studied in [DJO 08].

The multidimensional fractional diffusion-wave equation containing the source term was studied in [HAN 01, HAN 02a, HAN 02b] within the framework of an abstract Volterra equation. Namely, well-posedness of Cauchy problems for space-fractional, space–time-fractional and time-fractional diffusion-wave equations was the subject of analysis in [HAN 01, HAN 02a, HAN 02b]. The Cauchy problem for multidimensional time-fractional diffusion-wave equations, which was treated in [HAN 02b], reads

$$
{}_0\mathrm{D}_t^{\alpha} T(\boldsymbol{x},t) - \mathcal{D}\nabla^2 T(\boldsymbol{x},t) = F\delta(\boldsymbol{x})\,\delta(t), \quad \boldsymbol{x}\in\mathbb{R}^3, \ t\in[0,\infty), \qquad [6.6]
$$

where $\nabla^2$ denotes the Laplace space operator, $\delta(\boldsymbol{x})$, $\boldsymbol{x} \in \mathbb{R}^3$ and $\delta(t)$, $t \geq 0$, are space and time Dirac distributions, respectively, $\mathcal{D} > 0$ is the diffusivity coefficient and $F$ is a given constant. Equation [6.6] is subject to initial conditions

$$T(\boldsymbol{x},0) = T_0\delta(\boldsymbol{x}),\ \alpha \in (0,1],\quad \left.\begin{array}{l} T(\boldsymbol{x},0) = T_0\delta(\boldsymbol{x}) \\ \frac{\partial}{\partial t}T(\boldsymbol{x},0) = T_1\delta(\boldsymbol{x}) \end{array}\right\},\ \alpha \in (1,2],\quad \boldsymbol{x} \in \mathbb{R}^n, \qquad [6.7]$$

where $T_0$ and $T_1$ are given constants. Note that [6.6] and [6.7] have to be interpreted in the sense of distributions. Solutions are calculated by the use of the Laplace and Fourier transforms in cases of one, two and three space dimensions. We also mention [EID 04], where a fractional diffusion equation was studied with an elliptic second-order differential operator.

The classical heat conduction equation [6.2] is parabolic and has a physically unacceptable property: heat impulses propagate at an infinite speed in a bar in which heat is introduced at an arbitrary point (see [WID 75]). This property may be removed by assuming, as Cattaneo did in [CAT 48], that the thermal flux depends on the temperature gradient and on a special "mixed" functional (see [FIC 92]). The Cattaneo equation for heat flux $q$ is $\tau\frac{\partial}{\partial t}q + q = -\lambda\frac{\partial}{\partial x}T$, where $\tau$ is relaxation time and $\lambda$ is thermal conductivity. Heat flux $q$ may also be expressed as an integral of the history of the temperature gradient, as is given in [JOS 89]. When this constitutive assumption is made, the resulting partial differential equation, describing one-dimensional heat conduction becomes the telegraph equation

$$\tau\frac{\partial^2}{\partial t^2}T(x,t) + \frac{\partial}{\partial t}T(x,t) = \mathcal{D}\frac{\partial^2}{\partial x^2}T(x,t). \qquad [6.8]$$

This modification of the parabolic equation of heat conduction ensures that the heat impulses propagate with finite speed. It is shown in [SWE 78] that the longer the memory is, the more wave-like the solution of [6.8] is. Although often applied to the study of heat waves, the telegraph equation [6.8] is not compatible with the local equilibrium of entropy (in an isolated rigid body, entropy monotonically increases). In other words, the dissipation inequality $q\frac{\partial}{\partial x}T \leq 0$ is satisfied only if the term $\tau\frac{\partial}{\partial t}q\frac{\partial}{\partial x}T$ is sufficiently small (see [IGN 10, JOU 01, STR 11]). In the special case when the relaxation time is zero, the telegraph equation [6.8] becomes the classical heat conduction equation [6.2].

The one-dimensional fractional telegraph equation reads

$$\tau\,{}_0^C\mathrm{D}_t^{\alpha}T(x,t) + {}_0^C\mathrm{D}_t^{\beta}T(x,t) = \mathcal{D}\frac{\partial^2}{\partial x^2}T(x,t). \qquad [6.9]$$

It is studied in [ATA 07b], where it is assumed $0 < \beta < \alpha < 2$. Fractional telegraph equation in various forms, with time- and/or space-fractional derivatives, as well as with the source term, is analyzed in [CAM 09, CAM 08, LAN 06, ORS 03, SUN 09].

Further generalization of the diffusion-wave and telegraph equation is conducted by introducing the (time) distributed-order fractional derivative $\mathrm{D}_\phi T$, given by [2.91], so that we obtain the (time) distributed-order diffusion-wave equation in the form

$$\mathrm{D}_\phi T(x,t) = \int_\mu^\nu \phi(\gamma)\, {}_0\mathrm{D}_t^\gamma T(x,t)\,\mathrm{d}\gamma = \mathcal{D}\frac{\partial^2}{\partial x^2} T(x,t), \qquad [6.10]$$

where $\phi$ is the constitutive (weight) function or distribution, and $\mu = 0$, $\nu = 2$, or $\mu = 0$, $\nu = 1$, or $\mu = 1$, $\nu = 2$. We note that the diffusion-wave equation [6.4] and the fractional telegraph equation [6.9] can be obtained from [6.10] by choosing the constitutive distributions as $\phi(\gamma) = \delta(\gamma - \alpha)$ and $\phi(\gamma) = \tau\delta(\gamma - \alpha) + \delta(\gamma - \beta)$, respectively.

The distributed-order diffusion-wave equation was the subject of analysis in [ATA 09c, ATA 09d, ATA 09e]. The time and space distributed-order diffusion equation is also studied in [CHE 02] through the analysis of second moments of certain probability density functions. A similar approach can be found in [SOK 04]. Time distributed-order diffusion equations were solved in [NAB 04] for the Dirichlet, Neumann and Cauchy problems. Time distributed-order equations of relaxation are studied in [MAI 07a]. In [MAI 07a, MAI 07b], solutions are obtained by the use of the Mellin–Barnes integrals, and Laplace and Fourier transforms.

We note that results on time distributed-order diffusion-wave equations can also be found in [HAN 07, KOC 08, MAI 08]. In [HAN 07], equation [6.10] with $\mu = \alpha$, $\nu = 1$ was treated, while the weight function is assumed to be a right-continuous non-decreasing function on interval $[\alpha, 1]$ satisfying $\phi(\gamma) = 0$ for $\gamma < \alpha$. Asymptotic analysis of the solution was made in [HAN 07] by the use of regularly varying functions. An equation similar to [6.10] was treated in [KOC 08] with $\mu = 0$, $\nu = 1$ and $\phi \in C^2[0,1]$, where $\phi(\alpha) = \alpha^\nu \phi_1(\alpha)$, $\phi_1(\alpha) \geq \rho > 0$, $0 \leq \alpha \leq 1$, $\nu \geq 0$. The existence was proved by the use of asymptotic properties of the Fourier and Laplace transforms and their inversions. In [MAI 08], an equation similar to [6.10], with $\mu = 0$, $\nu = 1$ and with a general weight function $\phi$, was solved by the direct calculation, i.e. by the application of the Fourier and Laplace transforms. The special case $\phi(\alpha) = 1$ was considered as well.

In this chapter, we present results from [ATA 07b, ATA 09c, ATA 09d, ATA 09e, DJO 08].

## 6.1. Nonlinear fractional diffusion-wave and fractional Burgers/Korteweg–de Vries equations

Following [DJO 08], we consider fractional generalization of the nonlinear classical heat conduction equation [6.1], as well as of the Burgers and Korteweg–de Vries equations.

The nonlinear fractional diffusion-wave equation [6.5] will be considered in the case when the thermal diffusivity $\mathcal{D}$ is the function of temperature

$$\mathcal{D}(T) = k + mT^n, \tag{6.11}$$

where $k, m, n > 0$ are fixed constants. With the thermal diffusivity assumed in the form [6.11], the nonlinear fractional diffusion-wave equation [6.5] reads

$${}_0\mathrm{D}_t^{\alpha} T(x,t) = \frac{\partial}{\partial x}\left( (k + mT^n(x,t)) \frac{\partial}{\partial x} T(x,t) \right), \quad x \in [0,\infty), \; t > 0. \tag{6.12}$$

The fractional Burgers/Korteweg–de Vries equation is

$$\frac{\partial}{\partial t}\phi(x,t) + \phi \frac{\partial}{\partial x}\phi(x,t) = \nu \, {}_0\mathrm{D}_x^{\beta}\phi(x,t), \quad x \in [0,\infty), \; t > 0, \tag{6.13}$$

where $\beta \in (2,3)$ and $\nu > 0$. In the limiting case $\beta = 2$, we have that [6.13] reduces to the classical Burgers equation

$$\frac{\partial}{\partial t}\phi(x,t) + \phi(x,t) \frac{\partial}{\partial x}\phi(x,t) = \nu \frac{\partial^2}{\partial x^2}\phi(x,t), \quad x \in [0,\infty), \; t > 0,$$

see [WHI 74], while if $\beta = 3$, then [6.13] becomes the classical Korteweg–de Vries equation

$$\frac{\partial}{\partial t}\phi(x,t) + \phi(x,t) \frac{\partial}{\partial x}\phi(x,t) = \nu \frac{\partial^3}{\partial x^3}\phi(x,t), \quad x \in [0,\infty), \; t > 0.$$

We present similarity transformations, different from those used in [BUC 98, MAI 97b], that reduce the nonlinear fractional diffusion-wave equation [6.12] and the fractional Burgers/Korteweg–de Vries equation [6.13] to the ordinary nonlinear fractional-order differential equations. These equations are solved numerically. Our results will show qualitative agreement with the corresponding results for the integer values of $\alpha$ in [6.12] and $\beta$ in [6.13]. Also, we present two exact solutions to the nonlinear fractional diffusion-wave equation. One of these solutions has the form of a wave with the prescribed relation between its amplitude and the speed.

Results regarding [6.12] and [6.13] can be, in principle, extended by the use of the Lie-group symmetry analysis to obtain different similarity variables from the ones used here and corresponding reduced equations. Such an analysis may be performed along the lines given in [BLU 89, IBR 94, OLV 86] and presented in specific cases in [OZE 04, OZE 05, ROB 85]. However, the presence of non-integer derivatives in equations treated here makes such an analysis non-trivial.

### 6.1.1. *Nonlinear fractional diffusion-wave equation*

To prove that the equation [6.12] possesses similarity solutions, we perform the Lie-group scaling transformation (see [CAN 02]). We transform [6.12] by introducing new independent and dependent variables denoted by $\tilde{t}$, $\tilde{x}$, $\tilde{T}$ as

$$t = \lambda \tilde{t}, \quad x = \lambda^p \tilde{x}, \quad T = \lambda^q \tilde{T}, \qquad [6.14]$$

where $p$ and $q$ are parameters that are determined by requiring that [6.12] remains invariant under transformations [6.14]. We may easily verify that the transformed equation becomes

$$\lambda^{q-2p} k \frac{\partial^2}{\partial \tilde{x}^2} \tilde{T}\left(\tilde{x}, \tilde{t}\right) + \lambda^{q(1+n)-2p} m \frac{\partial}{\partial \tilde{x}} \left( \tilde{T}^n \left(\tilde{x}, \tilde{t}\right) \frac{\partial}{\partial \tilde{x}} \tilde{T}\left(\tilde{x}, \tilde{t}\right) \right)$$
$$= \lambda^{q-\alpha} {}_0\mathrm{D}_{\tilde{t}}^{\alpha} \tilde{T}\left(\tilde{x}, \tilde{t}\right).$$

If $k \neq 0$, the condition of invariance reads: $q - \alpha = q - 2p = q\,(1+n) - 2p$, from which we obtain $q = 0$ and $p = \frac{\alpha}{2}$. If $k = 0$, the condition of invariance simplifies to: $q - \alpha = q\,(1+n) - 2p$, yielding $q = \frac{2p-\alpha}{n}$ and $p$ arbitrary. Eliminating $\lambda$ from [6.14], we obtain $x^{-\frac{1}{p}} t = \tilde{x}^{-\frac{1}{p}} \tilde{t}$, $\frac{T}{\tilde{T}} = \left(\frac{x}{\tilde{x}}\right)^{\frac{q}{p}}$ and consequently $T$ is expressed as

$$T\left(x, t\right) = x^{\frac{q}{p}} U\left(x^{-\frac{1}{p}} t\right), \quad x \in [0, \infty), \; t > 0,$$

where $U$ satisfies an ordinary differential equation. This actually means that equation [6.12] possesses similarity solutions of the form

$$T\left(x, t\right) = x^a U\left(\xi\right), \quad \xi = x^b t, \quad x \in [0, \infty), \; t > 0, \qquad [6.15]$$

where $a = \frac{q}{p}$ and $b = -\frac{1}{p}$ are constants. Namely, if $k \neq 0$, we have $a = 0, b = -\frac{2}{\alpha}$, while if $k = 0$, $a$ is arbitrary, $b = \frac{an-2}{\alpha}$.

To derive the ordinary differential equation for $U$, we express ${}_0\mathrm{D}_t^\alpha T$, $\frac{\partial}{\partial x}T$, $\frac{\partial^2}{\partial x^2}T$ and $\frac{\partial}{\partial x}\left(T^n \frac{\partial}{\partial x}T\right)$ via $U$ using [6.15]. To find ${}_0\mathrm{D}_t^\alpha T$ via $U$, we write

$$\begin{aligned} {}_0\mathrm{D}_t^\alpha T(x,t) &= \frac{1}{\Gamma(1-\alpha)}\frac{\partial}{\partial t}\int_0^t \frac{T(x,\tau)}{(t-\tau)^\alpha}\mathrm{d}\tau \\ &= \frac{1}{\Gamma(1-\alpha)}x^b \frac{\mathrm{d}}{\mathrm{d}\xi}\int_0^\xi \frac{x^a U(y)\, x^{b\alpha}}{(\xi-y)^\alpha}x^{-b}\mathrm{d}y \\ &= x^{a+b\alpha}{}_0\mathrm{D}_\xi^\alpha U(\xi), \quad \xi = x^b t, \quad x\in[0,\infty), \, t>0. \end{aligned} \qquad [6.16]$$

Similarly, using [6.15], we calculate $\frac{\partial}{\partial x}T$, $\frac{\partial^2}{\partial x^2}T$ and $\frac{\partial}{\partial x}\left(T^n\frac{\partial}{\partial x}T\right)$ via $U$ as $(x \in [0,\infty), t>0, \xi\in[0,\infty))$

$$\frac{\partial}{\partial x}T(x,t) = x^{a-1}F_1(\xi), \quad \text{where} \quad F_1(\xi) = aU(\xi) + b\xi\frac{\mathrm{d}}{\mathrm{d}\xi}U(\xi),$$

$$\frac{\partial^2}{\partial x^2}T(x,t) = x^{a-2}F_2(\xi), \quad \text{where} \quad F_2(\xi) = (a-1)F_1(\xi) + b\xi\frac{\mathrm{d}}{\mathrm{d}\xi}F_1(\xi),$$

$$\begin{aligned} &\frac{\partial}{\partial x}\left(T^n(x,t)\frac{\partial}{\partial x}T(x,t)\right) \\ &= x^{a(1+n)-2}\left((a(1+n)-1)F(\xi) + b\xi\frac{\mathrm{d}}{\mathrm{d}\xi}F(\xi)\right), \quad \text{where} \\ &\qquad F(\xi) = U^n(\xi)\left(aU(\xi) + b\xi\frac{\mathrm{d}}{\mathrm{d}\xi}U(\xi)\right). \end{aligned} \qquad [6.17]$$

Substituting [6.16] and [6.17] in [6.12], we obtain the ordinary fractional-order differential equation for $U$. Since exponents of $x$ on the left- and right-hand sides of this equation must be equal, we verify that the conditions guaranteeing existence of similarity solution are $a = 0$, $b = -\frac{2}{\alpha}$ if $k \neq 0$ and $a$ arbitrary, and $b = \frac{an-2}{\alpha}$ if $k = 0$. Bearing both of these cases in mind, we state the following proposition.

PROPOSITION 6.1.– The nonlinear fractional diffusion equation [6.12] has the similarity solution

$$T(x,t) = x^a U(\xi), \quad \xi = x^b t, \quad x\in[0,\infty), \, t>0, \qquad [6.18]$$

where

$$a=0,\ \ b=-\frac{2}{\alpha},\ \ \text{if}\ \ k\neq 0,\ \ \text{or}\ \ a\ \text{arbitrary},\ b=\frac{an-2}{\alpha},\ \ \text{if}\ \ k=0.$$

Function $U$ is a solution to the ordinary nonlinear fractional equation

$$\begin{aligned}{}_{0}\mathrm{D}_{\xi}^{\alpha}U(\xi) &= k\left(a(a-1)U(\xi)+b(2a+b-1)\xi\frac{\mathrm{d}}{\mathrm{d}\xi}U(\xi)+b^{2}\xi^{2}\frac{\mathrm{d}^{2}}{\mathrm{d}\xi^{2}}U(\xi)\right)\\ &\quad +m\left((a(1+n)-1)F(\xi)+b\xi\frac{\mathrm{d}}{\mathrm{d}\xi}F(\xi)\right),\ \ \xi\in[0,\infty), \qquad [6.19]\end{aligned}$$

where $F(\xi)=U^{n}(\xi)\left(aU(\xi)+b\xi\frac{\mathrm{d}}{\mathrm{d}\xi}U(\xi)\right),\ \xi\in[0,\infty)$.

REMARK 6.1.– The similarity variable obtained in [BUC 98, MAI 97b] for the linear fractional diffusion-wave equation reads $z=\frac{x}{t^{\frac{\alpha}{2}}}$. Comparing this with $[6.18]_2$, we conclude that the relation between our similarity variable and the similarity variable of [BUC 98, MAI 97b] is $\xi=z^{-\frac{2}{\alpha}}$.

REMARK 6.2.– The classical diffusion equation is obtained by setting $\alpha=1, b=-2$, $m=0$, $k=1$. In this case, [6.19] becomes

$$\frac{\mathrm{d}}{\mathrm{d}\xi}U(\xi)=4\xi^{2}\frac{\mathrm{d}^{2}}{\mathrm{d}\xi^{2}}U(\xi)+6\xi\frac{\mathrm{d}}{\mathrm{d}\xi}U(\xi),\ \ \xi\in[0,\infty). \qquad [6.20]$$

The solution to [6.20] reads

$$U(\xi)=\int_{\frac{1}{2\sqrt{\xi}}}^{\infty}\mathrm{e}^{-s^{2}}\mathrm{d}s,\ \ \xi\in[0,\infty). \qquad [6.21]$$

Solution [6.21] agrees with the known classical case (see [WID 75, p. 10]).

### 6.1.2. *Fractional Burgers/Korteweg–de Vries equation*

Lie-group scaling transformation for the fractional Burgers/Korteweg–de Vries equation [6.13] is very similar to the corresponding analysis of the nonlinear fractional diffusion-wave equation [6.12]. Transformation of variables

$$t=\lambda\tilde{t},\quad x=\lambda^{p}\tilde{x},\quad \phi=\lambda^{q}\tilde{\phi} \qquad [6.22]$$

changes [6.13] into

$$\lambda^{q-1}\frac{\partial}{\partial\tilde{t}}\tilde{\phi}\left(\tilde{x},\tilde{t}\right)+\lambda^{2q-p}\tilde{\phi}\left(\tilde{x},\tilde{t}\right)\frac{\partial}{\partial\tilde{x}}\tilde{\phi}\left(\tilde{x},\tilde{t}\right)=\lambda^{q-\beta p}\nu\,{}_0\mathrm{D}_{\tilde{x}}^{\beta}\tilde{\phi}\left(\tilde{x},\tilde{t}\right).$$

Then, the invariance requires $q-1=2q-p=q-\beta p$, so that $p=\frac{1}{\beta}$, $q=\frac{1}{\beta}-1$. Elimination of $\lambda$ in [6.22] yields $t^{-p}x=\tilde{t}^{-p}\tilde{x}$ and $\frac{\phi}{\tilde{\phi}}=\left(\frac{t}{\tilde{t}}\right)^{q}$. This proves the existence of the similarity solution to equation [6.13] of the form

$$\phi\left(x,t\right)=t^{a}U\left(\xi\right),\;\;\xi=t^{b}x,\;\;x\in\left[0,\infty\right),\;t>0, \tag{6.23}$$

with $a=\frac{1}{\beta}-1$, $b=-\frac{1}{\beta}$.

To derive the ordinary differential equation for $U$, we express $\frac{\partial}{\partial t}\phi$, $\frac{\partial}{\partial x}\phi$, $\phi\frac{\partial}{\partial x}\phi$ and ${}_0\mathrm{D}_x^{\beta}\phi$ via $U$ using [6.23] as ($x\in\left[0,\infty\right)$, $t>0$, $\xi\in\left[0,\infty\right)$)

$$\begin{aligned}
&\frac{\partial}{\partial t}\phi\left(x,t\right)=t^{a-1}\left(aU\left(\xi\right)+b\xi\frac{\mathrm{d}}{\mathrm{d}\xi}U\left(\xi\right)\right),\\
&\frac{\partial}{\partial x}\phi\left(x,t\right)=t^{a+b}\frac{\mathrm{d}}{\mathrm{d}\xi}U\left(\xi\right),\\
&\phi\left(x,t\right)\frac{\partial}{\partial x}\phi\left(x,t\right)=t^{2a+b}U\left(\xi\right)\frac{\mathrm{d}}{\mathrm{d}\xi}U\left(\xi\right),
\end{aligned} \tag{6.24}$$

along with

$$\begin{aligned}
{}_0\mathrm{D}_x^{\beta}\phi\left(x,t\right)&=\frac{1}{\Gamma\left(2-\beta\right)}\frac{\partial^2}{\partial x^2}\int_0^x\frac{\phi\left(\zeta,t\right)}{\left(x-\zeta\right)^{\beta-1}}\mathrm{d}\zeta\\
&=t^{a+b\beta}\frac{1}{\Gamma\left(2-\beta\right)}\frac{\mathrm{d}^2}{\mathrm{d}\xi^2}\int_0^{\xi}\frac{U\left(\eta\right)}{\left(\xi-\eta\right)^{\beta-1}}\mathrm{d}\eta\\
&=t^{a+b\beta}{}_0\mathrm{D}_{\xi}^{\beta}U\left(\xi\right).
\end{aligned} \tag{6.25}$$

Substituting [6.23] and [6.25] in [6.13], we obtain the ordinary fractional-order differential equation for $U$. We state this in the following proposition.

PROPOSITION 6.2.– The fractional Burgers/Korteweg–de Vries equation [6.13] has the similarity solution

$$\phi\left(x,t\right)=t^{\frac{1}{\beta}-1}U\left(\xi\right),\;\;\xi=t^{-\frac{1}{\beta}}x,\;\;x\in\left[0,\infty\right),\;t>0, \tag{6.26}$$

where $U$ is a solution to the ordinary fractional differential equation

$$U(\xi)\frac{\mathrm{d}}{\mathrm{d}\xi}U(\xi)+\frac{1}{\beta}\left((1-\beta)U(\xi)-\xi\frac{\mathrm{d}}{\mathrm{d}\xi}U(\xi)\right)=\nu\,{}_0\mathrm{D}_{\xi}^{\beta}U(\xi),\quad \xi\in[0,\infty). \quad [6.27]$$

### 6.1.3. *Exact solutions of the nonlinear fractional diffusion-wave equation*

We present two exact solutions to the nonlinear fractional diffusion-wave equation [6.12]. Let $k=0$ in [6.12]. We seek the exact solution for $U$ to [6.19], which, using [6.18], will give the solution to [6.12] as well. Suppose the solution to [6.19] is in the form

$$U(\xi)=u\xi^{\beta},\quad \xi\in[0,\infty), \quad [6.28]$$

where $u$ and $\beta$ are constants. Then, for $\beta>-1$, [6.19] becomes ($\xi\in[0,\infty)$)

$$\begin{aligned} &u\frac{B(1-\alpha,1+\beta)}{\Gamma(1-\alpha)}(1-\alpha+\beta)\xi^{\beta-\alpha}\\ &\quad=m(a+b\beta)u^{n+1}(a(1+n)+b\beta(n+1)-1)\xi^{(n+1)\beta}, \end{aligned} \quad [6.29]$$

where $B(v,w)=\int_0^1 u^{v-1}(1-u)^{w-1}\mathrm{d}u$ is the beta function. From [6.29], it is clear that [6.28] is a solution to [6.19] if $\beta=-\frac{\alpha}{n}$, $n>\alpha$. Also, using some elementary properties of the beta and gamma functions and the fact that $b=\frac{an-2}{\alpha}$, from [6.29] we obtain

$$u=\left(\frac{n^2\Gamma\left(1-\frac{\alpha}{n}\right)}{2m(2+n)\Gamma\left(1-\alpha-\frac{\alpha}{n}\right)}\right)^{\frac{1}{n}}. \quad [6.30]$$

Note that the (arbitrary) value of $a$ does not affect the final result. Inserting [6.28] into [6.18], we also obtain a result that does not depend on $a$

$$T(x,t)=u\cdot\left(\frac{x}{t^{\frac{\alpha}{2}}}\right)^{\frac{2}{n}},\quad x\in[0,\infty),\ t>0. \quad [6.31]$$

Thus, [6.30] and [6.31] hold for $a=0$ too.

Let $k\neq 0$. Motivated by the fact that solution to [6.12] in the case when $k=0$, given by [6.31], does not depend on $a$, we seek the solution to [6.19] in the case when $k\neq 0$ again in the form [6.28]. Thus, assuming a solution to [6.19] with $k\neq 0$ in the form [6.28], we verify that it exists if $\beta = -\frac{\alpha}{n}$ for $n = 2$. The solution to [6.12], if $k \neq 0$, is given by [6.31] with $n = 2$.

Another exact solution to the nonlinear fractional diffusion-wave equation [6.12] has the form of a wave propagation along the $x$-axis with certain speed $c$ and it cannot be obtained from [6.19]. To obtain a solution to [6.12] with $k = 0$, we introduce the similarity transformation

$$T(x,t) = x^a U(\xi), \ \ \xi = \frac{ct}{x} - 1, \ \ x \in (0,\infty), \ t > 0, \tag{6.32}$$

where $a$ and $c$ are constants. With the solution assumed to be in the form of [6.32], [6.12] transforms into

$$\frac{c^\alpha x^{a-\alpha}}{\Gamma(1-\alpha)} \frac{\mathrm{d}}{\mathrm{d}\xi} \int_{-1}^{\xi} \frac{U(\eta)}{(\xi-\eta)^\alpha} \mathrm{d}\eta$$
$$= m x^{(n+1)a-2} \left( ((n+1)a - 1) F(\xi) - (1+\xi) \frac{\mathrm{d}}{\mathrm{d}\xi} F(\xi) \right), \ \ \xi \in \mathbb{R}, \tag{6.33}$$

where

$$F(\xi) = U^n(\xi) \left( aU(\xi) - (1+\xi) \frac{\mathrm{d}}{\mathrm{d}\xi} U(\xi) \right), \ \ \xi \in \mathbb{R}.$$

The condition for the existence of this similarity solution is $a - \alpha = (n+1)a - 2$ implying $a = \frac{2-\alpha}{n}$, so that [6.33] becomes

$$\frac{c^\alpha}{\Gamma(1-\alpha)} \frac{\mathrm{d}}{\mathrm{d}\xi} \int_{-1}^{\xi} \frac{U(\eta)}{(\xi-\eta)^\alpha} \mathrm{d}\eta$$
$$= m \left( ((n+1)a - 1) F(\xi) - (1+\xi) \frac{\mathrm{d}}{\mathrm{d}\xi} F(\xi) \right), \ \ \xi \in \mathbb{R}. \tag{6.34}$$

We look for the solution [6.34] in the form

$$U(\xi) = \begin{cases} u\xi^\beta, & \xi \geq 0, \ (x \leq ct), \\ 0, & \xi < 0, \ (x > ct), \end{cases} \tag{6.35}$$

where $u$ is a constant. Inserting [6.35] into [6.34], we have ($\xi \in \mathbb{R}$)

$$uc^{\alpha}\frac{\Gamma(1+\beta)}{\Gamma(2-\alpha+\beta)}(1-\alpha+\beta)\xi^{\beta-\alpha}$$
$$= mu^{n+1}(a-\beta)((n+1)(a-\beta)-1)\xi^{(n+1)\beta}$$
$$-2mu^{n+1}\beta(n+1)(a-\beta)\xi^{(n+1)\beta-1}+mu^{n+1}\beta((n+1)\beta-1)\xi^{(n+1)\beta-2}.$$

This equation can be satisfied by the unique choice $\beta = a = \frac{2-\alpha}{n}$, providing

$$u = \left(\frac{nc^{\alpha}\Gamma\left(1+\frac{2-\alpha}{n}\right)}{m(2-\alpha)\Gamma\left(2-\alpha+\frac{2-\alpha}{n}\right)}\right)^{\frac{1}{n}}. \quad [6.36]$$

Thus, solution [6.32] to [6.12] reads

$$T(x,t) = u(ct-x)^{\frac{2-\alpha}{n}}, \quad x \leq ct, \; t > 0. \quad [6.37]$$

Obviously, this solution represents a wave with amplitude $u$, which propagates along the $x$-axis with the speed $c$. The relation between the amplitude and the speed is determined by [6.36]. Solving [6.36] for $c$, we obtain

$$c = \left(\frac{mu^{n}\Gamma\left(2-\alpha+\frac{2-\alpha}{n}\right)}{n\Gamma\left(1+\frac{2-\alpha}{n}\right)}\right)^{\frac{1}{\alpha}}. \quad [6.38]$$

Inserting [6.38] into [6.37], we obtain

$$T(x,t) = u\left(t\left(\frac{mu^{n}(2-\alpha)\Gamma\left(2-\alpha+\frac{2-\alpha}{n}\right)}{n\Gamma\left(1+\frac{2-\alpha}{n}\right)}\right)^{\frac{1}{\alpha}}-x\right)^{\frac{2-\alpha}{n}}. \quad [6.39]$$

Note that in the literature, there are not many known relations of this type for fully nonlinear problems (see [WHI 74]). For $\alpha = 1$ (classical case), our results agree with the results presented in [TIK 77, p. 595].

We plot $T$ given by [6.37] in Figures 6.1–6.3 for the special case $m = 1$, $u = 1, \alpha = 0.5$ for different values of $n$ and $t$. When the exponent $n$ in [6.37] is increased to $n = 1.5$, the wave profiles are shown in Figure 6.2. Note that in this case, the dependence of $T$ on $x$ in [6.37] is linear. A further increase in $n$ to the value $n = 2$ makes the wave profiles as shown in Figure 6.3. Next, we examine the influence of the parameter $\alpha$ on the wave shape for fixed time instant. In particular, we take for $n = 1.6$, $t = 0.3$ and solve [6.37] for four different values of $\alpha$. The results are shown in Figure 6.4. It is seen that by decreasing the order of the derivative $\alpha$, the wave profile changes from concave ($\alpha = 0.8$) to convex ($\alpha = 0.3$).

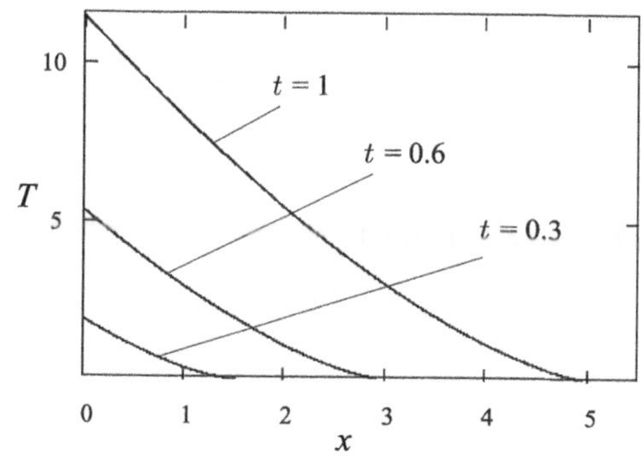


**Figure 6.1.** *Wave profiles for $u = 1$, $m = 1$, $\alpha = 0.5$, $n = 1.5$ and three different time instants*

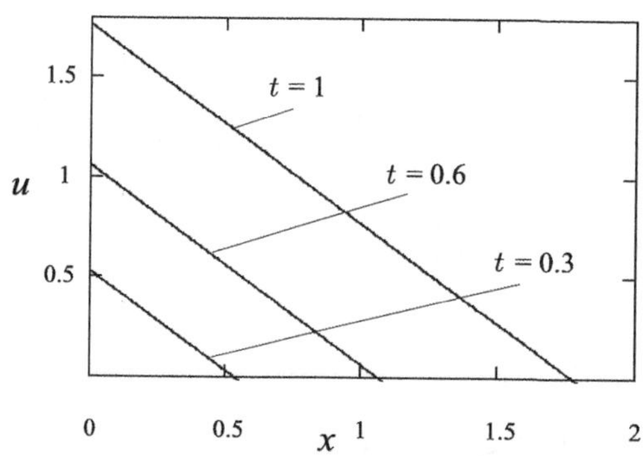


**Figure 6.2.** *Wave profiles for $u = 1$, $m = 1$, $\alpha = 0.5$, $n = 1.5$ and three different time instants*

### 6.1.4. *Numerical solutions to [6.19] and [6.27]*

We may use different numerical procedures (see, for example, [POD 99]) to solve [6.19] and [6.27]. Since we are interested in the solution valid for a large range of independent variables, especially well-suited methods are those that reduce fractional-order differential equations to a system of first-order equations. We use the expansion formula [2.31] in order to express the fractional derivative ${}_0\mathrm{D}_\xi^\alpha U$, so that [6.19] becomes a system

$$\begin{aligned}
&\mathcal{A}(N)\frac{U(\xi)}{\xi^\alpha}+\mathcal{B}(N)\xi^{1-\alpha}\frac{\mathrm{d}}{\mathrm{d}\xi}U(\xi)-\sum_{p=1}^{N}\mathcal{C}_{p-1}\frac{V_{p-1}(U)(\xi)}{\xi^{p+\alpha}}\\
&\quad=k\left(a\left(a-1\right)U\left(\xi\right)+b\left(2a+b-1\right)\xi\frac{\mathrm{d}}{\mathrm{d}\xi}U\left(\xi\right)+b^2\xi^2\frac{\mathrm{d}^2}{\mathrm{d}\xi^2}U\left(\xi\right)\right)\\
&\qquad+m\left(\left(a\left(1+n\right)-1\right)F\left(\xi\right)+b\xi\frac{\mathrm{d}}{\mathrm{d}\xi}F\left(\xi\right)\right),\quad\text{where}
\end{aligned}\qquad [6.40]$$

$$\begin{aligned}
&F\left(\xi\right)=U^n\left(\xi\right)\left(aU\left(\xi\right)+b\xi\frac{\mathrm{d}}{\mathrm{d}\xi}U\left(\xi\right)\right),\\
&\frac{\mathrm{d}}{\mathrm{d}\xi}V_{p-1}(U)(\xi)=\xi^{p-1}U(\xi),\quad p=2,3,\ldots,N,
\end{aligned}\qquad [6.41]$$

subject to

$$U\left(0\right)=x_0,\quad\frac{\mathrm{d}}{\mathrm{d}\xi}U\left(0\right)=v_0,\quad V_{p-1}\left(0\right)=0,\quad p=2,3,\ldots,N,\qquad [6.42]$$

where

$$\begin{aligned}
&\mathcal{A}(N)=\frac{\sin\left(\alpha\pi\right)}{\alpha\pi}\frac{\Gamma(N+\alpha)}{\Gamma(N)},\quad\mathcal{B}(N)=\frac{\Gamma(N+\alpha)}{\Gamma(\alpha)\Gamma(2-\alpha)\Gamma\left(N+1\right)},\\
&\mathcal{C}_{p-1}=\frac{\Gamma(p+\alpha)}{\Gamma(\alpha)\Gamma(1-\alpha)\Gamma(p)}.
\end{aligned}$$

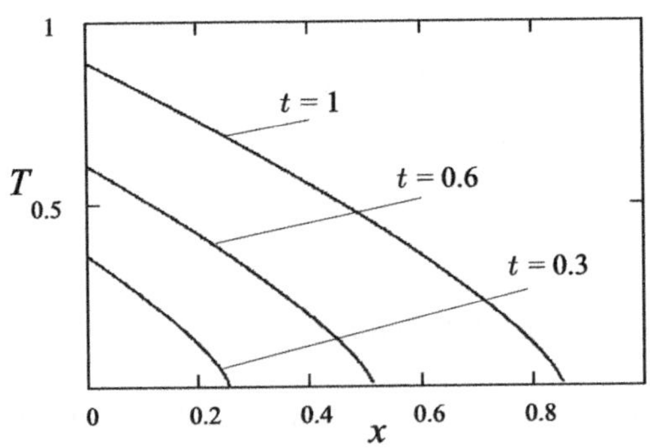


**Figure 6.3.** *Wave profiles for $u=1$, $m=1$, $\alpha=0.5$, $n=2$ and three different time instants*

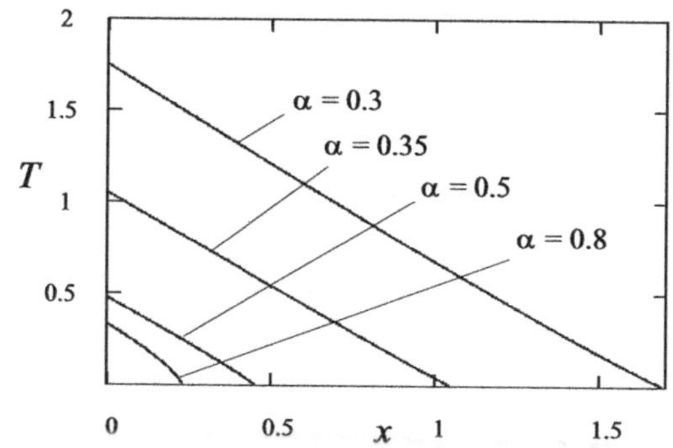


**Figure 6.4.** *Wave profiles for $u = 1$, $m = 1$, $n = 1.6$, $t = 0.3$ and different values of $\alpha$*

Systems [6.40] and [6.41], with [6.42], can be solved using the standard numerical procedures. A similar procedure is used to solve [6.27].

In the examples that follow we take $a = 0$ so that $b = -\frac{2}{\alpha}$. As a first example, we treat [6.19] with $m = 1$, $k = 1$, $n = 1$, $U(0) = 0$, $\lim_{\xi \to \infty} U(\xi) = 1$ for three different values of $\alpha \in \{0.5, 0.7, 1.0\}$. The results are shown in Figure 6.5. In the special case $\alpha = 1$, we obtain the example treated in [TIK 77].

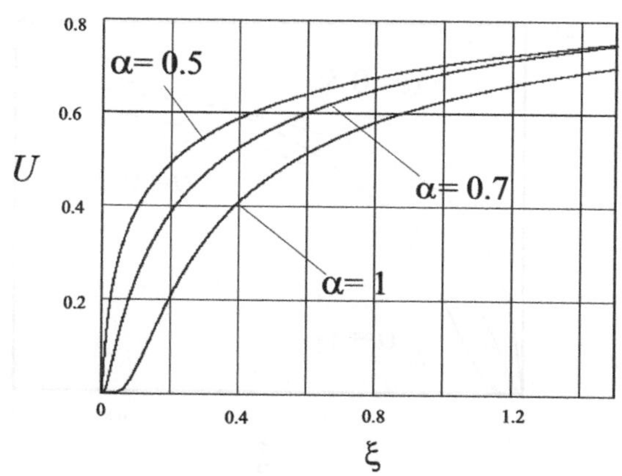


**Figure 6.5.** *Solution of [6.19] for $m = 1$, $k = 1$, $n = 1$, $U(0) = 0$, $\lim_{\xi \to \infty} U(\xi) = 1$ and three different values of $\alpha$*

We show in Figure 6.6 the behavior of $U$ for small $\xi$. The figure shows that decrease in $\alpha$ makes a solution to [6.19] steeper for small values of $\xi$. It also shows that all solutions have a part with a small slope. This part is shorter for smaller values of $\alpha$.

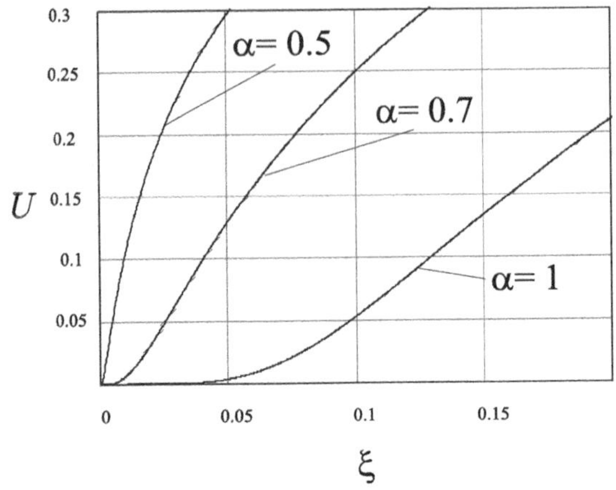


**Figure 6.6.** *Solution of* [6.19] *for* $m = 1, k = 1, n = 1, U(0) = 0,$ $\lim_{\xi\to\infty} U(\xi) = 1$ *and small value of* $\xi$

As a second example, we treat [6.19] with $m = 1, k = 1, n = 2$ for three different values of $\alpha \in \{0.5, 0.7, 1.0\}$. Again, we take $U(0) = 0$, $\lim_{\xi\to\infty} U(\xi) = 1$. The results are shown in Figure 6.7.

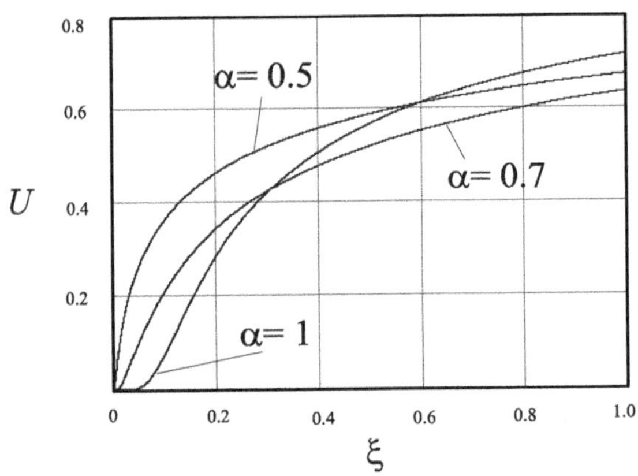


**Figure 6.7.** *Solution of [6.19] for* $m = 1, k = 1, n = 2, U(0) = 0,$ $\lim_{\xi\to\infty} U(\xi) = 1$ *and three different values of*

As a third example, we solve [6.27] for the value $\nu = 1$, $\beta = 2.9$. The results are shown in Figure 6.8. In the same figure, we show the solution to Burgers ($\beta = 2$) and Korteweg–de Vries equations ($\beta = 3$).

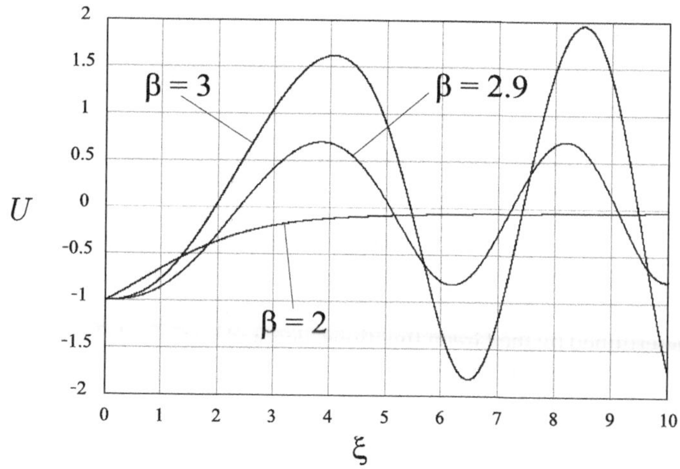


**Figure 6.8.** *Solution of* [6.27] *for* $\nu = 1$ *and three different values of* $\beta$

## 6.2. Fractional telegraph equation

We follow [ATA 07b] and treat the fractional generalization [6.9] of the telegraph equation [6.8]. The one-dimensional fractional telegraph equation is

$$\tau\, {}_0^C\mathrm{D}_t^{\alpha} T(x,t) + {}_0^C\mathrm{D}_t^{\beta} T(x,t) = \mathcal{D}\frac{\partial^2}{\partial x^2} T(x,t), \quad x \in I,\ t > 0, \qquad [6.43]$$

where $I$ can be $I = [0, l]$, or $I = [0, \infty)$, or $I = \mathbb{R}$ and $0 < \beta \leq \alpha \leq 2$. Thus, we are able to recover: the classical heat conduction equation ($\tau = 0$, $\beta = 1$), telegraph equation ($\alpha = 2$, $\beta = 1$) and classical wave equation ($\alpha = \beta = 2$).

We treat three initial-boundary value problems for [6.43].

1) Dirichlet problem

$$\begin{gathered} T(x, 0^+) = g(x), \quad \tfrac{\partial}{\partial t} T(x, 0^+) = 0, \quad x \in [0, l], \\ T(0, t) = T(l, t) = 0, \quad t > 0. \end{gathered} \qquad [6.44]$$

2) Signaling problem

$$\begin{gathered} T(x, 0^+) = 0, \quad \tfrac{\partial}{\partial t} T(x, 0^+) = 0, \quad x \in [0, \infty), \\ \lim_{x \to 0} T(x,t) = h(t), \quad \lim_{x \to \infty} T(x,t) = 0, \quad t > 0. \end{gathered} \qquad [6.45]$$

3) Cauchy problem

$$T(x,0^+) = g(x), \quad \frac{\partial}{\partial t}T(x,0^+) = 0, \;\; x \in \mathbb{R},$$
$$\lim_{x\to\pm\infty} T(x,t) = 0, \;\; t > 0. \qquad [6.46]$$

In [6.45] and [6.46], we use $h$ and $g$ to denote two given, sufficiently well-behaving function of time and space variables, respectively. We note that for the case when $h = \delta$ and $g = \delta$ where $\delta$ is the Dirac distribution, the solutions to [6.43], [6.45] and [6.43], [6.46] are determined by the Green functions, denoted by $\mathcal{G}_s$ and $\mathcal{G}_c$, respectively. With the Green functions known, the solution to problems [6.43], [6.45] and [6.43], [6.46] with arbitrary $h$ and $g$ read

$$T(x,t) = \int_0^t \mathcal{G}_s(x,t-\theta)\, h(\theta)\, d\theta \;\; \text{and}$$
$$T(x,t) = \int_{-\infty}^{\infty} \mathcal{G}_c(x-\xi,t)\, g(\xi)\, d\xi, \qquad [6.47]$$

respectively. Also note that for the case when $0 < \alpha < 1$, the initial condition $\frac{\partial}{\partial t}T(x,0)$ is redundant. To be able to compare results for the case $0 < \alpha < 1$, with the case $1 < \alpha < 2$, we set $\frac{\partial}{\partial t}T(x,0) = 0$.

### 6.2.1. *Dirichlet problem*

We present a solution to [6.43] and [6.44]. For the case $\tau = 0$, the problem was treated in [AGR 02b, ALK 05] by different methods. Assuming that a solution has a form

$$T(x,t) = X(x)\, U(t), \;\; x \in [0,l], \; t > 0,$$

and using the standard separation of variables procedure, we obtain ($x \in [0,l]$, $t > 0$)

$$T(x,t) = \sum_{k=1}^{\infty} C_k U_k(t) \sin(\lambda_k x), \quad C_k = \frac{2}{l}\int_0^l g(x) \sin(\lambda_k x)\, dx, \qquad [6.48]$$

where $\lambda_k = \frac{k\pi}{l}$, and $U_k$ satisfies

$$\tau\, {}_0^C D_t^{\alpha} U_k(t) + {}_0^C D_t^{\beta} U_k(t) + \lambda_k^2 \mathcal{D} U_k(t) = 0, \;\; t > 0, \qquad [6.49]$$
$$U_k(0^+) = 1, \quad \frac{d}{dt}U_k(0^+) = 0. \qquad [6.50]$$

Laplace transform of [6.49], with $\mathrm{Re}\, s > s_0$, which will be discussed later, leads to

$$\tilde{U}_k(s) = \frac{\tau s^{\alpha-1} + s^{\beta-1}}{\tau s^{\alpha} + s^{\beta} + \lambda_k^2 \mathcal{D}} = \frac{1}{s} - \frac{\lambda_k^2 \mathcal{D}}{s\left(\tau s^{\alpha} + s^{\beta} + \lambda_k^2 \mathcal{D}\right)}. \quad [6.51]$$

Expanding the right-hand side of [6.51] in a power series of $s$ (see [POD 99, p. 155]), we obtain

$$\tilde{U}_k(s) = \left(\tau s^{\alpha-1} + s^{\beta-1}\right) \frac{s^{-\beta}}{\tau s^{\alpha-\beta} + 1} \frac{1}{1 + \frac{\lambda_k^2 \mathcal{D} s^{-\beta}}{\tau s^{\alpha-\beta}+1}}. \quad [6.52]$$

Since $\frac{1}{1+\frac{\lambda_k^2 \mathcal{D} s^{-\beta}}{\tau s^{\alpha-\beta}+1}} = \sum_{j=0}^{\infty} (-1)^j \left(\frac{\lambda_k^2 \mathcal{D} s^{-\beta}}{\tau s^{\alpha-\beta}+1}\right)^j$ for $\left|\tau s^{\alpha} + s^{\beta}\right| > \lambda_k^2 \mathcal{D}$, we have

$$\tilde{U}_k(s) = \sum_{j=0}^{\infty} \frac{(-1)^j \left(\lambda_k^2 \mathcal{D}\right)^j \left(\tau s^{(\alpha-\beta)-(1+j\beta)} + s^{-(1+j\beta)}\right)}{\left(\tau s^{\alpha-\beta} + 1\right)^{1+j}}, \quad \mathrm{Re}\, s > s_0, [6.53]$$

where $s_0$ is determined so that $\left|\tau s^{\alpha} + s^{\beta}\right| > \lambda_k^2 \mathcal{D}$ holds. From [POD 99, p. 21], we have

$$\mathcal{L}^{-1}\left[\sum_{j=0}^{\infty} \frac{1}{j!} \frac{(-1)^j \left(\lambda_k^2 \mathcal{D}\right)^j}{\tau^j} \frac{j! s^{(\alpha-\beta)-(1+j\beta)}}{\left(s^{\alpha-\beta} + \frac{1}{\tau}\right)^{1+j}}\right](t)$$

$$= \sum_{j=0}^{\infty} 1 \frac{(-1)^j \left(\lambda_k^2 \mathcal{D}\right)^j}{\tau^j j!} t^{\alpha j} E^{(j)}_{\alpha-\beta,1+j\beta}\left(-\frac{t^{\alpha-\beta}}{\tau}\right), \quad t > 0,$$

$$\mathcal{L}^{-1}\left[\sum_{j=0}^{\infty} \frac{1}{j!} \frac{(-1)^j \left(\lambda_k^2 \mathcal{D}\right)^j}{\tau^{j+1}} \frac{j! s^{-(1+j\beta)}}{\left(s^{\alpha-\beta} + \frac{1}{\tau}\right)^{1+j}}\right](t)$$

$$= \sum_{j=0}^{\infty} \frac{(-1)^j \left(\lambda_k^2 \mathcal{D}\right)^j}{\tau^{j+1} j!} t^{(\alpha+1)j-\beta} E^{(j)}_{\alpha-\beta,\alpha+1+\beta(j-1)}\left(-\frac{t^{\alpha-\beta}}{\tau}\right), \quad t > 0.$$

So that the solution $U_k$ becomes

$$U_k(t) = \sum_{j=0}^{\infty} \frac{(-1)^j \left(\lambda_k^2 \mathcal{D}\right)^j}{\tau^j j!} t^{\alpha j} E^{(j)}_{\alpha-\beta,1+j\beta}\left(-\frac{t^{\alpha-\beta}}{\tau}\right)$$
$$+ \sum_{j=0}^{\infty} \frac{(-1)^j \left(\lambda_k^2 \mathcal{D}\right)^j}{\tau^{j+1} j!} t^{(\alpha+1)j-\beta} E^{(j)}_{\alpha-\beta,\alpha+1+\beta(j-1)}\left(-\frac{t^{\alpha-\beta}}{\tau}\right). \qquad [6.54]$$

Next, we express the inversion to [6.51] in terms of the contour integral. Thus, to find original $U_k$, we consider

$$U_k(t) = H(t) - \frac{1}{2\pi \mathrm{i}} \int_\gamma \psi(s) \mathrm{e}^{st} \mathrm{d}s, \quad t > 0, \qquad [6.55]$$

where $H$ is the Heaviside step function and

$$\psi(s) = \frac{\lambda_k^2 \mathcal{D}}{s\left(\tau s^\alpha + s^\beta + \lambda_k^2 \mathcal{D}\right)}, \quad \operatorname{Re} s > s_0.$$

Contour $\gamma$ is given by $\gamma = \{s \mid \operatorname{Re} s = \sigma,\ \sigma > \sigma_0 = 0\}$. Let $\gamma_0 = \overset{\frown}{AB} \cup \overset{\frown}{BD} \cup \overset{\frown}{DE} \cup \overset{\frown}{EF} \cup \overset{\frown}{FG} \cup \overset{\frown}{GA}$ be the closed contour shown in Figure 6.9.

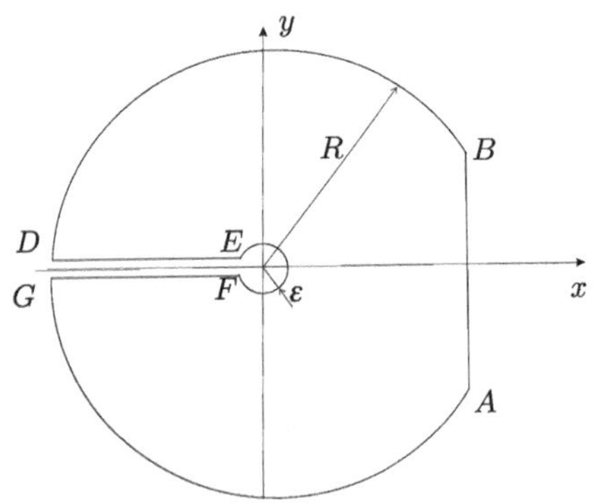


**Figure 6.9.** *Integration contour for inverse Laplace transform of [6.55]*

Note that function $\psi$ is analytic within $\gamma_0$, if $0 \leq \beta \leq \alpha \leq 1$. However, if $0 \leq \beta \leq \alpha \leq 2$, function $\psi$ is not analytic within $\gamma_0$. To prove this, suppose that there is a solution to equation

$$\tau s^{\alpha} + s^{\beta} + \lambda_k^2 \mathcal{D} = 0. \qquad [6.56]$$

Let $s = r\mathrm{e}^{\mathrm{i}\phi}$ be a solution to [6.56]. Separating the real and imaginary parts in [6.56], we obtain

$$\begin{aligned} &\tau r^{\alpha} \cos(\alpha\phi) + r^{\beta} \cos(\beta\phi) + \lambda_k^2 \mathcal{D} = 0, \\ &\tau r^{\alpha} \sin(\alpha\phi) + r^{\beta} \sin(\beta\phi) = 0. \end{aligned} \qquad [6.57]$$

It follows from $[6.57]_2$ that the solutions are complex conjugate, i.e. if $\phi$ is a solution so is $-\phi$. If $0 \leq \beta \leq \alpha \leq 1$, then $\sin(\alpha\phi)$ and $\sin(\beta\phi)$ are of the same sign, so that [6.57] cannot be satisfied. Thus, there are no solutions to [6.56]. If $0 \leq \beta \leq \alpha \leq 2$, then, by the argument principle, it can be shown that the function $\tau s^{\alpha} + s^{\beta} + \lambda_k^2 \mathcal{D}$ has exactly two (conjugate) zeros. Since, for $\operatorname{Re} s > 0$, by $[6.57]_2$, $\operatorname{Im}\left(\tau s^{\alpha} + s^{\beta} + \lambda_k^2 \mathcal{D}\right) > 0$, it follows that both zeros lie in the left part of the complex plane $\operatorname{Im} s < 0$.

We calculate the integral in [6.55] by the Cauchy residues theorem

$$\frac{1}{2\pi \mathrm{i}} \int_{\gamma_0} \psi(s)\, \mathrm{e}^{st} \mathrm{d}s = \operatorname{Res}\left(\psi(s)\, \mathrm{e}^{st}, s_0\right) + \operatorname{Res}\left(\psi(s)\, \mathrm{e}^{st}, \bar{s}_0\right), \qquad [6.58]$$

where $s_0$ and $\bar{s}_0$ are solutions of [6.56]. Integrals along the contours $\widehat{BD}$ and $\widehat{GA}$ tend to zero when $R \to \infty$, so they do not contribute to the left-hand side of [6.58]. Contributions to [6.58] come from the integrals along $\widehat{AB}$, $\widehat{DE}$, $\widehat{EF}$ and $\widehat{FG}$. To determine $U_k$, we let $s = r\mathrm{e}^{\mathrm{i}\pi}$ on $\widehat{DE}$ and $s = r\mathrm{e}^{-\mathrm{i}\pi}$ on $\widehat{FG}$, and $s = \varepsilon \mathrm{e}^{\mathrm{i}\theta}$, $\theta \in (-\pi, \pi)$, $\varepsilon > 0$ on $\widehat{EF}$. Then, we have

$$\begin{aligned} \int_{\widehat{DE}} \psi(s)\, \mathrm{e}^{st} \mathrm{d}s &= -\lambda_k^2 \mathcal{D} \int_{\varepsilon}^{R} \frac{1}{\tau r^{\alpha} \mathrm{e}^{\mathrm{i}\alpha\pi} + r^{\beta} \mathrm{e}^{\mathrm{i}\beta\pi} + \lambda_k^2 \mathcal{D}} \frac{\mathrm{e}^{-rt}}{r} \mathrm{d}r \\ \int_{\widehat{FG}} \psi(s)\, \mathrm{e}^{st} \mathrm{d}s &= \lambda_k^2 \mathcal{D} \int_{\varepsilon}^{R} \frac{1}{\tau r^{\alpha} \mathrm{e}^{-\mathrm{i}\alpha\pi} + r^{\beta} \mathrm{e}^{-\mathrm{i}\beta\pi} + \lambda_k^2 \mathcal{D}} \frac{\mathrm{e}^{-rt}}{r} \mathrm{d}r \\ \int_{\widehat{EF}} \psi(s)\, \mathrm{e}^{st} \mathrm{d}s &= -\mathrm{i}\lambda_k^2 \mathcal{D} \int_{-\pi}^{\pi} \frac{1}{\tau \varepsilon^{\alpha} \mathrm{e}^{\mathrm{i}\alpha\theta} + \varepsilon^{\beta} \mathrm{e}^{\mathrm{i}\beta\theta} + \lambda_k^2 \mathcal{D}} \mathrm{e}^{\varepsilon t(\cos\theta + \mathrm{i}\sin\theta)} \mathrm{d}\theta. \end{aligned}$$

$$[6.59]$$

Combining the terms in [6.59] and taking the limit when $R \to \infty$, $\varepsilon \to 0$, we obtain, for $t > 0$,

$$U_k(t) = \frac{\lambda_k^2 \mathcal{D}}{\pi} \int_0^\infty \frac{\tau r^\alpha \sin(\alpha\pi) + r^\beta \sin(\beta\pi)}{K_k(r)} \frac{\mathrm{e}^{-rt}}{r} \mathrm{d}r + \mathrm{Res}\left(\psi(s)\,\mathrm{e}^{st}, s_0\right) + \mathrm{Res}\left(\psi(s)\,\mathrm{e}^{st}, \bar{s}_0\right), \quad [6.60]$$

where, for $r > 0$,

$$K_k(r) = \tau^2 r^{2\alpha} + r^{2\beta} + \left(\lambda_k^2 \mathcal{D}\right)^2 + 2\lambda_k^2 \mathcal{D}\left(\tau r^\alpha \cos(\alpha\pi) + r^\beta \cos(\beta\pi)\right) + 2\tau r^{\alpha+\beta} \cos((\alpha-\beta)\pi). \quad [6.61]$$

With $U_k$ determined from [6.60], if $0 \leq \beta \leq \alpha \leq 1$ the complete solution of the problem becomes

$$T(x,t) = \frac{\mathcal{D}}{\pi} \sum_{k=0}^{\infty} C_k \lambda_k^2 \sin(\lambda_k x) \int_0^\infty \frac{\left(\tau r^\alpha \sin(\alpha\pi) + r^\beta \sin(\beta\pi)\right)}{K_k(r)} \frac{\mathrm{e}^{-rt}}{r} \mathrm{d}r. \quad [6.62]$$

### 6.2.2. *Signaling problem*

Applying the Laplace transform with respect to the time $t$ to [6.43] and [6.45], we obtain the following differential equation for the Laplace transform of the Green function

$$\frac{\partial^2}{\partial x^2} \tilde{\mathcal{G}}_s(x,s) - \frac{\tau s^\alpha + s^\beta}{\mathcal{D}} \tilde{\mathcal{G}}_s(x,s) = 0, \quad x \in [0,\infty), \ \mathrm{Re}\, s > 0, \quad [6.63]$$

with

$$\lim_{x\to\infty} \tilde{\mathcal{G}}_s(x,s) = 0, \quad \tilde{\mathcal{G}}_s(0,s) = 1, \quad \mathrm{Re}\, s > 0. \quad [6.64]$$

The solution to [6.63] and [6.64] is

$$\tilde{\mathcal{G}}_s(x,s) = \mathrm{e}^{-\frac{x}{\sqrt{\mathcal{D}}}\sqrt{\tau s^\alpha + s^\beta}}, \quad x \in [0,\infty), \ \mathrm{Re}\, s > 0. \quad [6.65]$$

We present the inversion of [6.65] in the form of series and in the integral form. From [6.65], we have

$$\begin{aligned}\tilde{\mathcal{G}}_s(x,s) &= \sum_{n=0}^{\infty}\frac{1}{n!}\left(-\frac{x}{\sqrt{\mathcal{D}}}\right)^n (\tau s^{\alpha})^{\frac{n}{2}}\left(1+\frac{1}{\tau s^{\alpha-\beta}}\right)^{\frac{n}{2}} \\ &= \sum_{n=0}^{\infty}\sum_{k=0}^{\infty}\left(-\frac{x}{\sqrt{\mathcal{D}}}\right)^n \frac{\tau^{\frac{n}{2}-k}}{n!k!}\frac{\Gamma\left(\frac{n}{2}+1\right)}{\Gamma\left(\frac{n}{2}+1-k\right)}\frac{1}{s^{(\alpha-\beta)k-\frac{\alpha n}{2}}}, \\ & \qquad x\in[0,\infty),\ \operatorname{Re}s>s_0,\end{aligned} \qquad [6.66]$$

where we used the following series representation $(1+z)^{\gamma}=\sum_{k=0}^{\infty}\frac{\Gamma(\gamma+1)}{k!\Gamma(\gamma+1-k)}z^k$, $\gamma>0$, $|z|<1$ and $s_0$ is determined so that $|s|>\left(\frac{1}{\tau}\right)^{\frac{1}{\alpha-\beta}}$. Term-by-term inversion of [6.66] leads to

$$\mathcal{G}_s(x,t)=\sum_{n=0}^{\infty}\sum_{k=0}^{\infty}\left(-\frac{x}{\sqrt{\mathcal{D}}}\right)^n \frac{\tau^{\frac{n}{2}-k}}{n!k!}\frac{\Gamma\left(\frac{n}{2}+1\right)}{\Gamma\left(\frac{n}{2}+1-k\right)}\frac{t^{(\alpha-\beta)k-\frac{\alpha n}{2}-1}}{\Gamma\left((\alpha-\beta)k-\frac{\alpha n}{2}\right)}. \qquad [6.67]$$

To express the solution in the integral form, we additionally assume $0<\alpha-\beta\leq 1$. We have

$$\mathcal{G}_s(x,t)=\mathcal{L}^{-1}\left[\tilde{\mathcal{G}}_s(x,s)\right](t)=\frac{1}{2\pi\mathrm{i}}\int_{\gamma}\mathrm{e}^{-\frac{x}{\sqrt{\mathcal{D}}}\sqrt{\tau s^{\alpha}+s^{\beta}}}\mathrm{e}^{st}\mathrm{d}s,\quad x\in[0,\infty),\ t>0,$$

where $\gamma$ is the same contour as the one shown in Figure 6.9. In case $0<\alpha-\beta\leq 1$, the only branch point of $\sqrt{\tau s^{\alpha}+s^{\beta}}$ is $s=0$. Therefore, the function $\mathrm{e}^{-\frac{x}{\sqrt{\mathcal{D}}}\sqrt{\tau s^{\alpha}+s^{\beta}}}\mathrm{e}^{st}$ is analytic within $\gamma_0$. Applying the same procedure as in the previous case, we obtain

$$\mathcal{G}_s(x,t)=\frac{1}{\pi}\int_0^{\infty}\sin\left(\frac{x}{\sqrt{\mathcal{D}}}\sqrt{R(p)}\sin\frac{\phi(p)}{2}\right)\mathrm{e}^{-\frac{x}{\sqrt{\mathcal{D}}}\sqrt{R(p)}\cos\frac{\phi(p)}{2}-pt}\mathrm{d}p,\quad t>0, \qquad [6.68]$$

where

$$\begin{aligned}R(p) &= \sqrt{\tau^2p^{2\alpha}+p^{2\beta}+2\tau p^{\alpha+\beta}\cos((\alpha-\beta)\pi)}, \\ \phi(p) &= \operatorname{arctg}\frac{\tau p^{\alpha-\beta}\sin\pi\alpha+\sin\pi\beta}{\tau p^{\alpha-\beta}\cos\pi\alpha+\cos\pi\beta}.\end{aligned} \qquad [6.69]$$

For arbitrary $h$, the solution is given by $[6.47]_1$. Note that [6.67] is valid for any $\alpha, \beta$ while [6.68] holds for $0 < \alpha - \beta \leq 1$. For the concrete values of $\alpha, \beta$, when applicable, we will compare numerical results obtained according to [6.67] and [6.68].

### 6.2.3. *Cauchy problem*

Finally, we consider [6.43] and [6.46]. For arbitrary $g$ in [6.46], the solution is given by $[6.47]_1$. Therefore, we determine the Green function $\mathcal{G}_c$. Applying the Laplace transform to [6.43], we obtain

$$\frac{\partial^2}{\partial x^2}\tilde{\mathcal{G}}_c(x,s) - \frac{\tau s^\alpha + s^\beta}{\mathcal{D}}\tilde{\mathcal{G}}_c(x,s) = -\frac{\tau s^{\alpha-1} + s^{\beta-1}}{\mathcal{D}}\delta(x), \quad x \in \mathbb{R}, \ \operatorname{Re} s > 0. \tag{6.70}$$

The differential equation [6.70] can be easily solved by using the same procedure as in [GOR 00]. Thus, we obtain

$$\tilde{\mathcal{G}}_c(x,s) = \frac{1}{2\sqrt{\mathcal{D}}}\frac{\sqrt{\tau s^\alpha + s^\beta}}{s}\mathrm{e}^{-\frac{|x|}{\sqrt{\mathcal{D}}}\sqrt{\tau s^\alpha + s^\beta}}, \quad x \in \mathbb{R}, \ \operatorname{Re} s > 0. \tag{6.71}$$

The inverse Laplace transform of [6.71] is obtained in the form of a power series. Thus, we expand [6.71] and obtain

$$\tilde{\mathcal{G}}_c(x,s) = \frac{1}{2\sqrt{\mathcal{D}}}\sum_{n=0}^{\infty}\sum_{k=0}^{\infty}\left(-\frac{|x|}{\sqrt{\mathcal{D}}}\right)^n \frac{\tau^{\frac{n+1}{2}-k}}{n!k!}\frac{\Gamma\left(\frac{n+3}{2}\right)}{\Gamma\left(\frac{n+3}{2}-k\right)}\frac{1}{s^{(\alpha-\beta)k-\alpha\frac{n+1}{2}+1}}, \quad x \in \mathbb{R}, \ \operatorname{Re} s > s_0, \tag{6.72}$$

where $s_0$ is determined so that $|s| > \left(\frac{1}{\tau}\right)^{\frac{1}{\alpha-\beta}}$. Term-by-term inversion leads to ($x \in \mathbb{R}$, $t > 0$)

$$\mathcal{G}_c(x,t) = \frac{1}{2\sqrt{\mathcal{D}}}\sum_{n=0}^{\infty}\sum_{k=0}^{\infty}\left(-\frac{|x|}{\sqrt{\mathcal{D}}}\right)^n \frac{\tau^{\frac{n+1}{2}-k}}{n!k!}\frac{\Gamma\left(\frac{n+3}{2}\right)}{\Gamma\left(\frac{n+3}{2}-k\right)} \times\frac{t^{(\alpha-\beta)k-\alpha\frac{n+1}{2}}}{\Gamma\left((\alpha-\beta)k-\alpha\frac{n+1}{2}+1\right)}. \tag{6.73}$$

To check the numerical accuracy of [6.73], we derive the connection between the solution of the Cauchy and signaling problems. The Laplace transforms corresponding to those two problems are given by [6.65] and [6.71]. By differentiating [6.65] with respect to $s$, we obtain

$$\frac{\mathrm{d}}{\mathrm{d}s}\tilde{\mathcal{G}}_s(x,s) = -\frac{x}{2\sqrt{\mathcal{D}}}\frac{\alpha\tau s^{\alpha-1}+\beta s^{\beta-1}}{\sqrt{\tau s^{\alpha}+s^{\beta}}}\mathrm{e}^{-\frac{x}{\sqrt{\mathcal{D}}}\sqrt{\tau s^{\alpha}+s^{\beta}}}, \quad x\in\mathbb{R},\ \mathrm{Re}\, s>0. \quad [6.74]$$

By comparing [6.71] and [6.74], we conclude that

$$x\tilde{\mathcal{G}}_c(x,s) = \frac{\alpha-\beta}{\alpha\beta}s\frac{s^{\alpha-\beta-1}}{s^{\alpha-\beta}+\frac{\beta}{\alpha\tau}}\frac{\mathrm{d}}{\mathrm{d}s}\tilde{\mathcal{G}}_s(x,s) - \frac{1}{\beta}\frac{\mathrm{d}}{\mathrm{d}s}\tilde{\mathcal{G}}_s(x,s), \quad x\in\mathbb{R},\ \mathrm{Re}\, s>0. \quad [6.75]$$

Using the following properties of the Laplace transform in [6.75], we obtain ($x\in\mathbb{R},\ t>0$)

$$x\mathcal{G}_c(x,t) = \frac{1}{\beta}t\mathcal{G}_s(x,t) - \frac{\alpha-\beta}{\alpha\beta}\left(\frac{\mathrm{d}}{\mathrm{d}t}(t\mathcal{G}_s(x,t))\right) * E_{\alpha-\beta}\left(-\frac{\beta}{\alpha\tau}t^{\alpha-\beta}\right), \quad [6.76]$$

where $E_\sigma$ is the one-parameter Mittag-Leffler function. Different grouping of terms in [6.75] leads to another relation between $\mathcal{G}_s$ and $\mathcal{G}_c$ given as ($x\in\mathbb{R},\ t>0$)

$$x\mathcal{G}_c(x,t) = \frac{1}{\beta}t\mathcal{G}_s(x,t) - \frac{\alpha-\beta}{\alpha\beta}\left(\frac{\mathrm{d}}{\mathrm{d}t}E_{\alpha-\beta}\left(-\frac{\beta}{\alpha\tau}t^{\alpha-\beta}\right)\right) * (t\mathcal{G}_s(x,t)). \quad [6.77]$$

It is known that in the case of the single fractional derivative, there is a relation between the solutions of the Cauchy and signaling problems (see [MAI 97a, p. 327]). We derived a generalization of this relation given by [6.76] and [6.77]. Note that for the case when $\alpha=\beta$, both relations lead to

$$x\mathcal{G}_c(x,t) = \frac{t}{\alpha}\mathcal{G}_s(x,t), \quad x\in\mathbb{R},\ t>0. \quad [6.78]$$

Equation [6.78] is given in [MAI 97a].

### 6.2.4. *Numerical results*

#### 6.2.4.1. *Dirichlet problem*

We present numerical results for [6.62] in the special case $\alpha=0.9$ and for three values of $\beta\in\{0.2,0.5,0.9\}$. We used the integral representation in [6.62] and the following special values of parameters: $\tau=0.1$, $C_k=2$ for $k=1$ and all other

$C_k = 0$. Also, we used $\lambda_1 = 1$, $\mathcal{D} = 1$ and $x = \frac{\pi}{2}$. The results are shown in Figure 6.10. We observe from this figure that, in the beginning of the process, the solution $T$ decreases more rapidly as $\beta$ decreases. However, as $t \to \infty$, the solution tends to zero more rapidly when $\beta$ is larger.

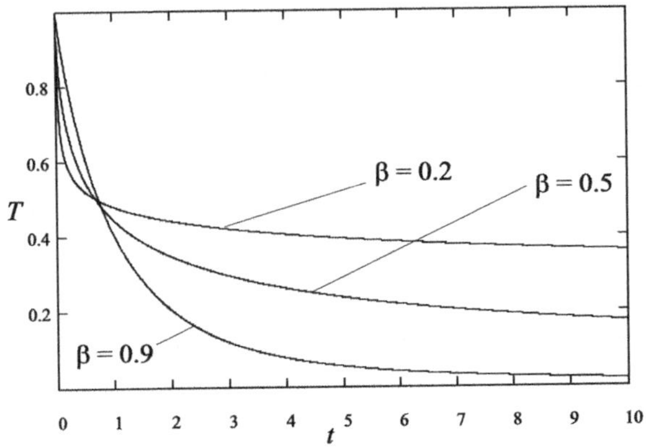


**Figure 6.10.** *Solution given by [6.62] for* $\alpha = 0.9$ *and* $\beta \in \{0.2, 0.5, 0.9\}$

6.2.4.2. *Signaling problem*

First, we consider the case $\alpha = 1.8$, $\beta = 0.9$, $\tau = 0.01$, $\mathcal{D} = 1$. The function $\mathcal{G}_s$ calculated by using [6.67] and [6.68] is shown in Figure 6.11 for three different time instants. The numerical results show that for fixed $\alpha$ and $\beta$, the maximum of $\mathcal{G}_s$ decreases with the increase of time $t$.

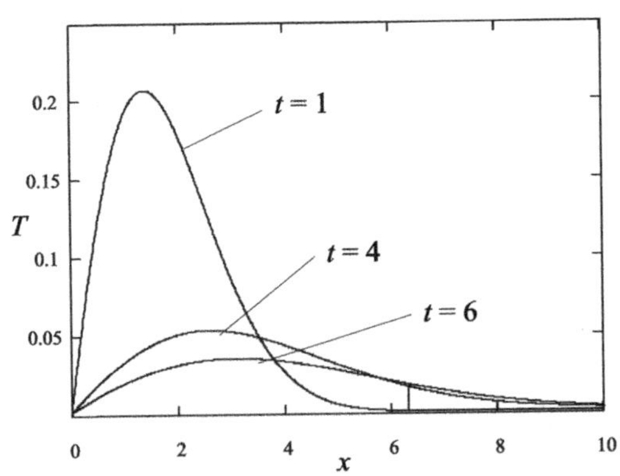


**Figure 6.11.** *Solution given by [6.68] for* $\alpha = 1.8$, $\beta = 0.9$, $\tau = 0.01$, $\mathcal{D} = 1$

6.2.4.3. *Cauchy problem*

In Figure 6.12, we present the solution $\mathcal{G}_c$ obtained by using [6.73] for special values of parameters, given as: $\mathcal{D} = 1$, $\tau = 0.001$, $\alpha = 2$, $\beta = 1.25$, $t = 2$. In Figure 6.13, we present the solution $\mathcal{G}_c$ obtained by [6.73] for $\mathcal{D} = 1$, $\tau = 0.001$, $\alpha = 2$, $\beta = 1.5$, $t = 2$. Finally, in Figure 6.14, we present the solution $\mathcal{G}_c$ obtained by using [6.73] for $\mathcal{D} = 1$, $\tau = 0.001$, $\alpha = 2$, $\beta = 1.9$, $t = 2$.

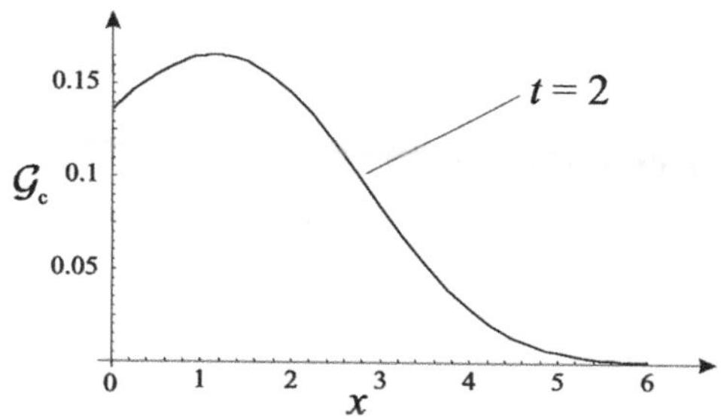


**Figure 6.12.** *Solution given by [6.73] for* $\alpha = 2$, $\beta = 1.25$, $\tau = 0.001$, $\mathcal{D} = 1$, $t = 2$

We note that for the case $\beta = 2$, the classical wave equation is obtained and Figures 6.12–6.14 represent an impulse in the form of the Dirac distribution, traveling with the speed $c = \sqrt{\frac{\mathcal{D}}{1+\tau}}$. In figures, it is seen that $\mathcal{G}_c$ changes from

$$\mathcal{G}_c^{(\text{diff. eq.})}(x,t) = \frac{1}{2\sqrt{\pi t}} e^{-\frac{x^2}{4t}} \qquad [6.79]$$

corresponding to the diffusion equation to

$$\mathcal{G}_c^{(\text{wave eq.})}(x,t) = \delta\left(x - \sqrt{\frac{\mathcal{D}}{1+\tau}} t\right) \qquad [6.80]$$

corresponding to the wave equation.

In trying to come as close as possible to solution [6.80], we solved [6.43] and [6.46] for $\mathcal{D} = 1 + \tau$, $\alpha = \beta = 1.995$ and $t = 1$. We could not get the convergence, for this set of parameters, in the series [6.73] and we used the integral form of solution [6.77]. The result is shown in Figure 6.15. This figure shows the solution to [6.77] that is closest to [6.80].

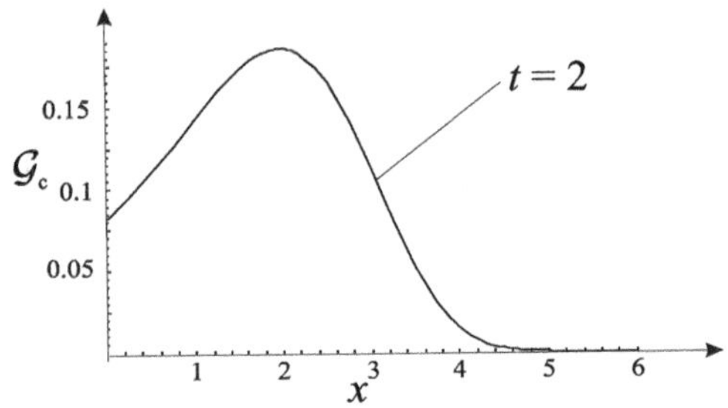


**Figure 6.13.** *Solution given by [6.73] for* $\alpha = 2$, $\beta = 1.5$, $\tau = 0.001$, $\mathcal{D} = 1$, $t = 2$

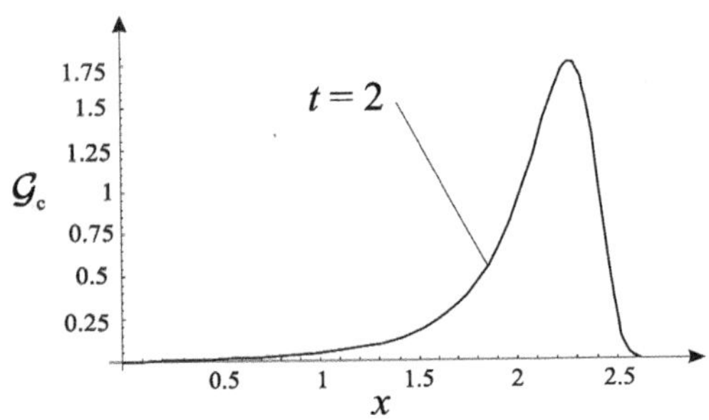


**Figure 6.14.** *Solution given by [6.73] for* $\alpha = 2$, $\beta = 1.9$, $\tau = 0.001$, $\mathcal{D} = 1$, $t = 2$

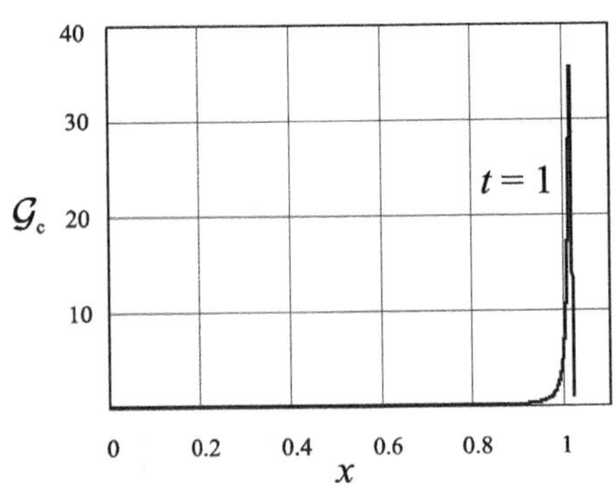


**Figure 6.15.** *Solution given by [6.73] for* $\mathcal{D} = 1 + \tau$, $\alpha = \beta = 1.995$ *and* $t = 1$

### 6.3. Distributed-order diffusion-wave equation

We consider a slight generalization of time distributed-order diffusion-wave equation [6.10], where the distributed-order fractional derivative $\mathrm{D}_\phi$ is interpreted in the sense of distributions. Also, instead of the second-order partial derivative we use linear differential operator $\mathcal{A}$ in $\mathbb{R}^n$ and we add the source term. Thus, we generalized [6.10] as

$$ {}_D\mathrm{D}_\phi u(\boldsymbol{x},t) - \mathcal{A}u(\boldsymbol{x},t) = F(\boldsymbol{x},t), \quad \boldsymbol{x}\in\mathbb{R}^n,\ t>0, \qquad [6.81]$$

where, by definition 2.7,

$$ {}_D\mathrm{D}_\phi u(\boldsymbol{x},t) = \int_{\operatorname{supp}\phi} \phi(\alpha)\, {}_D\mathrm{D}_t^\alpha u(\boldsymbol{x},t)\,\mathrm{d}\alpha, \quad \boldsymbol{x}\in\mathbb{R}^n,\ t>0,$$

with ${}_D\mathrm{D}_t^\alpha u$ given by [2.82]. Equation [6.81] is subject to initial conditions

$$\begin{aligned} &u(\boldsymbol{x},0) = u_0(\boldsymbol{x}), && \operatorname{supp}\phi\subset[0,1],\\ &u(\boldsymbol{x},0) = u_0(\boldsymbol{x}),\ \tfrac{\partial}{\partial t}u(\boldsymbol{x},0) = v_0(\boldsymbol{x}), && \operatorname{supp}\phi\subset[0,2]. \end{aligned} \qquad [6.82]$$

We prove the existence and uniqueness of a solution to [6.81] and [6.82] with

$$\phi(\alpha) = \sum_{j=0}^{k} a_j\delta(\alpha-\alpha_j), \quad a\in\mathbb{R}_+,$$

where $\operatorname{supp}\phi\subset[0,2]$. Then, [6.81] becomes

$$\sum_{j=0}^{k} a_j\, {}_0\mathrm{D}_t^{\alpha_j} u(\boldsymbol{x},t) - \mathcal{A}u(\boldsymbol{x},t) = F(\boldsymbol{x},t), \quad \boldsymbol{x}\in\mathbb{R}^n,\ t>0. \qquad [6.83]$$

In the presentation that follows, we need the following definition.

DEFINITION 6.1.– *Let $X$ be a Banach space and let $\mathcal{A}$ be a linear unbounded operator defined on a dense subset $X_\mathcal{A}$ of $X$. Then, $u$ is a solution to [6.83] and [6.82] in the case* $\operatorname{supp}\phi\subset[0,1]$, *respectively* $\operatorname{supp}\phi\subset[0,2]$, *if, with appropriate assumptions on $F$ and appropriate assumptions on $u_0$, respectively $u_0$ and $v_0$, $u$ belongs to $AC^1_{loc}([0,\infty), X_\mathcal{A})$, respectively to $AC^2_{loc}([0,\infty), X_\mathcal{A})$, and that $u$ satisfies [6.83] and [6.82].*

The existence and uniqueness with this specific choice of a weight distribution will be proven by reducing the problem to an abstract Volterra equation. In a special case, when $\mathcal{A} = \nabla^2$, we refer to equation [6.81] as a time distributed-order diffusion-wave equation. Then, $X = L^2(\mathbb{R}^n)$ and $X_{\nabla^2} = W^{2,2}(\mathbb{R}^n)$. Throughout this section, we use notation $J = [0,T]$, $T > 0$.

Concerning the notation, we call [6.81]

– a distributed diffusion-wave equation if $[a,b] \subset \operatorname{supp}\phi \subset (0,2]$, where $a < 1 < b < 2$ and $\mathcal{A} = \nabla^2$;

– a distributed diffusion equation if $\operatorname{supp}\phi \subset (0,1]$ and $\mathcal{A} = \nabla^2$;

– a distributed wave equation if $\operatorname{supp}\phi \subset (1,2]$ and $\mathcal{A} = \nabla^2$.

In section 6.3.1.5, we will separately consider solutions to the Cauchy problem ([6.81] and [6.82]) and to the Volterra-type equation

$$u(\boldsymbol{x},t) = w(\boldsymbol{x},t) + a(t) * \mathcal{A}u(\boldsymbol{x},t), \quad \boldsymbol{x} \in \mathbb{R}^n, \ t > 0, \qquad [6.84]$$

where

$$w(\boldsymbol{x},t) = u_0(\boldsymbol{x}) + [v_0(\boldsymbol{x}) \cdot t * B_v(t) + F(\boldsymbol{x},t)] * a(t), \quad \boldsymbol{x} \in \mathbb{R}^n, \ t > 0, \qquad [6.85]$$

if $\operatorname{supp}\phi \subset [0,2]$, or $\phi \in C([0,2])$, or

$$w(\boldsymbol{x},t) = u_0(\boldsymbol{x}) + F(\boldsymbol{x},t) * a(t), \quad \boldsymbol{x} \in \mathbb{R}^n, \ t > 0, \qquad [6.86]$$

if $\operatorname{supp}\phi \subset [0,1]$, or $\phi \in C([0,1])$. Here, $a$, $B_u$ and $B_v$ are defined via Laplace transforms as follows:

$$\begin{aligned} \tilde{a}(s) &= \frac{1}{\tilde{B}_u(s)}, \quad s \in \mathbb{C}_+ = \{s \mid \operatorname{Re} s > 0\}, \\ \tilde{B}_u(s) &= \int_{\alpha\in[0,2]} \phi(\alpha)\, s^\alpha \mathrm{d}\alpha, \quad \operatorname{supp}\phi \subset [0,2], \\ \tilde{B}_v(s) &= \int_{\alpha\in(1,2]} \phi(\alpha)\, s^\alpha \mathrm{d}\alpha. \end{aligned} \qquad [6.87]$$

Explanations for [6.84, 6.85, 6.86 and 6.87] will be given in section 6.3.1.1.

A separate analysis of Volterra equations ([6.84]–[6.87]) and Cauchy problems ([6.81] and [6.82]) is necessary, since a strong solution to Volterra equation [6.84] belongs to $C(J, X_{\mathcal{A}})$ and may not satisfy [6.81] and [6.82], while a solution to the Cauchy problem ([6.81] and [6.82]) belongs to $AC^1(J, X_{\mathcal{A}})$ or to $AC^2(J, X_{\mathcal{A}})$.

Let us summarize the main results of this section.

1) Volterra equations ([6.84] and [6.86]) with assumptions $\operatorname{supp}\phi \subset [0,1]$, $u_0 \in X_{\mathcal{A}}$ and $F \in W^{1,1}(J, X_{\mathcal{A}})$ have a unique solution $u \in C(J, X_{\mathcal{A}})$.

2) The Cauchy problems ([6.81] and [6.82]) with assumptions $\operatorname{supp}\phi \subset [0,1]$ and

a) $F \in AC^2(J, X_{\mathcal{A}})$, $u_0 \equiv 0$ have a unique solution $u \in AC^1(J, X_{\mathcal{A}})$;

b) $\mathcal{A} = \nabla^2$, $F \in C_0^{0,\alpha}(J, W^{2,2}(\mathbb{R}^n))$, where $\alpha > 0$, $u_0 \equiv 0$ has a unique solution $u \in AC^1(J, W^{2,2}(\mathbb{R}^n))$.

3) Volterra equations ([6.84] and [6.85]) with assumptions $\operatorname{supp}\phi \subset [0,2]$, $u_0, v_0 \in X_{\mathcal{A}}$, $a(t) \in AC^1(J)$ and $F \in L^1_{loc}(J, X_{\mathcal{A}})$ have $u \in C(J, X_{\mathcal{A}})$ as a unique solution.

4) The Cauchy problems ([6.81] and [6.82]) with assumptions $\operatorname{supp}\phi \subset [0,2]$, $u_0 = v_0 \equiv 0$ and $F \in AC^3(J, X_{\mathcal{A}})$, $F(\boldsymbol{x},0) = \frac{\partial}{\partial t}F(\boldsymbol{x},0) = \frac{\partial^2}{\partial t^2}F(\boldsymbol{x},0) = 0$, $\boldsymbol{x} \in \mathbb{R}^n$ have $u \in AC^2(J, X_{\mathcal{A}})$ as a unique solution.

### 6.3.1. *Existence of a solution to Cauchy problems ([6.81] and [6.82])*

Our goal is to prove the existence of a solution to Cauchy problems ([6.81] and [6.82]), which is a vector-valued function or distribution $t \mapsto u(\cdot, t)$. Because of this, the next conclusion is essential.

*All the calculations for functions, distributions and regular distributions, defined on the real line, can be applied to* $t \mapsto \langle u(\boldsymbol{x},t), \psi(\boldsymbol{x})\rangle$, $\psi \in S(\mathbb{R}^n)$, *where* $t \mapsto u(\cdot,t) \in S'(\mathbb{R}^n)$ *is a distribution-valued function, or its regularization* $\bar{u}(\cdot,t) = u(\cdot,t)H(t)$, $t \in \mathbb{R}$, *or distribution-valued distribution, defined as* $\langle\langle u(\boldsymbol{x},t), \psi(\boldsymbol{x})\rangle, \varphi(t)\rangle$, $\psi \in S(\mathbb{R}^n)$, $\varphi \in S(\mathbb{R})$.

*Here, we implicitly use the Schwartz kernel theorem and the generalized Fubini theorem*

$$\langle\langle u(\boldsymbol{x},t), \psi(\boldsymbol{x})\rangle, \varphi(t)\rangle = \langle\langle u(\boldsymbol{x},t), \varphi(t)\rangle, \psi(\boldsymbol{x})\rangle = \langle u(\boldsymbol{x},t), \psi(\boldsymbol{x})\varphi(t)\rangle,$$

*for every* $\psi \in S(\mathbb{R}^n)$, $\varphi \in S(\mathbb{R})$.

6.3.1.1. *Reduction to an abstract Volterra equation*

We apply the Laplace transform to [6.81] and [6.82], understood in the sense of distributions. Let $\phi$ belong to $\mathcal{E}'(\mathbb{R})$, $\operatorname{supp}\phi \subset [0,2]$ or $\operatorname{supp}\phi \subset [0,1]$. Then, we use the Laplace transform of distributed-order derivative given by [2.102] or [2.103]. Let $X$ be a Banach space of functions on $\mathbb{R}^n$ with the norm $|\cdot|$, for example $X = L^2(\mathbb{R}^n)$. We will use the notation $u(\boldsymbol{x},t)$, $\boldsymbol{x} \in \mathbb{R}^n$, $t > 0$, for the mapping $\mathbb{R} \mapsto X$, $t \mapsto u(\boldsymbol{x},t)$, $\boldsymbol{x} \in \mathbb{R}^n$. Let $u \in L^1_{loc+}(\mathbb{R}, X) \cap L(\mathcal{S}, X)$. Then, the Laplace transform of $u$ with respect to $t$ is given by

$$\tilde{u}(\boldsymbol{x},s) = \mathcal{L}[u(\boldsymbol{x},t)](s), \quad \boldsymbol{x} \in \mathbb{R}^n, \ s \in \mathbb{C}_+.$$

This is an analytic function in $\mathbb{C}_+$ with values in $X$. We will assume for $F$ the same as for $u$ in order to perform the Laplace transform. So, we assume $F \in L^1_{loc+}(\mathbb{R}, X) \cap L(\mathcal{S}, X)$ and later we will give additional assumptions on $F$.

Thus, by the use of the Laplace transform, we obtain

$$\begin{aligned}\tilde{u}(\boldsymbol{x},s)\int_{\alpha\in[0,2]}\phi(\alpha)\, s^{\alpha}\mathrm{d}\alpha - u_0(\boldsymbol{x})\frac{1}{s}\int_{\alpha\in[0,2]}\phi(\alpha)\, s^{\alpha}\mathrm{d}\alpha \\ - v_0(\boldsymbol{x})\frac{1}{s^2}\int_{\alpha\in(1,2]}\phi(\alpha)\, s^{\alpha}\mathrm{d}\alpha - \mathcal{A}\tilde{u}(\boldsymbol{x},s) = \tilde{F}(\boldsymbol{x},s), \quad \text{in } X, \ s \in \mathbb{C}_+,\end{aligned} \tag{6.88}$$

if $\operatorname{supp}\phi \subset [0,2]$. According to remark 2.8, if $\operatorname{supp}\phi \subset [0,1]$, then

$$\begin{aligned}\tilde{u}(\boldsymbol{x},s)\int_{\alpha\in[0,1]}\phi(\alpha)\, s^{\alpha}\mathrm{d}\alpha - u_0(\boldsymbol{x})\frac{1}{s}\int_{\alpha\in[0,1]}\phi(\alpha)\, s^{\alpha}\mathrm{d}\alpha - \mathcal{A}\tilde{u}(\boldsymbol{x},s) = \tilde{F}(\boldsymbol{x},s), \\ \text{in } X, \ s \in \mathbb{C}_+.\end{aligned} \tag{6.89}$$

Let

$$\begin{aligned}\tilde{B}_u(s) &= \int_{\alpha\in[0,2]}\phi(\alpha)\, s^{\alpha}\mathrm{d}\alpha, \quad \operatorname{supp}\phi \subset [0,2], \\ \tilde{B}_v(s) &= \int_{\alpha\in(1,2]}\phi(\alpha)\, s^{\alpha}\mathrm{d}\alpha, \quad s \in \mathbb{C}_+.\end{aligned}$$

Both are analytic functions in $\mathbb{C}_+$. Further, assume that $\tilde{B}_u$ and $\tilde{B}_v$ satisfy

$$\tilde{B}_u(s) \neq 0, \quad \tilde{B}_v(s) \neq 0, \quad s \in \mathbb{C}_+.$$

Equation [6.88] becomes

$$\tilde{u}(\boldsymbol{x},s)\tilde{B}_u(s) - u_0(\boldsymbol{x})\frac{1}{s}\tilde{B}_u(s) - v_0(\boldsymbol{x})\frac{1}{s^2}\tilde{B}_v(s) - \mathcal{A}\tilde{u}(\boldsymbol{x},s) = \tilde{F}(\boldsymbol{x},s),$$
$$\boldsymbol{x}\in\mathbb{R}^n,\ s\in\mathbb{C}_+, \qquad [6.90]$$

while [6.89] transforms to

$$\tilde{u}(\boldsymbol{x},s)\tilde{B}_u(s) - u_0(\boldsymbol{x})\frac{1}{s}\tilde{B}_u(s) - \mathcal{A}\tilde{u}(\boldsymbol{x},s) = \tilde{F}(\boldsymbol{x},s),\quad \boldsymbol{x}\in\mathbb{R}^n,\ s\in\mathbb{C}_+. \qquad [6.91]$$

Next, we use [6.90], respectively [6.91], to obtain an abstract Volterra equation

$$\begin{aligned}\tilde{u}(\boldsymbol{x},s) &= u_0(\boldsymbol{x})\frac{1}{s} + v_0(\boldsymbol{x})\frac{1}{s^2}\tilde{B}_v(s)\frac{1}{\tilde{B}_u(s)}\\ &\quad +\tilde{F}(\boldsymbol{x},s)\frac{1}{\tilde{B}_u(s)} + \frac{1}{\tilde{B}_u(s)}\mathcal{A}\tilde{u}(\boldsymbol{x},s),\quad \boldsymbol{x}\in\mathbb{R}^n,\ s\in\mathbb{C}_+, \qquad [6.92]\end{aligned}$$

respectively

$$\tilde{u}(\boldsymbol{x},s) = u_0(\boldsymbol{x})\frac{1}{s} + \tilde{F}(\boldsymbol{x},s)\frac{1}{\tilde{B}_u(s)} + \frac{1}{\tilde{B}_u(s)}\mathcal{A}\tilde{u}(\boldsymbol{x},s),\quad \boldsymbol{x}\in\mathbb{R}^n,\ s\in\mathbb{C}_+. [6.93]$$

The application of the inverse Laplace transforms to [6.92], respectively to [6.93], implies a new form of [6.81]

$$\begin{aligned}u(\boldsymbol{x},t) &= u_0(\boldsymbol{x}) + v_0(\boldsymbol{x})\cdot t * B_v(t) * a(t)\\ &\quad +F(\boldsymbol{x},t) * a(t) + a(t) * \mathcal{A}u(\boldsymbol{x},t),\quad \boldsymbol{x}\in\mathbb{R}^n,\ t>0, \qquad [6.94]\end{aligned}$$

if $\operatorname{supp}\phi \subset [0,2]$, respectively

$$u(\boldsymbol{x},t) = u_0(\boldsymbol{x}) + F(\boldsymbol{x},t) * a(t) + a(t) * \mathcal{A}u(\boldsymbol{x},t),\quad \boldsymbol{x}\in\mathbb{R}^n,\ t>0, \qquad [6.95]$$

if $\operatorname{supp}\phi \subset [0,1]$, where

$$a(t) = \mathcal{L}^{-1}\left[\frac{1}{\tilde{B}_u(s)}\right](t),\quad t>0, \qquad [6.96]$$

is called scalar kernel. It belongs to $L^1_{loc+}(\mathbb{R})$.

Equation [6.81] in the form of [6.94], respectively [6.95], is actually of the form of an abstract Volterra equation

$$u(t) = w(t) + a(t) * \mathcal{A}u(t), \quad t \in J,$$

where $w \in C(J, X)$ (see [PRÜ 93, p. 30]). However, we need to prescribe a stronger condition for $w$, since a solution to [6.81] and [6.82] should belong to $AC^1(J, X)$ if $\operatorname{supp}\phi \subset [0,1]$ or to $AC^2(J, X)$ if $\operatorname{supp}\phi \subset [0,2]$. This will be explained in section 6.3.1.5. We assume that $\mathcal{A}$ is an unbounded operator defined on a dense subset $X_{\mathcal{A}}$ equipped with the graph norm $|\cdot|_{\mathcal{A}} : |x|_{\mathcal{A}} = |x| + |\mathcal{A}x|$.

Note that if $\operatorname{supp}\phi \subset [0,2]$, then $w$ is given by

$$w(\boldsymbol{x},t) = u_0(\boldsymbol{x}) + [v_0(\boldsymbol{x}) \cdot t * B_v(t) + F(\boldsymbol{x},t)] * a(t), \quad \boldsymbol{x} \in \mathbb{R}^n, \ t > 0. \qquad [6.97]$$

If $\operatorname{supp}\phi \subset [0,1]$, then $w$ is given by

$$w(\boldsymbol{x},t) = u_0(\boldsymbol{x}) + F(\boldsymbol{x},t) * a(t), \quad \boldsymbol{x} \in \mathbb{R}^n, \ t > 0. \qquad [6.98]$$

Thus, we have obtained equations [6.84], [6.85], [6.86] and [6.87].

6.3.1.2. *Specific form of $\phi$ in [6.81]*

We consider

$$\phi(\alpha) = \sum_{j=0}^{k} a_j \delta(\alpha - \alpha_j), \qquad [6.99]$$

where we assume

condition I: $a_j \in \mathbb{R}_+, \ \alpha \in [0,1], \ 0 \leq \alpha_k \leq \alpha_j \leq \alpha_0 \leq 1;$
condition II: $a_j \in \mathbb{R}_+, \ \alpha \in [0,2], \ 0 \leq \alpha_k \leq \alpha_j \leq \alpha_0 \leq 2, \ \alpha_0 > 1.$

Note that condition I actually means that $\operatorname{supp}\phi \subset [0,1]$, while condition II that $\operatorname{supp}\phi \subset [0,2]$.

Distribution $\phi$, in the form [6.99], has its importance in viscoelasticity theory (see [MAI 97a, p. 300]). Physically, coefficients $a_j$, $j \in \{0, \ldots, k\}$, may be interpreted as the relaxation times and therefore $a_j \in \mathbb{R}_+$, $j \in \{0, \ldots, k\}$. We will show in section 6.3.1.3 that this condition is in accordance with the existence of a resolvent. Functions $\tilde{B}_u$ and $\tilde{B}_v$ are now given as follows:

1) If condition I holds, then

$$\tilde{B}_u(s) = \sum_{j=0}^{k} a_j s^{\alpha_j}, \quad \alpha_j \in [0,1], \ s \in \mathbb{C}_+.$$

2) If condition II holds, then

$$\begin{aligned} \tilde{B}_u(s) &= \sum_{j=0}^{k} a_j s^{\alpha_j}, \quad \alpha_j \in [0,2], \ s \in \mathbb{C}_+, \\ \tilde{B}_v(s) &= \sum_{j=0}^{l} a_j s^{\alpha_j}, \quad \alpha_j \in (1,2], \ \alpha_l > 1, \ s \in \mathbb{C}_+. \end{aligned}$$

Note that $B_v$ is a singular distribution, but $t * B_v(t) \in L^1_{loc+}(\mathbb{R})$.

Now, scalar kernel [6.96] becomes

$$a(t) = \mathcal{L}^{-1}[\tilde{a}(s)](t), \quad t > 0, \quad \tilde{a}(s) = \left(\sum_{j=0}^{k} a_j s^{\alpha_j}\right)^{-1}, \quad s \in \mathbb{C}_+, \qquad [6.100]$$

with the same conditions as before. These functions are studied in [ATA 09b], where the following lemma was stated.

LEMMA 6.1.– Let $a(t) = \mathcal{L}^{-1}[\tilde{a}(s)](t)$, $t > 0$, be defined by [6.100]. Assume $\sum_{j=0}^{k} a_j s^{\alpha_j} \neq 0$, $s \in \mathbb{C}_+$. Then

1) $a$ is locally integrable function in $\mathbb{R}$ with $a(t) = 0, t < 0$;

2) $a$ is absolutely continuous in $J$, if $\alpha_0 - \alpha_k > 1$.

The inversion of $\tilde{a}$, defined by [6.100], is given by [ATA 09b] in its integral representation, while by [POD 99], it is given in terms of derivatives of a two-parameter Mittag-Leffler function.

In the following, we will investigate equation [6.81] in the form of [6.94] if $\mathcal{A}$ is a general linear differential operator and in the form of [6.95] if $\mathcal{A} = \nabla^2$. Thus, the following equations are considered:

– Volterra time distributed-order equation with linear differential operator $\mathcal{A}$

$$u(\boldsymbol{x}, t) = w(\boldsymbol{x}, t) + a(t) * \mathcal{A}u(\boldsymbol{x}, t), \quad t \in J \qquad [6.101]$$

on a Banach space $X$.

– Volterra time distributed-order diffusion-wave equation

$$u(\boldsymbol{x},t) = w(\boldsymbol{x},t) + a(t) * \nabla^2 u(\boldsymbol{x},t), \quad t \in J \qquad [6.102]$$

on $L^2(\mathbb{R}^n)$.

Recall that the scalar kernel $a$ is given by [6.100] and distribution $w$ either by [6.97], for $\operatorname{supp}\phi \subset [0,2]$, or by [6.98], for $\operatorname{supp}\phi \subset [0,1]$.

#### 6.3.1.3. *Existence of resolvent in a general case of operator $\mathcal{A}$*

In the sequel, we will give assertions assuming separately that scalar kernel $a$, defined by [6.100], satisfies condition I or condition II. First, we state some properties of scalar kernel $a$.

Recall definition 4.5 of [PRÜ 93], where the notion of a completely positive function is introduced.

PROPOSITION 6.3.– Let condition I hold for $a$. Then, it is a completely positive function.

PROOF.– According to proposition 4.5 of [PRÜ 93], in order for $a$ to be a completely positive function, we should have that $a \in BV_{loc}(\mathbb{R}_+)$ as well as $\tilde{a}(s) > 0$ for $s > 0$, which is clear in our case, since $a$ is continuous for $t > 0$ and coefficients in $\tilde{a}$ are positive. Moreover, we should have that $\varphi(s) = \frac{1}{\tilde{a}(s)} = \sum_{j=0}^{k} a_j s^{\alpha_j}$, $s > 0$, as a Bernstein function. By definition 4.3 of [PRÜ 93], a smooth function $\varphi(s)$ is a Bernstein function if it is positive for $s > 0$ and $f = \varphi'$ is completely monotonic. Let us prove this. Clearly, the assumptions $\alpha_j \in [0,1]$ and $a_j \in \mathbb{R}_+$, $j \in \{0,\ldots,k\}$ imply

$$f(s) = \sum_{j=0}^{k} a_j \alpha_j s^{\alpha_j - 1} \geq 0, \quad f^{(1)}(s) = -\sum_{j=0}^{k} a_j \alpha_j (1-\alpha_j) s^{\alpha_j - 2} \leq 0$$

for $s > 0$ and in general, for $n \in \mathbb{N}$

$$f^{(n)}(s) = (-1)^n \sum_{j=0}^{k} a_j \alpha_j (1-\alpha_j) \ldots (n-\alpha_j) s^{\alpha_j - 1 - n}, \quad s > 0.$$

Thus, $f$ satisfies $(-1)^n f^{(n)}(s) \geq 0$ for $s > 0$, $\alpha_j \in [0,1]$ and $a_j \in \mathbb{R}_+$, $j \in \{0,\ldots,k\}$. ∎

A simple result is the following corollary.

COROLLARY 6.1.– Let condition II hold for $a$. Suppose that

$$\frac{1}{\tilde{a}(s)} = \sum_{j=0}^{k} a_j s^{\alpha_j} = \left(\sum_{p=0}^{l} b_p s^{\beta_p}\right)^2, \quad s \in \mathbb{C}_+, \qquad [6.103]$$

where $\beta_p \in [0,1)$, $b_p \in \mathbb{R}_+$, $p \in \{0, \ldots, l\}$ $(2\beta_0 = \alpha_0 \in (1,2))$. Then, $a$ can be written as $a = c * c$, where

$$c(t) = \mathcal{L}^{-1}\left[\left(\sum_{p=0}^{l} b_p s^{\beta_p}\right)^{-1}\right](t), \quad t > 0,$$

is a completely positive function.

The theorems that are to follow state the existence of a resolvent, the basic result for this section.

THEOREM 6.1.– Let condition I hold for $a$. Let $\mathcal{A}$ be a linear differential operator on a Banach space $X$ and let it generate an exponentially bounded $C_0$ semi-group $T$ in $X$, such that $|T(t)| \leq M\mathrm{e}^{\omega t}$, $t > 0$, for $M > 0$, $\omega \geq 0$. Then, the Volterra time distributed-order equation with linear differential operator $\mathcal{A}$, [6.101], admits exponentially bounded resolvent $S$.

Moreover, the growth bound $\omega_a$ is defined as

$$\omega_a = \begin{cases} \frac{1}{\tilde{a}\left(\frac{1}{\omega}\right)}, & \tilde{a}(0) > \frac{1}{\omega}, \\ 0, & \text{otherwise.} \end{cases} \qquad [6.104]$$

If $\omega = 0$, then $\omega_a = 0$ (because $\tilde{a}(0) > \frac{1}{\omega}$ does not hold). If $\omega > 0$ and $\omega\tilde{a}(0) \neq 1$, then we have two cases determined by [6.104]. Resolvent $S$ is of type $(M_a, \omega_a)$. If $\omega > 0$ and $\omega\tilde{a}(0) = 1$, then resolvent is of type $(M_\varepsilon, \varepsilon)$ for every $\varepsilon > 0$.

PROOF.– Proof follows from theorem 4.2 of [PRÜ 93], since, from proposition 6.3, $a$ is a completely positive function. ■

THEOREM 6.2.– Let condition II and [6.103] hold for $a$. Let $\mathcal{A}$ be a linear differential operator on a Banach space $X$ and let it generate an exponentially bounded cosine family $T$ in $X$, such that $|T(t)| \leq M\mathrm{e}^{\omega t}$, $t > 0$, for some $M > 0$, $\omega \geq 0$. Then, the Volterra time distributed-order equation [6.101] with linear differential operator $\mathcal{A}$ admits an exponentially bounded resolvent $S$.

Moreover, the growth bound $\omega_a$ is defined as

$$\omega_a = \begin{cases} \frac{1}{\sqrt{\tilde{a}\left(\frac{1}{\omega}\right)}}, & \tilde{a}(0) > \frac{1}{\omega^2}, \\ 0, & \text{otherwise.} \end{cases} \qquad [6.105]$$

PROOF.– Proof follows from theorem 4.3 (iii) of [PRÜ 93], since $a = c * c$, where $c$ is a completely positive function. ■

COROLLARY 6.2.– Let $X = L^2(\mathbb{R}^n)$ and $\mathcal{A} = \nabla^2$. The Volterra time distributed-order diffusion-wave equation [6.102] admits an exponentially bounded resolvent $S$ as follows:

1) If condition I holds, then resolvent $S$ is of type $\left(M_a, \sum_{j=0}^{k} \frac{a_j}{\omega^{\alpha_j}}\right)$ if $\alpha_k > 0$ or if $\alpha_k = 0$ and $a_k < \omega$; or of type $(M_a, 0)$ if $\alpha_k = 0$ and $a_k > \omega$; or of type $(M_\varepsilon, \varepsilon)$ for every $\varepsilon > 0$ if $\alpha_k = 0$ and $a_k = \omega$.

2) If condition II holds, then resolvent $S$ has the growth bound $\omega_a = \sqrt{\sum_{j=0}^{k} \frac{a_j}{\omega^{\alpha_j}}}$ if $\alpha_k > 0$ or if $\alpha_k = 0$ and $a_k < \omega^2$; or $\omega_a = 0$ if $\alpha_k = 0$ and $a_k > \omega^2$.

PROOF.– Let us first show (1). It is known that the Laplacian generates a bounded holomorphic $C_0$ semi-group of angle $\frac{\pi}{2}$ on $L^2(\mathbb{R}^n)$ (example 3.7.6 of [ARE 01]). Therefore, by proposition 3.7.2 of [ARE 01], it generates $C_0$ semi-group of type $(M, \omega)$, $M > 0, \omega > 0$, on $L^2(\mathbb{R}^n)$. From [6.100], we see

$$\frac{1}{\tilde{a}(0)} = \begin{cases} a_k, & \alpha_k = 0, \\ 0, & \alpha_k > 0. \end{cases}$$

Let $\alpha_k = 0$. Since $\omega > 0$, from theorem 6.1, [6.104], if $a_k < \omega$, then $\omega_a = \sum_{j=0}^{k} \frac{a_j}{\omega^{\alpha_j}}$, by [6.100], and if $a_k > \omega$, then $\omega_a = 0$. Also if $a_k = \omega$, then every $\varepsilon > 0$ is the growth bound. Let $\alpha_k > 0$, then $\omega > 0$ implies that $\tilde{a}(0) > \frac{1}{\omega}$ is satisfied in [6.104] and therefore $\omega_a = \sum_{j=0}^{k} \frac{a_j}{\omega^{\alpha_j}}$.

We now prove (2). From theorem 8.3.12 of [ARE 01], the Laplacian generates a cosine family on $L^2(\mathbb{R}^n)$. Let $\alpha_k = 0$. Since $\omega > 0$, from theorem 6.2, [6.105], if $a_k < \omega^2$, then $\omega_a = \sqrt{\sum_{j=0}^{k} \frac{a_j}{\omega^{\alpha_j}}}$, by [6.100], and if $a_k > \omega^2$, then $\omega_a = 0$. Let $\alpha_k > 0$, then $\omega > 0$ implies that $\tilde{a}(0) > \frac{1}{\omega^2}$ is satisfied in [6.105] and therefore $\omega_a = \sqrt{\sum_{j=0}^{k} \frac{a_j}{\omega^{\alpha_j}}}$. ■

6.3.1.4. *Additional properties if $\mathcal{A} = \nabla^2$ and condition I holds*

In this case, we have additional properties of [6.102].

PROPOSITION 6.4.– Let condition I hold for $a$. Then, the Volterra time distributed-order diffusion-wave equation [6.102] is parabolic.

PROOF.– According to definition 3.1 of [PRÜ 93], equation is parabolic if scalar kernel $a$ satisfies:

(P1) $\tilde{a}(s) \neq 0$ and $\frac{1}{\tilde{a}(s)} \in \rho\left(\nabla^2\right), s \in \mathbb{C}_+$;

(P2) there exists a constant $M \geq 1$ such that $H(s) = \frac{1}{s}\left(I - \tilde{a}(s)\nabla^2\right)^{-1}, s \in \mathbb{C}_+$, satisfies

$$|H(s)| \leq \frac{M}{|s|}, \quad s \in \mathbb{C}_+.$$

We first prove (P1). By writing $s = r\mathrm{e}^{\mathrm{i}\varphi}$, the Laplace transform of $a$ is

$$\tilde{a}(s) = \left(\sum_{j=0}^{k} a_j s^{\alpha_j}\right)^{-1} = \left(\sum_{j=0}^{k} a_j r^{\alpha_j}\cos(\alpha_j\varphi) + \mathrm{i}\sum_{j=0}^{k} a_j r^{\alpha_j}\sin(\alpha_j\varphi)\right)^{-1}, \quad s \in \mathbb{C}_+.$$

It is different from zero for $s \in \mathbb{C}_+$, since $\cos(\alpha_j\varphi) > 0$ for $\varphi \in \left(-\frac{\pi}{2}, \frac{\pi}{2}\right)$ and $\alpha_j \in [0,1]$. Next, we prove that $\frac{1}{\tilde{a}(s)} \in \rho\left(\nabla^2\right)$, $s \in \mathbb{C}_+$. The resolvent set of the Laplacian is $\rho\left(\nabla^2\right) = \mathbb{C}\setminus(-\infty, 0]$ and therefore we should have

$$\operatorname{Re}\frac{1}{\tilde{a}(s)} = \sum_{j=0}^{k} a_j r^{\alpha_j}\cos(\alpha_j\varphi) \not\leq 0, \quad |\varphi| < \frac{\pi}{2}.$$

It holds since $\operatorname{Re}\frac{1}{\tilde{a}(s)} > 0$, $s \in \mathbb{C}_+$ for $\alpha_j \in [0,1]$, $a_j \in \mathbb{R}_+$, $j \in \{0, \ldots, k\}$.

To prove (P2), we use the Fourier transform for $Hv$, where $v$ is any function $v \in L^2(\mathbb{R}^n)$, and the Parseval formula. Thus

$$\begin{aligned}\|\mathcal{F}[H(s)v(\boldsymbol{x})](\boldsymbol{\omega})\|_{L^2(\mathbb{R}^n)} &= \left\|\frac{1}{s\tilde{a}(s)}\mathcal{F}\left[\left(\frac{I}{\tilde{a}(s)}-\nabla^2\right)^{-1}v(\boldsymbol{x})\right](\boldsymbol{\omega})\right\|_{L^2(\mathbb{R}^n)} \\ &= \frac{1}{|s|}\left\|\frac{1}{1+\tilde{a}(s)\omega^2}\hat{v}(\boldsymbol{\omega})\right\|_{L^2(\mathbb{R}^n)}, \quad s\in\mathbb{C}_+. \qquad [6.106]\end{aligned}$$

We have

$$\begin{aligned}\left|1+\tilde{a}(s)\omega^2\right| &= \left|1+\omega^2\left(\sum_{j=0}^{k}a_j s^{\alpha_j}\right)^{-1}\right| \\ &= \left|1+\left(\omega^2\sum_{j=0}^{k}a_j r^{\alpha_j}\cos(\alpha_j\varphi)\right.\right. \\ &\quad \left.\left. -\mathrm{i}\omega^2\sum_{j=0}^{k}a_j r^{\alpha_j}\sin(\alpha_j\varphi)\right)|\tilde{a}(s)|^{-1}\right|,\end{aligned}$$

i.e.

$$\left|1+\tilde{a}(s)\omega^2\right|\geq 1, \quad s\in\mathbb{C}_+.$$

Therefore, [6.106] leads to

$$\|\mathcal{F}[H(s)v(\boldsymbol{x})](\boldsymbol{\omega})\|_{L^2(\mathbb{R}^n)} \leq \frac{1}{|s|}\|\hat{v}(\boldsymbol{\omega})\|_{L^2(\mathbb{R}^n)}, \quad s\in\mathbb{C}_+,$$

and by the Parseval formula, we have

$$H(s)\leq\frac{1}{|s|}, \quad s\in\mathbb{C}_+.$$

Thus, (P2) is satisfied with $M=1$. ■

Recall definition 3.3 of Prüss [PRÜ 93], where the notion of an $r$-regular, $r \in \mathbb{N}$, function is introduced.

PROPOSITION 6.5.– Let condition I hold for $a$. Then, it is $r$-regular, $r \in \mathbb{N}$.

PROOF.– According to proposition 3.2 of [PRÜ 93], in order to prove that $a$ is an $r$-regular kernel, $r \in \mathbb{N}$, we have to prove that $\tilde{a}(s)$, $s \in \mathbb{C}_+$, admits analytic extension to sector $\Sigma(0,\phi) = \{s \in \mathbb{C} \mid |\arg s| < \phi\}$, $\phi > \frac{\pi}{2}$ and that there exists $\theta \in (0,\infty)$, such that $|\arg \tilde{a}(s)| < \theta$, $s \in \Sigma(0,\phi)$. By taking $\phi = \pi$, $\tilde{a}$ admits analytic extension to $\Sigma(0,\pi)$, since $\bar{\partial}_s \tilde{a}(s) = \left(\frac{\partial}{\partial x} + \mathrm{i}\frac{\partial}{\partial y}\right)\tilde{a}(s) = 0$, $s \in \mathbb{C}_+$. Next, by taking $\theta = \pi$, we have $0 < \arg s^{\alpha_j} < \pi$, since $\alpha_j \in [0,1]$ and $a_j \in \mathbb{R}_+$, $j \in \{0,\ldots,k\}$. Therefore, $0 < \arg\left(\sum_{j=0}^{k} a_j s^{\alpha_j}\right) < \pi$ holds as well. Similar arguments hold in the case $-\pi < \arg s^{\alpha_j} < 0$, so it can be concluded that $\left|\arg\left(\sum_{j=0}^{k} a_j s^{\alpha_j}\right)\right| < \pi$, i.e. $|\arg \tilde{a}(s)| < \pi$. ■

THEOREM 6.3.– Let condition I hold for $a$. Then, the Volterra time distributed-order diffusion-wave equation [6.102] admits resolvent $S \in C^\infty\left((0,\infty), \mathcal{B}\left(L^2(\mathbb{R}^n)\right)\right)$.

Moreover, for every $r \in \mathbb{N}$, there is $M_r > 0$ such that

$$|t^r S^r(t)| \leq M_r, \quad t \geq 0.$$

PROOF.– Note that the Laplacian is a closed linear operator in $L^2(\mathbb{R}^n)$ with the dense domain $\mathcal{D}(\nabla^2) = W^{2,2}(\mathbb{R}^n)$. Moreover, from proposition 6.4, equation [6.102] is parabolic and $a$, given by [6.100], is an $r$-regular, $r \in \mathbb{N}$, scalar kernel, as is shown in proposition 6.5. Thus, theorem 6.3 follows from theorem 3.1 of [PRÜ 93]. ■

Moreover, it can be shown that the Volterra time distributed-order diffusion-wave equation [6.102] admits an analytic resolvent as stated in the following theorem. Recall that a sector in the complex plane is defined as $\Sigma(\omega,\theta) = \{\lambda \in \mathbb{C} \mid |\arg(\lambda - \omega)| < \theta\}$.

THEOREM 6.4.– Let condition I hold for $a$ and let $\alpha_0 < 1$. Then, the Volterra time distributed-order diffusion-wave equation [6.102] admits an analytic resolvent $S$ of analyticity type $(0,\theta_0)$, $\theta_0 < \frac{\pi}{2}\left(\frac{1}{\alpha_0} - 1\right)$.

Moreover, for every $\omega > 0$ and $\theta < \theta_0$, there exists $M = M(\omega,\theta)$ such that

$$\left|S^{(n)}(t)\right| \leq M \frac{n! \mathrm{e}^{\omega t(1+\sin\theta)}}{(t\sin\theta)^n}, \quad t > 0,\ n \in \mathbb{N}.$$

PROOF.– Let us show that $a$ satisfies conditions A1–A3 (see [PRÜ 93, theorem 2.1]).

(A1) $\tilde{a}$ has meromorphic extension to $\Sigma\left(0, \frac{\pi}{2}+\theta_0\right)$. It is clear that $\tilde{a}$ can only have poles in $\Sigma\left(0, \frac{\pi}{2}+\theta_0\right)$.

(A2) $\tilde{a}(s) \neq 0$ and $\frac{1}{\tilde{a}(s)} \in \rho\left(\nabla^2\right)$, for all $s \in \Sigma\left(0, \frac{\pi}{2}+\theta_0\right)$. First, we determine $\theta_0$ so that $\arg s^{\alpha_j}|_{s=r\exp\left[\mathrm{i}\left(\frac{\pi}{2}+\theta_0\right)\right]} < \frac{\pi}{2}, j \in \{0,\ldots,k\}$. Since $\alpha_0$ is the greatest, the condition reduces to $\alpha_0\left(\frac{\pi}{2}+\theta_0\right) < \frac{\pi}{2}$, i.e. $\theta_0 < \frac{\pi}{2}\left(\frac{1}{\alpha_0}-1\right)$. Then, $\frac{1}{\tilde{a}(s)} \in \rho\left(\nabla^2\right)$ holds since $\frac{1}{\tilde{a}(s)} = \sum_{j=0}^{k} a_j s^{\alpha_j} \in \mathbb{C}\setminus(-\infty, 0]$, $s \in \Sigma\left(0, \frac{\pi}{2}+\theta_0\right)$, where $\theta_0$ is determined so that $\arg s^{\alpha_j} < \frac{\pi}{2}, j \in \{0,\ldots,k\}$.

(A3) To prove that there exists $C = C(\xi, \theta)$ such that for all $\theta < \theta_0$ and $\xi > 0$ operator $H(s) = \frac{1}{s}\left(I - \tilde{a}(s)\nabla^2\right)^{-1}$ satisfies

$$|H(s)| \leq \frac{C}{|s-\xi|}, \quad s \in \Sigma\left(\xi, \frac{\pi}{2}+\theta\right), \qquad [6.107]$$

we start with [6.106]

$$\left\|\mathcal{F}\left[H(s)v(\boldsymbol{x})\right](\boldsymbol{\omega})\right\|_{L^2(\mathbb{R}^n)} = \frac{1}{|s|}\left\|\frac{1}{1+\tilde{a}(s)\omega^2}\hat{v}(\boldsymbol{\omega})\right\|_{L^2(\mathbb{R}^n)}, \quad s \in \mathbb{C}_+. \qquad [6.108]$$

Let us show that $\left|1+\tilde{a}(s)\omega^2\right| > 1$, $s \in \Sigma\left(0, \frac{\pi}{2}+\theta_0\right)$. We have

$$\operatorname{Re}\tilde{a}(s)\omega^2 = \frac{\omega^2 \sum\limits_{j=0}^{k} a_j r^{\alpha_j}\cos(\alpha_j\varphi)}{|\tilde{a}(s)|}, \quad |\varphi| < \frac{\pi}{2}+\theta_0 = \frac{\pi}{2\alpha_0}.$$

For $\alpha_j \leq \alpha_0 < 1$ and $|\varphi| < \frac{\pi}{2}+\theta_0$, it holds that $\cos(\alpha_j\varphi) > 0$. Therefore, $\omega^2\sum_{j=0}^{k} a_j r^{\alpha_j}\cos(\alpha_j\varphi) > 0$. Thus, we proved that $\left|1+\tilde{a}(s)\omega^2\right| > 1$, $s \in \Sigma\left(0, \frac{\pi}{2}+\theta_0\right)$ and [6.108] holds with $s \in \Sigma\left(0, \frac{\pi}{2}+\theta_0\right)$. This gives

$$\left\|\mathcal{F}\left[H(s)v(\boldsymbol{x})\right](\boldsymbol{\omega})\right\|_{L^2(\mathbb{R}^n)} \leq \frac{1}{|s|}\left\|\hat{v}(\boldsymbol{\omega})\right\|_{L^2(\mathbb{R}^n)}, \quad s \in \Sigma\left(0, \frac{\pi}{2}+\theta_0\right).$$

Now, we will prove that [6.107] holds for every $\theta < \theta_0$ and $\xi > 0$. Fix $\theta < \theta_0$ and $\xi > 0$ and let define circles $K_1$ and $K_2$ as: $K_1 = \{s \in \mathbb{C} \mid |s| < |\xi|\}$ and $K_2 = \{s \in \mathbb{C} \mid |s-\xi| < |\xi|\}$. Assume that $s \in \mathbb{C}$ is out of both the circles, i.e. $|s| \geq |\xi|$ and $|s-\xi| \geq |\xi|$. We have

$$|s| = |s-\xi+\xi| \geq |s-\xi| - |\xi|, \quad 2|s| \geq |s-\xi| + |s| - |\xi| \geq |s-\xi|,$$

and this implies that

$$\frac{1}{|s|} \leq \frac{2}{|s-\xi|}.$$

Next, assume $s \in \mathbb{C}$ is inside both the circles. Since $s$ also belongs to $\Sigma\left(\xi, \frac{\pi}{2}+\theta\right)$, it holds that $r_0 \leq |s| < |\xi|$ and $|s-\xi| < |\xi|$. We have

$$\frac{|s|}{|s-\xi|} \geq \frac{r_0}{|s-\xi|} \geq \frac{r_0}{|\xi|}, \text{ which implies } \frac{1}{|s|} \leq \frac{r_0}{|\xi|}\frac{1}{|s-\xi|}.$$

We conclude that

$$\frac{1}{|s|} \leq \frac{C}{|s-\xi|}, \quad s \in \Sigma\left(\xi, \frac{\pi}{2}+\theta\right), \ \theta < \theta_0, \ \xi > 0,$$

and therefore, for $s \in \Sigma\left(\xi, \frac{\pi}{2}+\theta\right)$, $\theta < \theta_0$, $\xi > 0$, we have

$$\left\|\mathcal{F}\left[H(s)\, v(\boldsymbol{x})\right](\boldsymbol{\omega})\right\|_{L^2(\mathbb{R}^n)} \leq \frac{1}{|s|}\left\|\hat{v}(\boldsymbol{\omega})\right\|_{L^2(\mathbb{R}^n)} \leq \frac{C}{|s-\xi|}\left\|\hat{v}(\boldsymbol{\omega})\right\|_{L^2(\mathbb{R}^n)}.$$

Thus, from the Parseval formula, for every $v \in L^2(\mathbb{R}^n)$ [6.107] holds.

Since the Laplacian is an unbounded, closed and linear operator in $L^2(\mathbb{R}^n)$ with the dense domain $\mathcal{D}\left(\nabla^2\right) = W^{2,2}(\mathbb{R}^n)$, the assertion follows from theorem 2.1 and corollary 2.1 of [PRÜ 93]. ■

6.3.1.5. *Solution to [6.81] and [6.82] via resolvent*

Theorems 6.1 and 6.2 state that there exists a resolvent to the Volterra time distributed-order equation [6.101] with the linear differential operator $\mathcal{A}$, in the cases when $a$, given by [6.100], satisfies condition I or condition II. Moreover, for the Volterra time distributed-order diffusion-wave equation [6.102] with the same scalar kernel $a$, theorems 6.3 and 6.4 and corollary 6.2 state the existence of the resolvent. Therefore, it is possible to construct a solution to [6.81] and [6.82] by the use of the resolvent.

Formally, applying the Laplace transform to the Volterra time distributed-order equation [6.101] with linear differential operator $\mathcal{A}$ and by solving an algebraic equation with respect to $\tilde{u}$, we obtain

$$\tilde{u}(\boldsymbol{x}, s) = s\tilde{S}(s)\, \tilde{w}(\boldsymbol{x}, s), \quad \boldsymbol{x} \in \mathbb{R}^n, \ s \in \mathbb{C}_+, \qquad [6.109]$$

where the resolvent $S$ is given by its Laplace transform

$$\tilde{S}(s) \equiv H(s) = \frac{1}{s}\left(I - \tilde{a}(s)\mathcal{A}\right)^{-1}, \quad s \in \mathbb{C}_{+}.$$

Applying the inverse Laplace transform to [6.109], the solution is expressed via the resolvent operator by

$$u(\boldsymbol{x},t) = \frac{\partial}{\partial t}\left[S(t) * w(\boldsymbol{x},t)\right], \quad \boldsymbol{x} \in \mathbb{R}^n, \ t > 0, \qquad [6.110]$$

where $w$ belongs to an appropriate space. The analysis of the formal expression [6.110] will be the subject of the next two sections. We will consider strong solutions to the Volterra-type equation and solutions to the Cauchy problem ([6.81] and [6.82]). In the first case, solutions belong to $C(J, X_{\mathcal{A}})$ and in the second case, solutions belong to an appropriate space so that distributed-order derivative can be applied to them. Recall that $w$ is given by [6.98] if $\operatorname{supp}\phi \subset [0,1]$, i.e. if condition I holds, and by [6.97] if $\operatorname{supp}\phi \subset [0,2]$, i.e. if condition II holds. Now, we consider these two cases separately.

*Condition I holds.* In this section, we assume that $\mathcal{A}$ generates an exponentially bounded $C_0$ semi-group. We have proved that $\mathcal{A}$ also generates an exponentially bounded resolvent (corollary 6.2). Moreover, if $\mathcal{A} = \nabla^2$, then it generates a smooth ($C^\infty$) resolvent (theorem 6.3) if $\operatorname{supp}\phi \subset [0,1]$ and an analytic resolvent, with appropriate bounds on its derivatives (theorem 6.4), if $\operatorname{supp}\phi \subset [0,1)$.

Now by proposition 2.1 of [PRÜ 93], we have the next theorem.

THEOREM 6.5.– Assume that $\mathcal{A}$ generates an exponentially bounded $C_0$ semi-group. Let $u_0 \in X_{\mathcal{A}}$ and $F \in W^{1,1}(J, X_{\mathcal{A}})$. Then

$$u(\boldsymbol{x},t) = S(t)\,u_0(\boldsymbol{x}) + S(t) * \frac{\partial}{\partial t}\left[a(t) * F(\boldsymbol{x},t)\right], \quad \boldsymbol{x} \in \mathbb{R}^n, \ t \in J,$$

is a unique strong solution to [6.101], where $S$ is the resolvent of [6.101]. It belongs to $C(J, X_{\mathcal{A}})$.

However, in order to have $u \in AC^1(J, X_{\mathcal{A}})$, we need additional assumptions.

THEOREM 6.6.– Assume that $\mathcal{A}$ generates an exponentially bounded $C_0$ semi-group. Let $u_0 \equiv 0$ and $F \in AC^2(J, X_{\mathcal{A}})$, $F(\boldsymbol{x},0) = 0$, $\boldsymbol{x} \in \mathbb{R}^n$. Then

$$u(\boldsymbol{x},t) = S(t) * a(t) * \frac{\partial}{\partial t}F(\boldsymbol{x},t), \quad \boldsymbol{x} \in \mathbb{R}^n, \ t \in J$$

is a unique strong solution to [6.81] belonging to $AC^1(J, X_{\mathcal{A}})$.

THEOREM 6.7.– Assume $\mathcal{A} = \nabla^2$. Let $u_0 \equiv 0$ and $F \in C_0^{0,\alpha}(J, W^{2,2}(\mathbb{R}^n))$ for some $\alpha > 0$. Then

$$u(\boldsymbol{x}, t) = a(t) * F(\boldsymbol{x}, t) + \frac{\mathrm{d}}{\mathrm{d}t} S(t) * a(t) * F(\boldsymbol{x}, t), \quad \boldsymbol{x} \in \mathbb{R}^n, \ t \in J$$

is a unique strong solution to [6.81]. Moreover, $u \in AC^1(J, W^{2,2}(\mathbb{R}^n))$.

PROOF.– Assumption $F \in C_0^{0,\alpha}(J, W^{2,2}(\mathbb{R}^n))$ implies that $a * F \in C_0^{0,\alpha}(J, W^{2,2}(\mathbb{R}^n))$. Now, from theorem 2.4 (iii) of [PRÜ 93] condition $a * F \in C_0^{0,\alpha}(J, W^{2,2}(\mathbb{R}^n))$ implies $u \in AC^1(J, W^{2,2}(\mathbb{R}^n))$. ■

*Condition II holds.* Note that

$$t \mapsto t * B_v(t) = \sum_{j=0}^{l} \frac{a_j}{\Gamma(2 - \alpha_j) t^{\alpha_j - 1}}, \quad t > 0, \ 1 < \alpha_l \le \alpha_j \le \alpha_0 \le 2,$$

(see [2.101]) is locally integrable on $[0, \infty)$.

To have an exponentially bounded resolvent $S$, $\mathcal{A}$ must generate an exponentially bounded cosine family, as we have proved in theorem 6.2 and corollary 6.2. So with this, we have the next theorem.

THEOREM 6.8.– Assume that $a = c * c$ satisfies [6.103], with $\alpha_0 - \alpha_k > 1$ (see lemma 6.1) and that $\mathcal{A}$ generates an exponentially bounded cosine family (see theorem 6.2). Let $u_0, v_0 \in X_{\mathcal{A}}$, $a \in AC^1(J)$ (see lemma 6.1) and $F \in L^1_{loc}(J, X_{\mathcal{A}})$. Then

$$\begin{aligned} u(\boldsymbol{x}, t) \quad &= \quad S(t) u_0(\boldsymbol{x}) + S(t) * \frac{\mathrm{d}}{\mathrm{d}t} a(t) * [v_0(\boldsymbol{x}) \cdot t * B_v(t) + F(\boldsymbol{x}, t)], \\ &\qquad \boldsymbol{x} \in \mathbb{R}^n, \ t \in J \end{aligned}$$

is a unique strong solution to [6.101], where $S$ is the resolvent of [6.101] with the properties of theorem 6.2.

PROOF.– Bearing in mind theorem 6.2, this theorem is a result of proposition 1.2 (iv) of [PRÜ 93]. ■

To have $u \in AC^2(J, X_{\mathcal{A}})$, we have to assume additional conditions. Since, in theorem 6.8, we cannot extend assumptions on $a$ (again see lemma 6.1), we have to make additional assumptions on the resolvent and on $F$.

THEOREM 6.9.– Assume that $a = c * c$ satisfies [6.103] and that $\mathcal{A}$ generates an exponentially bounded cosine family (see theorem 6.2). Let $u_0 = v_0 \equiv 0$ and $F \in AC^3(J, X_{\mathcal{A}}),\ F(\boldsymbol{x},0) = \frac{\partial}{\partial t}F(\boldsymbol{x},0) = \frac{\partial^2}{\partial t^2}F(\boldsymbol{x},0) = 0,\ \boldsymbol{x} \in \mathbb{R}^n$. Then

$$u(\boldsymbol{x},t) = S(t) * a(t) * \frac{\partial}{\partial t}F(\boldsymbol{x},t), \quad \boldsymbol{x} \in \mathbb{R}^n,\ t \in J$$

is a unique strong solution to [6.81] and it belongs to $AC^2(J, X_{\mathcal{A}})$.

Moreover, if $\mathcal{A} = \nabla^2$ and $X_{\mathcal{A}} = W^{2,2}(\mathbb{R}^n)$, then $u$ is a strong unique solution to [6.81] and it belongs to $AC^2\left(J, W^{2,2}(\mathbb{R}^n)\right)$.

### 6.3.2. *Solution to the Cauchy problem*

We treat the Cauchy problem

$$_D\mathrm{D}_\phi u(\boldsymbol{x},t) - \nabla^2 u(\boldsymbol{x},t) = F(\boldsymbol{x},t), \quad \boldsymbol{x} \in \mathbb{R}^n,\ t > 0, \qquad [6.111]$$

$$u(\boldsymbol{x},0) = u_0(\boldsymbol{x}), \operatorname{supp}\phi \subset [0,1], \left.\begin{array}{l} u(\boldsymbol{x},0) = u_0(\boldsymbol{x}) \\ \frac{\partial}{\partial t}u(\boldsymbol{x},0) = v_0(\boldsymbol{x}) \end{array}\right\}, \operatorname{supp}\phi \subset [0,2], \boldsymbol{x} \in \mathbb{R}^n, \qquad [6.112]$$

where we assumed $u_0, v_0 \in \mathcal{S}'(\mathbb{R}^n)$ and $F \in C([0,\infty), \mathcal{S}'(\mathbb{R}^n))$, interpreted in $\mathcal{S}'_+(\mathbb{R}^{n+1})$ as

$$\int_{\alpha\in[0,2]} \phi(\alpha)\, {}_D\mathrm{D}_t^\alpha \Big(u(\boldsymbol{x},t)\, H(t)\Big)\mathrm{d}\alpha - u_0(\boldsymbol{x}) \int_{\alpha\in[0,2]} \phi(\alpha)\, f_{1-\alpha}(t)\,\mathrm{d}\alpha$$
$$-v_0(\boldsymbol{x}) \int_{\alpha\in(1,2]} \phi(\alpha)\, f_{2-\alpha}(t)\,\mathrm{d}\alpha - \nabla^2 u(\boldsymbol{x},t)\, H(t) = F(\boldsymbol{x},t)\, H(t). \qquad [6.113]$$

We reinterpret the Cauchy problem ([6.111] and [6.112]) within the space of distributions. Let $u \in AC^2_{loc}([0,\infty), \mathcal{S}'(\mathbb{R}^n))$, $F \in C([0,\infty), \mathcal{S}'(\mathbb{R}^n))$ and $\phi \in \mathcal{E}'(\mathbb{R}), \operatorname{supp}\phi \subset [0,2]$. Then, for every $\psi \in \mathcal{S}(\mathbb{R}^n)$, [6.111] reads

$$\langle \mathrm{D}_\phi u(\boldsymbol{x},t), \psi(\boldsymbol{x})\rangle - \langle \nabla^2 u(\boldsymbol{x},t), \psi(\boldsymbol{x})\rangle = \langle F(\boldsymbol{x},t), \psi(\boldsymbol{x})\rangle, \quad t > 0, \qquad [6.114]$$

and therefore

$$\langle \mathrm{D}_\phi u(\boldsymbol{x},t), \psi(\boldsymbol{x})\rangle = \int_0^2 \phi(\alpha) \langle {}_0\mathrm{D}_t^\alpha u(\boldsymbol{x},t), \psi(\boldsymbol{x})\rangle \, \mathrm{d}\alpha, \;\; t>0, \; \psi \in \mathcal{S}(\mathbb{R}^n),$$

or

$$\langle \mathrm{D}_\phi u(\boldsymbol{x},t), \psi(\boldsymbol{x})\rangle = \sum_{j=0}^{k} \alpha_j \left\langle {}_0\mathrm{D}_t^{\alpha_j} u(\boldsymbol{x},t), \psi(\boldsymbol{x})\right\rangle \mathrm{d}\alpha, \;\; t>0, \; \psi \in \mathcal{S}(\mathbb{R}^n).$$

We can see that terms $\langle \mathrm{D}_\phi u(\boldsymbol{x},t), \psi(\boldsymbol{x})\rangle$, $\left\langle u(\boldsymbol{x},t), \nabla^2\psi(\boldsymbol{x})\right\rangle$ and $\langle F(\boldsymbol{x},t), \psi(\boldsymbol{x})\rangle$ in [6.114] can be interpreted as the functions solely of variable $t>0$. Hence, for every $\psi \in \mathcal{S}(\mathbb{R}^n)$, regular distributions $u_{reg}$, $U_{reg}$ and $F_{reg}$ are defined by ($t \in \mathbb{R}$)

$$\begin{aligned} u_{reg}(t) &= \langle u(\boldsymbol{x},t) H(t), \psi(\boldsymbol{x})\rangle, \;\; U_{reg}(t) = \left\langle u(\boldsymbol{x},t) H(t), \nabla^2\psi(\boldsymbol{x})\right\rangle, \\ & F_{reg}(t) = \langle F(\boldsymbol{x},t) H(t), \psi(\boldsymbol{x})\rangle. \end{aligned} \quad [6.115]$$

Then $u_{reg}, U_{reg}, F_{reg} \in \mathcal{S}'_+(\mathbb{R})$. With [6.115] and

$$u_0 = \langle u_0(\boldsymbol{x}), \psi(\boldsymbol{x})\rangle, \;\; v_0 = \langle v_0(\boldsymbol{x}), \psi(\boldsymbol{x})\rangle, \;\; \psi \in \mathcal{S}(\mathbb{R}^n), \quad [6.116]$$

we reinterpret the Cauchy problem ([6.111] and [6.112]) in the sense of distributions, rewriting [6.114] as

$$\begin{aligned} &\int_{\alpha\in[0,2]} \phi(\alpha)\, {}_D\mathrm{D}_t^\alpha u_{reg}(t)\, \mathrm{d}\alpha - u_0 \int_{\alpha\in[0,2]} \phi(\alpha) f_{1-\alpha}(t)\, \mathrm{d}\alpha \\ &\quad -v_0 \int_{\alpha\in(1,2]} \phi(\alpha) f_{2-\alpha}(t)\, \mathrm{d}\alpha - U_{reg}(t) = F_{reg}(t), \quad \text{in } \mathcal{S}'_+(\mathbb{R}). \end{aligned} \quad [6.117]$$

Actually, [6.117] (in $\mathcal{S}'_+(\mathbb{R}^{n+1})$) has the form

$$\begin{aligned} &\int_{\alpha\in[0,2]} \phi(\alpha)\, {}_D\mathrm{D}_t^\alpha [u(\boldsymbol{x},t) H(t)]\, \mathrm{d}\alpha - u_0(\boldsymbol{x}) \int_{\alpha\in[0,2]} \phi(\alpha) f_{1-\alpha}(t)\, \mathrm{d}\alpha \\ &\quad -v_0(\boldsymbol{x}) \int_{\alpha\in(1,2]} \phi(\alpha) f_{2-\alpha}(t)\, \mathrm{d}\alpha - \nabla^2 u(\boldsymbol{x},t) H(t) = F(\boldsymbol{x},t) H(t), \end{aligned}$$

[6.118]

and this equation will be the subject of the further analysis, as noted in the introduction.

Applying the Laplace transform to [6.117], we obtain

$$\tilde{u}_{reg}(s)\int_{\alpha\in[0,2]}\phi(\alpha)\,s^{\alpha}\mathrm{d}\alpha-u_0\frac{1}{s}\int_{\alpha\in[0,2]}\phi(\alpha)\,s^{\alpha}\mathrm{d}\alpha$$
$$-v_0\frac{1}{s^2}\int_{\alpha\in(1,2]}\phi(\alpha)\,s^{\alpha}\mathrm{d}\alpha-\tilde{U}_{reg}(s)=\tilde{F}_{reg}(s),\quad s\in\mathbb{C}_+, \qquad [6.119]$$

where $\tilde{u}_{reg}$, $\tilde{U}_{reg}$ and $\tilde{F}_{reg}$ is used to denote the Laplace transforms of $u_{reg}$, $U_{reg}$ and $F_{reg}$. Let

$$\tilde{B}_u(s)=\int_{\alpha\in[0,2]}\phi(\alpha)\,s^{\alpha}\mathrm{d}\alpha,\quad \tilde{B}_v(s)=\int_{\alpha\in(1,2]}\phi(\alpha)\,s^{\alpha}\mathrm{d}\alpha,\quad s\in\mathbb{C}_+. \qquad [6.120]$$

These functions are analytic in $\mathbb{C}_+$. Moreover, assume

$$\tilde{B}_u(s)\neq 0,\quad \tilde{B}_v(s)\neq 0,\quad s\in\mathbb{C}_+.$$

Then, equation [6.119] becomes

$$\tilde{u}_{reg}(s)\,\tilde{B}_u(s)-u_0\frac{1}{s}\tilde{B}_u(s)-v_0\frac{1}{s^2}\tilde{B}_v(s)-\tilde{U}_{reg}(s)=\tilde{F}_{reg}(s),\quad s\in\mathbb{C}_+. [6.121]$$

Note that the Fourier transform of $g\in\mathcal{S}'(\mathbb{R}^n)$, with respect to the spatial variable, is defined by

$$\langle\mathcal{F}[g],\psi\rangle=\langle g,\mathcal{F}[\psi]\rangle,\quad \psi\in\mathcal{S}(\mathbb{R}^n),$$

where

$$\hat{\psi}(\boldsymbol{\omega})\equiv\mathcal{F}[\psi(\boldsymbol{x})](\boldsymbol{\omega})=\int_{\mathbb{R}^n}\psi(\boldsymbol{x})\,\mathrm{e}^{-\mathrm{i}\boldsymbol{x}\cdot\boldsymbol{\omega}}\mathrm{d}^n\boldsymbol{x},\quad \boldsymbol{\omega}\in\mathbb{R}^n.$$

We will use the identity $\langle g,\psi\rangle=\left\langle\hat{g}(\boldsymbol{\omega}),\hat{\Psi}(\boldsymbol{\omega})\right\rangle$, $\psi\in\mathcal{S}(\mathbb{R}^n)$, where $\hat{\Psi}(\boldsymbol{\omega})=\frac{1}{(2\pi)^n}\hat{\psi}(-\boldsymbol{\omega})\in\mathcal{S}(\mathbb{R}^n)$. Applying the Laplace transform to [6.115], [6.116] and

$$\mathcal{F}\left[\nabla^2\psi(\boldsymbol{x})\right](\boldsymbol{\omega})=-|\boldsymbol{\omega}|^2\,\psi(\boldsymbol{\omega}),\quad \boldsymbol{\omega}\in\mathbb{R}^n,$$

we obtain that for every $\hat{\Psi} \in \mathcal{S}(\mathbb{R}^n)$, $s \in \mathbb{C}_+$ and $t \in \mathbb{R}$

$$u_{reg}(t) = \left\langle \hat{u}(\boldsymbol{\omega}, t) H(t), \hat{\Psi}(\boldsymbol{\omega}) \right\rangle, \quad \tilde{u}_{reg}(s) = \left\langle \widehat{\tilde{u}}(\boldsymbol{\omega}, s), \hat{\Psi}(\boldsymbol{\omega}) \right\rangle,$$

$$U_{reg}(t) = \left\langle -|\boldsymbol{\omega}|^2 \hat{u}(\boldsymbol{\omega}, t) H(t), \hat{\Psi}(\boldsymbol{\omega}) \right\rangle, \quad \tilde{U}_{reg}(s) = \left\langle -|\boldsymbol{\omega}|^2 \widehat{\tilde{u}}(\boldsymbol{\omega}, s), \hat{\Psi}(\boldsymbol{\omega}) \right\rangle,$$

$$F_{reg}(t) = \left\langle \hat{F}(\boldsymbol{\omega}, t) H(t), \hat{\Psi}(\boldsymbol{\omega}) \right\rangle, \quad \tilde{F}_{reg}(s) = \left\langle \widehat{\tilde{F}}(\boldsymbol{\omega}, s), \hat{\Psi}(\boldsymbol{\omega}) \right\rangle, \qquad [6.122]$$

$$u_0 = \left\langle \hat{u}_0(\boldsymbol{\omega}), \hat{\Psi}(\boldsymbol{\omega}) \right\rangle, \quad v_0 = \left\langle \hat{v}_0(\boldsymbol{\omega}), \hat{\Psi}(\boldsymbol{\omega}) \right\rangle. \qquad [6.123]$$

Now, [6.121] with [6.122] and [6.123], written in $\mathcal{S}'_+(\mathbb{R}^n)$, becomes

$$\widehat{\tilde{u}}(\boldsymbol{\omega}, s) \tilde{B}_u(s) - \hat{u}_0(\boldsymbol{\omega}) \frac{1}{s} \tilde{B}_u(s) - \hat{v}_0(\boldsymbol{\omega}) \frac{1}{s^2} \tilde{B}_v(s) + |\boldsymbol{\omega}|^2 \widehat{\tilde{u}}(\boldsymbol{\omega}, s) = \widehat{\tilde{F}}(\boldsymbol{\omega}, s),$$

$$\boldsymbol{\omega} \in \mathbb{R}^n, \; s \in \mathbb{C}_+.$$

Finally, by solving the resulting equation, we obtain

$$\widehat{\tilde{u}}(\boldsymbol{\omega}, s) = \frac{\widehat{\tilde{F}}(\boldsymbol{\omega}, s) + \hat{u}_0(\boldsymbol{\omega}) \frac{1}{s} \tilde{B}_u(s) + \hat{v}_0(\boldsymbol{\omega}) \frac{1}{s^2} \tilde{B}_v(s)}{\tilde{B}_u(s) + |\boldsymbol{\omega}|^2}, \quad \boldsymbol{\omega} \in \mathbb{R}^n, \; s \in \mathbb{C}_+. \qquad [6.124]$$

Thus, [6.124] represents the Fourier and Laplace transforms of a solution to the distributional equation [6.118].

We introduce the solution kernel via its Fourier and Laplace transforms as

$$\widehat{\tilde{g}}(\omega, s) = \frac{1}{\tilde{B}_u(s) + \omega^2}, \quad \omega = |\boldsymbol{\omega}|, \; \boldsymbol{\omega} \in \mathbb{R}^n, \; s \in \mathbb{C}_+. \qquad [6.125]$$

This formula has to be understood in the sense of distributions, since $s \mapsto \widehat{\tilde{g}}(\cdot, s)$ is an analytic function in $\mathbb{C}_+$ having values in $\mathcal{S}'(\mathbb{R}^n)$. After making the Fourier inversion of [6.124], we obtain (in $\mathcal{S}'(\mathbb{R}^n)$)

$$\tilde{u}(\boldsymbol{x}, s) = \left( \tilde{F}(\boldsymbol{x}, s) + u_0(\boldsymbol{x}) \frac{1}{s} \tilde{B}_u(s) + v_0(\boldsymbol{x}) \frac{1}{s^2} \tilde{B}_v(s) \right) *_{\boldsymbol{x}} \tilde{g}(|\boldsymbol{x}|, s),$$

$$s \in \mathbb{C}_+. \qquad [6.126]$$

Note that in the case when $\operatorname{supp} \phi \subset [0, 1]$, the distributed diffusion equation is obtained by [6.126] by setting $v_0 \equiv 0$, since the initial condition $\frac{\partial}{\partial t} u(\cdot, 0)$ does not appear. After applying the Laplace inversion to [6.126], we obtain the solution for the distributed diffusion equation ($v_0 = 0$) as being

$$u(\boldsymbol{x},t) = F(\boldsymbol{x},t) *_{\boldsymbol{x},t} g(|\boldsymbol{x}|,t) + u_0(\boldsymbol{x}) *_{\boldsymbol{x}} [H(t) *_t B_u(t) *_t g(|\boldsymbol{x}|,t)], \quad [6.127]$$

as well as a solution for distributed diffusion-wave equation

$$\begin{aligned} u(x,t) &= F(\boldsymbol{x},t) *_{\boldsymbol{x},t} g(|\boldsymbol{x}|,t) + u_0(\boldsymbol{x}) *_{\boldsymbol{x}} [H(t) *_t B_u(t) *_t g(|\boldsymbol{x}|,t)] \\ &\quad + v_0(\boldsymbol{x}) *_{\boldsymbol{x}} [t *_t B_v(t) *_t g(|\boldsymbol{x}|,t)]. \end{aligned} \quad [6.128]$$

Distribution $u$ that solves the Cauchy problem [6.118] is given by [6.127] or by [6.128], where the explicit form of a weight function $\phi$ has to be given in order to obtain the solution kernel $g$ and functions $B_u$ and $B_v$. In the following sections, we determine $g$ in both distributed diffusion and distributed diffusion-wave equations.

6.3.2.1. *Solution kernel of distributed diffusion equation*

Similarly to [CHE 02, KOC 08], rewrite [6.125] in the form

$$\widehat{\tilde{g}}(\omega,s) = \int_0^\infty \mathrm{e}^{-p\left(\tilde{B}_u(s)+\omega^2\right)}\mathrm{d}p, \quad \omega \geq 0, \ \operatorname{Re}\tilde{B}_u(s) > 0, \ s \in \mathbb{C}_+.$$

This leads to ($x = |\boldsymbol{x}|$, $\boldsymbol{x} \in \mathbb{R}^n$, $s \in \mathbb{C}_+$)

$$\tilde{g}(x,s) = \mathcal{F}^{-1}\left[\widehat{\tilde{g}}(\omega,s)\right](x) = \frac{1}{(2\pi)^n}\int_{\mathbb{R}^n} \widehat{\tilde{g}}(\omega,s)\,\mathrm{e}^{\mathrm{i}\boldsymbol{x}\cdot\boldsymbol{\omega}}\mathrm{d}^n\boldsymbol{\omega},$$

and to ($x = |\boldsymbol{x}|$, $\boldsymbol{x} \in \mathbb{R}^n$, $s \in \mathbb{C}_+$)

$$\tilde{g}(x,s) = \int_0^\infty I_n(x,p)\,\mathrm{e}^{-p\tilde{B}_u(s)}\mathrm{d}p, \quad I_n(x,p) = \frac{1}{(2\pi)^n}\int_{\mathbb{R}^n} \mathrm{e}^{-p\omega^2+\mathrm{i}\boldsymbol{x}\cdot\boldsymbol{\omega}}\mathrm{d}^n\boldsymbol{\omega}. \quad [6.129]$$

Note that for $p = 0$, it yields $I_n(\boldsymbol{x},0) = \delta(\boldsymbol{x})$, $\boldsymbol{x} \in \mathbb{R}^n$. Distribution $I_n$ in [6.129] is obtained after reducing the exponent to the binomial square and integrating over $\mathbb{R}^n$ as

$$I_n(x,p) = \frac{1}{\left(\sqrt{4\pi p}\right)^n}\mathrm{e}^{-\frac{x^2}{4p}}, \quad x \geq 0, \ p > 0. \quad [6.130]$$

Inverse Laplace transform in [6.129] gives

$$g(x,t) = \mathcal{L}^{-1}[\tilde{g}(x,s)](t) = \frac{1}{2\pi\mathrm{i}}\int_{\gamma} \tilde{g}(x,s)\,\mathrm{e}^{st}\mathrm{d}s, \quad x \geq 0,\ t > 0,$$

where $\gamma = \{s \in \mathbb{C} \mid \operatorname{Re} s = s_0,\ \operatorname{Im} s \in (-\infty, \infty)\}$, with suitably chosen $s_0$, is a Bromwich contour. Thus, we obtain

$$g(x,t) = \int_0^{\infty} I_n(x,p)\, I(t,p)\,\mathrm{d}p, \quad I(t,p) = \frac{1}{2\pi\mathrm{i}}\int_{\gamma} \mathrm{e}^{-p\tilde{B}_u(s)+st}\mathrm{d}s, \quad x \geq 0,\ t > 0. \tag{6.131}$$

CONDITION 6.1.– Suppose that $\phi$ satisfies:

1) $\operatorname{Re}\tilde{B}_u(s) > 0,\ s \in \mathbb{C}_+$;

2) $J(s,t,p) = \mathrm{e}^{-p\tilde{B}_u(s)+st}, \quad s \in \mathbb{C}_+,\ t > 0,\ p > 0,$ [6.132]

has $s = 0$ and $s = \infty$ as the only branch points and all its singularities lie in the half plane $\operatorname{Re} s < 0$.

Then, integral $I$ can be evaluated by the Cauchy residue theorem

$$\oint_{\hat{\Gamma}} J(s,t,p)\,\mathrm{d}s = 2\pi\mathrm{i}\sum_j \operatorname{Res}(J(s,t,p), s_j), \quad t > 0,\ p > 0,$$

where $\hat{\Gamma} = \Gamma_1 \cup \Gamma_2 \cup \Gamma_\varepsilon \cup \Gamma_3 \cup \Gamma_4 \cup \gamma_0$ is contour given in Figure 6.16.

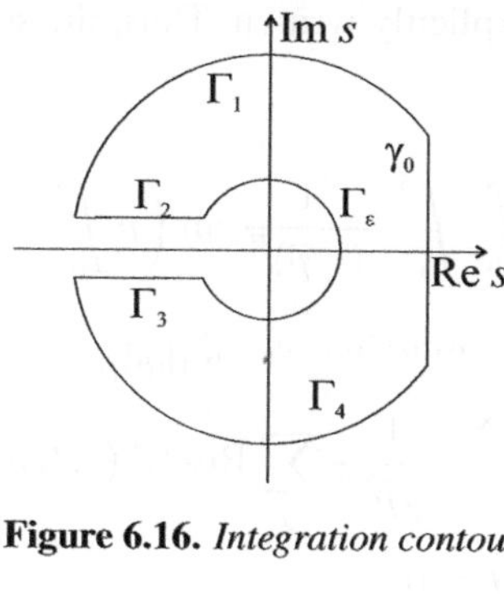


**Figure 6.16.** *Integration contour* $\hat{\Gamma}$

Since $\operatorname{Re}\tilde{B}_u(s) > 0$ for $\operatorname{supp}\phi \subset [0,1]$, in the limit $R \to \infty$, $\varepsilon \to 0$, integrals along contours $\Gamma_1$, $\Gamma_\varepsilon$ and $\Gamma_4$ tend to zero, while integrals along contours $\Gamma_2$, $\Gamma_3$ and $\gamma_0$ give ($t > 0,\ p > 0$)

$$\lim_{R\to\infty,\ \varepsilon\to 0} \int_{\Gamma_2} J(s,t,p)\,\mathrm{d}s = \int_0^\infty \mathrm{e}^{-p\tilde{B}_u\left(q\mathrm{e}^{\mathrm{i}\pi}\right)} \mathrm{e}^{-qt}\mathrm{d}q,$$

$$\lim_{R\to\infty,\ \varepsilon\to 0} \int_{\Gamma_3} J(s,t,p)\,\mathrm{d}s = -\int_0^\infty \mathrm{e}^{-p\tilde{B}_u\left(q\mathrm{e}^{-\mathrm{i}\pi}\right)} \mathrm{e}^{-qt}\mathrm{d}q,$$

$$\int_\gamma J(s,t,p)\,\mathrm{d}s = 2\pi\mathrm{i} I(t,p).$$

Thus $I$, given by [6.131], is of the form

$$I(t,p) = \frac{1}{2\pi\mathrm{i}} \int_0^\infty \left(\mathrm{e}^{-p\tilde{B}_u\left(q\mathrm{e}^{-\mathrm{i}\pi}\right)} - \mathrm{e}^{-p\tilde{B}_u\left(q\mathrm{e}^{\mathrm{i}\pi}\right)}\right) \mathrm{e}^{-qt}\mathrm{d}q + \sum_j \operatorname{Res}\left(J(s,t,p), s_j\right),$$
$$t > 0,\ p > 0,$$

which can be rewritten as

$$I(t,p) = \frac{1}{\pi} \int_0^\infty \sin\left(p\int_0^1 \phi(\alpha)\, q^\alpha \sin(\alpha\pi)\,\mathrm{d}\alpha\right) \mathrm{e}^{-p\int_0^1 \phi(\alpha) q^\alpha \cos(\alpha\pi)\mathrm{d}\alpha - qt}\mathrm{d}q$$
$$+\sum_j \operatorname{Res}\left(J(s,t,p), s_j\right), \quad t>0,\ p>0, \qquad [6.133]$$

see [6.120] for $\tilde{B}_u$.

Solution kernel $g$, given by [6.131], where $I_n$ and $I$ are given by [6.130] and [6.133], respectively, can be explicitly written. Thus, the solution to [6.125] is given by

$$g(x,t) = \frac{1}{\left(\sqrt{4\pi}\right)^n \pi} \int_0^\infty \int_0^\infty \frac{1}{\left(\sqrt{p}\right)^n} \sin\left(p\int_0^1 \phi(\alpha)\, q^\alpha \sin(\alpha\pi)\,\mathrm{d}\alpha\right)$$
$$\times \mathrm{e}^{-p\int_0^1 \phi(\alpha) q^\alpha \cos(\alpha\pi)\mathrm{d}\alpha - qt - \frac{x^2}{4p}}\mathrm{d}q\mathrm{d}p$$
$$+\frac{1}{\left(\sqrt{4\pi}\right)^n} \int_0^\infty \frac{1}{\left(\sqrt{p}\right)^n} \sum_j \operatorname{Res}\left(J(s,t,p), s_j\right) \mathrm{e}^{-\frac{x^2}{4p}}\mathrm{d}p,$$
$$x \geq 0,\ t > 0. \qquad [6.134]$$

After the weight function $\phi$ is specified, the explicit form of $g$ follows from [6.134]. Note that for fixed $t > 0$, $g$ is a distribution in $\mathcal{S}'(\mathbb{R}^n)$.

6.3.2.2. *Solution kernel of distributed diffusion-wave equation*

Similarly, as was done in [HAN 02b, KOC 08, MAI 08], we use [6.125] to obtain the solution kernel. Three cases can be distinguished, according to the number of spatial dimensions.

*One-dimensional case*. Inverse Fourier transform of [6.125] for $n = 1$ yields

$$\tilde{g}_1(x,s) = \frac{1}{2\pi}\int_{-\infty}^{\infty} \frac{\mathrm{e}^{\mathrm{i}x\omega}}{\tilde{B}_u(s) + \omega^2}\mathrm{d}\omega, \quad x \in \mathbb{R}, \ s \in \mathbb{C}_+.$$

The function $\tilde{g}_1$ is evaluated by integration in complex plane, using the Cauchy residue theorem

$$\oint_{\bar{\Gamma}} \frac{\mathrm{e}^{\mathrm{i}xz}}{\tilde{B}_u(s) + z^2}\mathrm{d}z = 2\pi\mathrm{i}\sum_j \operatorname{Res}\left(\frac{\mathrm{e}^{\mathrm{i}xz}}{\tilde{B}_u(s) + z^2}, z_j\right), \quad x \in \mathbb{R}, \ s \in \mathbb{C}_+, \qquad [6.135]$$

where contour $\bar{\Gamma}$ is either $\bar{\Gamma} = \Gamma_1 \cup \gamma_1$ for $x > 0$, or $\bar{\Gamma} = \Gamma_2 \cup \gamma_2$ for $x < 0$, as shown in Figure 6.17. In writing [6.135], we assumed that $\tilde{B}_u(s) + z^2$ has no zeros on the real axis for any $z$, i.e. if $z_{1,2}$ are solutions of $\tilde{B}_u(s) + z^2 = 0$, then $\operatorname{Im} z_1 = \operatorname{Re}\sqrt{\tilde{B}_u(s)} > 0$ and $\operatorname{Im} z_2 = -\operatorname{Re}\sqrt{\tilde{B}_u(s)} < 0$. Let $\gamma = \{s \in \mathbb{C} \mid \operatorname{Re} s = s_0 > 0, \ \operatorname{Im} s \in (-\infty, \infty)\}$, be the Bromwich contour in the Laplace transform. Cauchy residue theorem [6.135] determines $\tilde{g}_1$ as

$$\tilde{g}_1(x,s) = \frac{\mathrm{e}^{-|x|\sqrt{\tilde{B}_u(s)}}}{2\sqrt{\tilde{B}_u(s)}}, \quad x \in \mathbb{R}, \ s \in \mathbb{C}_+. \qquad [6.136]$$

CONDITION 6.2.– Suppose that $\phi$ satisfies:

1) $\operatorname{Re}\sqrt{\tilde{B}_u(s)} > 0, \ s \in \mathbb{C}_+$;

2) function $\tilde{g}_1$ given by [6.136] has $s = 0$ and $s = \infty$ as the only branch points and all its singularities lie in the half plane $\operatorname{Re} s < 0$.

Then, the inverse Laplace transform of [6.136], or the solution kernel itself, is evaluated similarly as in section 6.3.2.1. The Cauchy residue theorem applied to $\tilde{g}_1$ yields

$$\oint_{\hat{\Gamma}} \tilde{g}_1(x,s)\,\mathrm{e}^{st}\mathrm{d}s = 2\pi\mathrm{i}\sum_j \operatorname{Res}\left(\tilde{g}_1(x,s), s_j\right), \quad x \in \mathbb{R}, \ t > 0, \qquad [6.137]$$

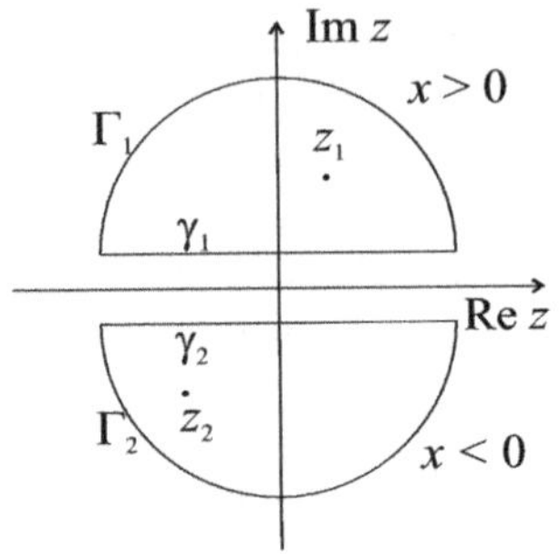


**Figure 6.17.** *Integration contour* $\bar{\Gamma}$

where $\hat{\Gamma} = \Gamma_1 \cup \Gamma_2 \cup \Gamma_\varepsilon \cup \Gamma_3 \cup \Gamma_4 \cup \gamma_0$ is the contour given in Figure 6.16.

In the limit $R \to \infty$, $\varepsilon \to 0$ integrals along contours $\Gamma_1$, $\Gamma_\varepsilon$ and $\Gamma_4$ tend to zero for all $\alpha \in (0,2)$, while integrals along contours $\Gamma_2, \Gamma_3$ give ($x \in \mathbb{R},\ t > 0$)

$$\begin{aligned}
\lim_{R\to\infty,\ \varepsilon\to 0} \int_{\Gamma_2} \tilde{g}_1(x,s)\, \mathrm{e}^{st} \mathrm{d}s &= \frac{1}{2} \int_0^\infty \frac{\mathrm{e}^{-|x|\sqrt{\tilde{B}_u(q\mathrm{e}^{\mathrm{i}\pi})}}}{\sqrt{\tilde{B}_u(q\mathrm{e}^{\mathrm{i}\pi})}} \mathrm{e}^{-qt} \mathrm{d}q, \\
\lim_{R\to\infty,\ \varepsilon\to 0} \int_{\Gamma_3} \tilde{g}_1(x,s)\, \mathrm{e}^{st} \mathrm{d}s &= -\frac{1}{2} \int_0^\infty \frac{\mathrm{e}^{-|x|\sqrt{\tilde{B}_u(q\mathrm{e}^{-\mathrm{i}\pi})}}}{\sqrt{\tilde{B}_u(q\mathrm{e}^{-\mathrm{i}\pi})}} \mathrm{e}^{-qt} \mathrm{d}q.
\end{aligned}$$

Since

$$g_1(x,t) = \frac{1}{2\pi\mathrm{i}} \int_\gamma \tilde{g}_1(x,s)\, \mathrm{d}s, \quad x \in \mathbb{R},\ t > 0,$$

the explicit form of the solution kernel $g_1$ follows from [6.137] and can be written as

$$\begin{aligned}
g_1(x,t) &= \frac{1}{4\pi\mathrm{i}} \int_0^\infty \left( \frac{\mathrm{e}^{-|x|\sqrt{\tilde{B}_u(q\mathrm{e}^{-\mathrm{i}\pi})}}}{\sqrt{\tilde{B}_u(q\mathrm{e}^{-\mathrm{i}\pi})}} - \frac{\mathrm{e}^{-|x|\sqrt{\tilde{B}_u(q\mathrm{e}^{\mathrm{i}\pi})}}}{\sqrt{\tilde{B}_u(q\mathrm{e}^{\mathrm{i}\pi})}} \right) \mathrm{e}^{-qt} \mathrm{d}q \\
&\quad + \sum_j \mathrm{Res}\left(\tilde{g}_1(x,s), s_j\right), \quad x \in \mathbb{R},\ t > 0. \qquad [6.138]
\end{aligned}$$

Note that for fixed $t > 0$, $g_1$ is a distribution in $\mathcal{S}'(\mathbb{R}^n)$.

*Two-dimensional case*. Inverse Fourier transform of the solution kernel [6.125], with $n = 2$, is obtained in polar coordinates. Let $\boldsymbol{\omega} = \omega_1 \boldsymbol{e}_1 + \omega_2 \boldsymbol{e}_2$. By pointing unit vector $\boldsymbol{e}_1$ along $\boldsymbol{x}$, the scalar product $\boldsymbol{x} \cdot \boldsymbol{\omega}$ becomes $\boldsymbol{x} \cdot \boldsymbol{\omega} = x\omega \cos\varphi$, where $x = |\boldsymbol{x}|$, $\omega = |\boldsymbol{\omega}|$ and $\varphi$ is a polar angle. Therefore, function $\tilde{g}_2$ is of the form

$$\tilde{g}_2(x,s) = \frac{1}{(2\pi)^2} \int_0^\infty \widehat{\tilde{g}}_2(\omega,s)\,\omega \mathrm{d}\omega \int_0^{2\pi} \mathrm{e}^{\mathrm{i}x\omega\cos\varphi} \mathrm{d}\varphi, \quad x \geq 0, \; s \in \mathbb{C}_+. \qquad [6.139]$$

Since $\int_0^{2\pi} \mathrm{e}^{\mathrm{i}x\omega\cos\varphi} \mathrm{d}\varphi = 2\pi J_0(x\omega)$, where $J_0(z) = \sum_{k=0}^\infty \frac{(-1)^k}{(2^k k!)^2} z^{2k}$ is the Bessel function of the first kind, and $\int_0^\infty \frac{y}{y^2+w^2} J_0(cy)\,\mathrm{d}y = K_0(cw)$, where $K_0(z)$ is the modified Bessel function of the second kind (also known as the Macdonald function), relation [6.139] is transformed into

$$\tilde{g}_2(x,s) = \frac{1}{2\pi} K_0\left(x\sqrt{\tilde{B}_u(s)}\right), \quad x \geq 0, \; s \in \mathbb{C}_+. \qquad [6.140]$$

Therefore, the solution kernel in two dimensions reads

$$g_2(x,t) = \frac{1}{(2\pi)^2 \mathrm{i}} \int_\gamma K_0\left(x\sqrt{\tilde{B}_u(s)}\right) \mathrm{e}^{st} \mathrm{d}s, \quad x \geq 0, \; t > 0, \qquad [6.141]$$

where $\gamma = \{s \in \mathbb{C} \mid \operatorname{Re} s = s_0 > 0, \; \operatorname{Im} s \in (-\infty, \infty)\}$ is the Bromwich contour.

*Three-dimensional case*. Let $n = 3$. The inverse Fourier transform of the solution kernel [6.125] is obtained in spherical coordinates. Again if $\boldsymbol{\omega} = \omega_1 \boldsymbol{e}_1 + \omega_2 \boldsymbol{e}_2 + \omega_3 \boldsymbol{e}_3$ and by choosing $\boldsymbol{e}_3$ so that $\boldsymbol{x} = x\boldsymbol{e}_3$, the scalar product $\boldsymbol{x} \cdot \boldsymbol{\omega}$ becomes $\boldsymbol{x} \cdot \boldsymbol{\omega} = x\omega \cos\theta$, where $x = |\boldsymbol{x}|$, $\omega = |\boldsymbol{\omega}|$ and $\theta$ is the angle between $\boldsymbol{e}_3$ and vector $\boldsymbol{\omega}$. Therefore, function $\tilde{g}_3$ is of the form

$$\tilde{g}_3(x,s) = \frac{1}{(2\pi)^3} \int_0^{2\pi} \mathrm{d}\varphi \int_0^\infty \widehat{\tilde{g}}_3(\omega,s)\,\omega^2 \mathrm{d}\omega \int_0^\pi \sin\theta\, \mathrm{e}^{\mathrm{i}x\omega\cos\theta} \mathrm{d}\theta, \quad x \geq 0, \; s \in \mathbb{C}_+.$$

Integrating over angles $\varphi$ and $\theta$ and by the substitution $\omega = -\omega$, the following is obtained

$$\tilde{g}_3(x,s) = \frac{1}{(2\pi)^2 \mathrm{i}x} \int_{-\infty}^\infty \frac{\mathrm{e}^{\mathrm{i}x\omega}}{\tilde{B}_u(s) + \omega^2} \omega \mathrm{d}\omega, \quad x \geq 0, \; s \in \mathbb{C}_+,$$

or by noting ($x \geq 0,\ s \in \mathbb{C}_+$)

$$\int_{-\infty}^{\infty} \frac{\mathrm{e}^{\mathrm{i}x\omega}}{\tilde{B}_u(s)+\omega^2}\omega \mathrm{d}\omega = -\mathrm{i}\frac{\partial}{\partial x}\int_{-\infty}^{\infty} \frac{\mathrm{e}^{\mathrm{i}x\omega}}{\tilde{B}_u(s)+\omega^2}\mathrm{d}\omega = -2\pi\mathrm{i}\frac{\partial}{\partial x}\tilde{g}_1(x,s),$$

$\tilde{g}_3$ becomes

$$\tilde{g}_3(x,s) = -\frac{1}{2\pi x}\frac{\partial}{\partial x}\tilde{g}_1(x,s),\quad x\geq 0,\ s\in\mathbb{C}_+. \tag{6.142}$$

The inverse Laplace transform of [6.142], or the solution kernel itself, is

$$g_3(x,t) = -\frac{1}{2\pi x}\frac{\partial}{\partial x}g_1(x,t),\quad x\geq 0,\ t>0.$$

By taking into account [6.138], the solution kernel finally becomes

$$\begin{aligned} g_3(x,t) &= \frac{1}{8\pi^2\mathrm{i}x}\int_0^{\infty}\left(\mathrm{e}^{-x\sqrt{\tilde{B}_u(q\mathrm{e}^{-\mathrm{i}\pi})}} - \mathrm{e}^{-x\sqrt{\tilde{B}_u(q\mathrm{e}^{\mathrm{i}\pi})}}\right)\mathrm{e}^{-qt}\mathrm{d}q \\ &\quad -\frac{1}{2\pi x}\frac{\partial}{\partial x}\sum_j \mathrm{Res}\left(\tilde{g}_1(x,s),s_j\right),\quad x\geq 0,\ t>0. \end{aligned} \tag{6.143}$$

6.3.2.3. *Special cases of weight function*

By specifying the form of the weight function $\phi$, it is possible not only to recover known equations and their solutions, but also to construct new solutions for the generalization of previously treated equations.

*Case* $\phi(\mu) = \delta(\mu-\alpha)$. Let $\alpha \in (0,2)$. The wave-diffusion equation, which was thoroughly studied in [MAI 97a] for the one-dimensional and in [HAN 02a, HAN 02b] for the $d$-dimensional case, $d \in \{1,2,3\}$, is obtained by setting: $\phi(\mu) = \delta(\mu-\alpha)$, $\mu = 0$, $\eta = 2$, $\mathcal{A} = \nabla^2$, $F = 0$, $v_0 = 0$ and $u_0 = 1$, in [6.111] and [6.112].

First, we use [6.120] to obtain function $\tilde{B}_u$ and then calculate the function $B_u = \mathcal{L}^{-1}\left[\tilde{B}_u\right]$ (for $\alpha \in (0,2)$) as

$$\begin{aligned} \tilde{B}_u(s) &= \int_0^2 \delta(\mu-\alpha)\, s^{\mu}\mathrm{d}\mu = s^{\alpha},\quad s\in\mathbb{C}_+, \\ B_u(t) &= \frac{1}{\Gamma(-\alpha)}\frac{1}{t^{\alpha+1}},\quad t>0. \end{aligned}$$

The explicit form of solutions can be constructed as follows.

In the case when $\alpha \in (0,1)$, the solution kernel $g$, given by [6.134], can be written as

$$\begin{aligned} g(x,t) &= \frac{1}{\left(\sqrt{4\pi}\right)^n \pi} \int_0^\infty \int_0^\infty \frac{1}{\left(\sqrt{p}\right)^n} \sin\left(pq^\alpha \sin(\alpha\pi)\right) \\ &\quad \times \mathrm{e}^{-pq^\alpha \cos(\alpha\pi) - qt - \frac{x^2}{4p}} \mathrm{d}q\mathrm{d}p, \quad x = |\boldsymbol{x}|, \ \boldsymbol{x} \in \mathbb{R}^n, \ t > 0. \end{aligned}$$

From [6.132], we calculate $J(s,t,p) = \mathrm{e}^{-ps^\alpha + st}$, $s \in \mathbb{C}_+, t > 0, \ p \geq 0$. Since it satisfies condition 6.1 (its only branch points are $s = 0$ and $s = \infty$ and it has no singularities in the right complex plane), the solution itself is given by [6.127] and it reads

$$\begin{aligned} u(x,t) &= \frac{1}{\left(\sqrt{4\pi}\right)^n \pi\Gamma(-\alpha)} \cdot H(t) *_t \frac{1}{t^{\alpha+1}} *_t \int_0^\infty \int_0^\infty \frac{1}{\left(\sqrt{p}\right)^n} \sin\left(pq^\alpha \sin(\alpha\pi)\right) \\ &\quad \times \mathrm{e}^{-pq^\alpha \cos(\alpha\pi) - qt - \frac{x^2}{4p}} \mathrm{d}q\mathrm{d}p, \quad x = |\boldsymbol{x}|, \ \boldsymbol{x} \in \mathbb{R}^n, \ t > 0. \end{aligned} \qquad [6.144]$$

Let $\alpha \in (0,2)$. For the one-dimensional case ($n = 1$), the function $\tilde{g}_1$, calculated by [6.136], reads

$$\tilde{g}_1(x,s) = \frac{\mathrm{e}^{-|x|\sqrt{s^\alpha}}}{2\sqrt{s^\alpha}}, \quad x \in \mathbb{R}, \ s \in \mathbb{C}_+.$$

It has neither branch points different from $s = 0$ and $s = \infty$, nor singularities in the right complex plane, i.e. it satisfies condition 6.2. Therefore, $g_1$, given by [6.138], becomes

$$\begin{aligned} g_1(x,t) &= \frac{1}{4\pi\mathrm{i}} \int_0^\infty \left( \frac{\mathrm{e}^{-|x|\sqrt{q^\alpha \mathrm{e}^{-\mathrm{i}\alpha\pi}}}}{\sqrt{q^\alpha \mathrm{e}^{-\mathrm{i}\alpha\pi}}} - \frac{\mathrm{e}^{-|x|\sqrt{q^\alpha \mathrm{e}^{\mathrm{i}\alpha\pi}}}}{\sqrt{q^\alpha \mathrm{e}^{\mathrm{i}\alpha\pi}}} \right) \mathrm{e}^{-qt} \mathrm{d}q, \\ &\quad x \in \mathbb{R}, \ t > 0. \end{aligned} \qquad [6.145]$$

Now, we will consider the case $\alpha \in (0,1)$. Then, $\mathbb{R} \ni t \mapsto g_1(\cdot,t)$ is a smooth function with the values in $\mathcal{S}'(\mathbb{R}^n)$. Let us write [6.145] for $t = 0$ and make the change of variables as $p = |x|^{\frac{2}{\alpha}} q$. Then, [6.145] reads

$$\begin{aligned} g_1(x,0) &= \frac{1}{4\pi \mathrm{i}} \frac{1}{|x|^{\frac{2}{\alpha}-1}} \int_0^\infty \left( \frac{\mathrm{e}^{-p^{\frac{\alpha}{2}} \mathrm{e}^{-\mathrm{i}\frac{\alpha\pi}{2}}}}{\mathrm{e}^{-\mathrm{i}\frac{\alpha\pi}{2}}} - \frac{\mathrm{e}^{-p^{\frac{\alpha}{2}} \mathrm{e}^{\mathrm{i}\frac{\alpha\pi}{2}}}}{\mathrm{e}^{\mathrm{i}\frac{\alpha\pi}{2}}} \right) \frac{\mathrm{d}p}{p^{\frac{\alpha}{2}}}, \quad x \in \mathbb{R}, \\ &\sim \frac{C}{|x|^{\frac{2}{\alpha}-1}}, \quad x \to \infty. \end{aligned}$$

The first time derivative of [6.145] in $t = 0$ reads

$$\frac{\partial}{\partial t} g_1(x,0) = -\frac{1}{4\pi \mathrm{i}} \int_0^\infty \left( \frac{\mathrm{e}^{-|x|\sqrt{q^\alpha \mathrm{e}^{-\mathrm{i}\alpha\pi}}}}{\sqrt{q^\alpha \mathrm{e}^{-\mathrm{i}\alpha\pi}}} - \frac{\mathrm{e}^{-|x|\sqrt{q^\alpha \mathrm{e}^{\mathrm{i}\alpha\pi}}}}{\sqrt{q^\alpha \mathrm{e}^{\mathrm{i}\alpha\pi}}} \right) q \mathrm{d}q, \quad x \in \mathbb{R},$$

and with the same change of variables as before, we obtain

$$\begin{aligned} \frac{\partial}{\partial t} g_1(x,0) &= -\frac{1}{4\pi \mathrm{i}} \frac{1}{|x|^{\frac{4}{\alpha}-1}} \\ &\times \int_0^\infty \left( \frac{\mathrm{e}^{-p^{\frac{\alpha}{2}} \mathrm{e}^{-\mathrm{i}\frac{\alpha\pi}{2}}}}{\mathrm{e}^{-\mathrm{i}\frac{\alpha\pi}{2}}} - \frac{\mathrm{e}^{-p^{\frac{\alpha}{2}} \mathrm{e}^{\mathrm{i}\frac{\alpha\pi}{2}}}}{\mathrm{e}^{\mathrm{i}\frac{\alpha\pi}{2}}} \right) p^{1-\frac{\alpha}{2}} \mathrm{d}p, \quad x \in \mathbb{R}, \\ &\sim \frac{C}{|x|^{\frac{4}{\alpha}-1}}. \end{aligned}$$

Note that for $\alpha \in [1,2]$, $g_1$ is a distribution with respect to time as well.

The solution can now be calculated by using [6.127] ($x \in \mathbb{R}$, $t > 0$)

$$\begin{aligned} u(x,t) &= \frac{1}{4\pi \mathrm{i} \Gamma(-\alpha)} \cdot H(t) *_t \frac{1}{t^{\alpha+1}} *_t \\ &\times \int_0^\infty \left( \frac{\mathrm{e}^{-|x|\sqrt{q^\alpha \mathrm{e}^{-\mathrm{i}\alpha\pi}}}}{\sqrt{q^\alpha \mathrm{e}^{-\mathrm{i}\alpha\pi}}} - \frac{\mathrm{e}^{-|x|\sqrt{q^\alpha \mathrm{e}^{\mathrm{i}\alpha\pi}}}}{\sqrt{q^\alpha \mathrm{e}^{\mathrm{i}\alpha\pi}}} \right) \mathrm{e}^{-qt} \mathrm{d}q. \end{aligned} \qquad [6.146]$$

As an example, we set $x = 1.23$, $t \in (0,5)$, $\alpha = 0.45$ in [6.146] and plot the resulting expression in Figure 6.18 as a solid line. Also in Figure 6.18, we present a plot of $u(1.23, t)$ (denoted by dots) obtained by using the solution given in [MAI 96]

$$u_{Mai}(x,t) = \frac{t}{2\pi \mathrm{i} \alpha |x|} \int_{\hat{\Gamma}} \mathrm{e}^{st - \frac{|x|}{\sqrt{D}} s^{\frac{\alpha}{2}}} \mathrm{d}s, \quad x \in \mathbb{R}, \ t > 0.$$

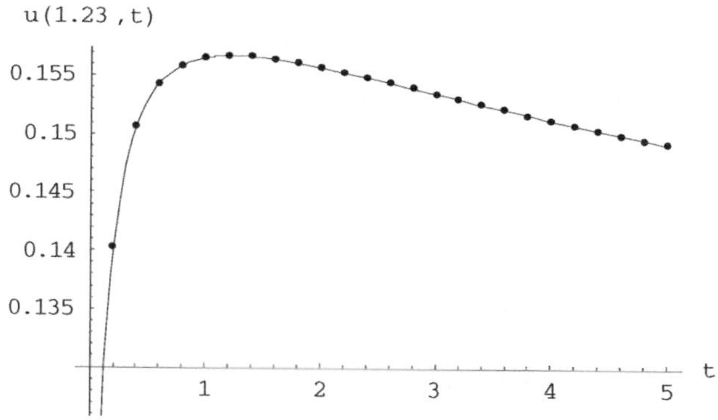


**Figure 6.18.** *Plots of the solutions to the diffusion-wave equation*

As could be seen from Figure 6.18, for the chosen set of parameters, both solutions agree well.

*Case* $\phi(\mu) = \tau\delta(\mu-\alpha) + \delta(\mu-\beta)$. The wave-diffusion equation with two fractional derivatives of different order, or the fractional telegraph equation, which was studied in [ATA 07b], is obtained by setting $\mu = 0$, $\eta = 2$, $\phi(\mu) = \tau\delta(\mu-\alpha) + \delta(\mu-\beta)$ for $0 < \beta < \alpha < 2$, $\mathcal{A} = \frac{\partial^2}{\partial x^2}$, $F = 0$, $v_0 = 0$ and $u_0 = 1$ in [6.111] and [6.112].

We use [6.120] to obtain the function $\tilde{B}_u$ and then calculate $B_u = \mathcal{L}^{-1}\left[\tilde{B}_u\right]$

$$\begin{aligned}
\tilde{B}_u(s) &= \int_0^2 (\tau\delta(\mu-\alpha) + \delta(\mu-\beta))\, s^{\mu}\mathrm{d}\mu = \tau s^{\alpha} + s^{\beta}, \quad s \in \mathbb{C}_+, \\
B_u(t) &= \tau\frac{1}{\Gamma(-\alpha)}\frac{1}{t^{\alpha+1}} + \frac{1}{\Gamma(-\beta)}\frac{1}{t^{\beta+1}}, \quad t > 0.
\end{aligned}$$

The explicit form of solutions can also be constructed. In the case when $0 < \beta < \alpha < 1$, the solution kernel $g$ given by [6.134] has the form

$$\begin{aligned}
g(x,t) &= \frac{1}{\left(\sqrt{4\pi}\right)^n \pi}\int_0^\infty\int_0^\infty \frac{1}{\left(\sqrt{p}\right)^n}\sin\left(p\left(\tau q^{\alpha}\sin(\alpha\pi) + q^{\beta}\sin(\beta\pi)\right)\right) \\
&\quad \times \mathrm{e}^{-p\left(\tau q^{\alpha}\cos(\alpha\pi) + q^{\beta}\cos(\beta\pi)\right) - qt - \frac{x^2}{4p}}\mathrm{d}q\mathrm{d}p, \quad x = |\boldsymbol{x}|, \ \boldsymbol{x} \in \mathbb{R}^n, \ t > 0,
\end{aligned}$$

where we used that $J(s,t,p) = \mathrm{e}^{-p(\tau s^{\alpha}+s^{\beta})+st}$, $s \in \mathbb{C}_+$, $t > 0$, $p \geq 0$, calculated by [6.132], has no singularities in the right complex plane and its only branch points are $s = 0$ and $s = \infty$, i.e. it satisfies condition 6.1. Thus, using [6.128], we obtain

$$\begin{aligned} u(x,t) &= \frac{1}{\left(\sqrt{4\pi}\right)^n \pi} \cdot H(t) *_t \left(\tau \frac{1}{\Gamma(-\alpha)} \frac{1}{t^{\alpha+1}} + \frac{1}{\Gamma(-\beta)} \frac{1}{t^{\beta+1}}\right) \\ &*_t \int_0^\infty \int_0^\infty \frac{1}{\left(\sqrt{p}\right)^n} \sin\left(p\left(\tau q^{\alpha} \sin(\alpha\pi) + q^{\beta} \sin(\beta\pi)\right)\right) \\ &\times \mathrm{e}^{-p\left(\tau q^{\alpha} \cos(\alpha\pi) + q^{\beta} \cos(\beta\pi)\right) - qt - \frac{x^2}{4p}} \mathrm{d}q \mathrm{d}p, \quad x = |\boldsymbol{x}|, \ \boldsymbol{x} \in \mathbb{R}^n, \ t > 0. \end{aligned} \quad [6.147]$$

The solution kernel for the same case and $n = 1$ is given via [6.138] by ($x \in \mathbb{R}$, $t > 0$)

$$g_1(x,t) = \frac{1}{4\pi \mathrm{i}} \int_0^\infty \left( \frac{\mathrm{e}^{-|x|\sqrt{\tau q^{\alpha} \mathrm{e}^{-\mathrm{i}\alpha\pi} + q^{\beta} \mathrm{e}^{-\mathrm{i}\beta\pi}}}}{\sqrt{\tau q^{\alpha} \mathrm{e}^{-\mathrm{i}\alpha\pi} + q^{\beta} \mathrm{e}^{-\mathrm{i}\beta\pi}}} - \frac{\mathrm{e}^{-|x|\sqrt{\tau q^{\alpha} \mathrm{e}^{\mathrm{i}\alpha\pi} + q^{\beta} \mathrm{e}^{\mathrm{i}\beta\pi}}}}{\sqrt{\tau q^{\alpha} \mathrm{e}^{\mathrm{i}\alpha\pi} + q^{\beta} \mathrm{e}^{\mathrm{i}\beta\pi}}} \right) \mathrm{e}^{-qt} \mathrm{d}q. \quad [6.148]$$

Note that the function $\tilde{g}_1(x,s) = \frac{\mathrm{e}^{-|x|\sqrt{\tau s^{\alpha}+s^{\beta}}}}{2\sqrt{\tau s^{\alpha}+s^{\beta}}}$, $x \in \mathbb{R}$, $s \in \mathbb{C}_+$, calculated by [6.136], has no singularities in the right complex plane and no branch points when $0 < \beta < \alpha < 1$, i.e. it satisfies condition 6.2. A solution can now be computed by using [6.128]

$$\begin{aligned} u(x,t) &= \frac{1}{4\pi \mathrm{i}} \cdot H(t) *_t \left(\tau \frac{1}{\Gamma(-\alpha)} \frac{1}{t^{\alpha+1}} + \frac{1}{\Gamma(-\beta)} \frac{1}{t^{\beta+1}}\right) \\ &*_t \int_0^\infty \left( \frac{\mathrm{e}^{-|x|\sqrt{\tau q^{\alpha} \mathrm{e}^{-\mathrm{i}\alpha\pi} + q^{\beta} \mathrm{e}^{-\mathrm{i}\beta\pi}}}}{\sqrt{\tau q^{\alpha} \mathrm{e}^{-\mathrm{i}\alpha\pi} + q^{\beta} \mathrm{e}^{-\mathrm{i}\beta\pi}}} - \frac{\mathrm{e}^{-|x|\sqrt{\tau q^{\alpha} \mathrm{e}^{\mathrm{i}\alpha\pi} + q^{\beta} \mathrm{e}^{\mathrm{i}\beta\pi}}}}{\sqrt{\tau q^{\alpha} \mathrm{e}^{\mathrm{i}\alpha\pi} + q^{\beta} \mathrm{e}^{\mathrm{i}\beta\pi}}} \right) \mathrm{e}^{-qt} \mathrm{d}q, \\ & x \in \mathbb{R}, \ t > 0. \end{aligned} \quad [6.149]$$

For $x = 1.232$, $t \in (0,8)$, $\alpha = 0.9$, $\beta = 0.56$, $\tau = 10^{-3}$, the plot of equation [6.149] is shown in Figure 6.19 by the solid line. In the same figure, we show

$u(1.232, t)$, denoted by dots, obtained by using solution ($x \in \mathbb{R}$, $t > 0$)

$$u_{APZ}(x,t) = \frac{1}{2\sqrt{\mathcal{D}}}\sum_{n=0}^{\infty}\sum_{k=0}^{\infty}\left(-\frac{|x|}{\sqrt{\mathcal{D}}}\right)^n \tau^k \frac{\Gamma\left(\frac{n+3}{2}\right)}{n!k!\Gamma\left(\frac{n+3}{2}-k\right)} \times \frac{t^{-\left((\alpha-\beta)k+\beta\frac{n+1}{2}\right)}}{\Gamma\left(-\left[(\alpha-\beta)k+\beta\frac{n+1}{2}-1\right]\right)},$$

which is given in [ATA 07b].

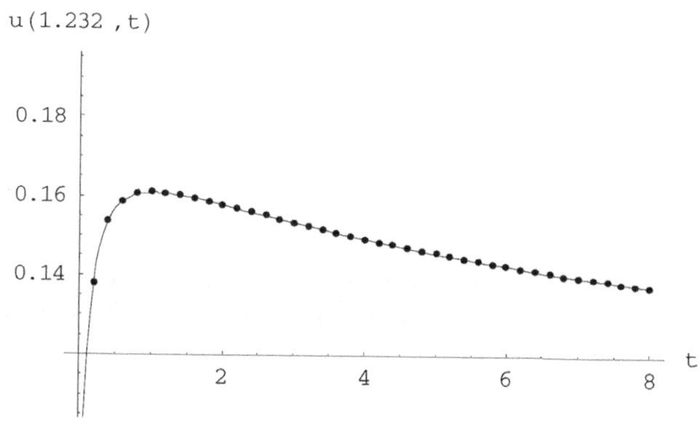


**Figure 6.19.** *Plots of the solutions to the fractional telegraph equation*

Again, for the selected set of parameters, both solutions agree well.

*Case* $\phi(\alpha) = \tau^\alpha$. This is the simplest case guaranteeing the dimensional homogeneity. Parameter $\tau > 0$ can be physically interpreted as the relaxation time.

Functions $\tilde{B}_u$, for $\operatorname{supp}\phi \subset (0,1)$ and $\operatorname{supp}\phi \subset (0,2)$, and $\tilde{B}_v$ can now be computed by using [6.120] as

$$\tilde{B}_u(s) = \int_0^1 (\tau s)^\alpha \, d\alpha = \frac{\tau s - 1}{\ln(\tau s)}, \quad s \in \mathbb{C}_+,$$

$$\tilde{B}_u(s) = \int_0^2 (\tau s)^\alpha \, d\alpha = \frac{(\tau s)^2 - 1}{\ln(\tau s)}, \quad \tilde{B}_v(s) = \int_1^2 (\tau s)^\alpha \, d\alpha = \frac{(\tau s)^2 - \tau s}{\ln(\tau s)}. \qquad [6.150]$$

Since $\varphi(t) = \mathcal{L}^{-1}\left[\frac{1}{\ln(\tau s)}\right](t) = \int_0^\infty \frac{t^{u-1}\mathrm{e}^{-u\ln\tau}}{\Gamma(u)}du$, $t > 0$ (see [ATA 05a]), Laplace inversions of $\tilde{B}_u$ and $\tilde{B}_v$ give

$$B_u(t) = \tau\frac{\mathrm{d}}{\mathrm{d}t}\varphi(t) - \varphi(t), \quad \alpha \in (0,1), \; t > 0,$$

$$B_u(t) = \tau^2\frac{\mathrm{d}^2}{\mathrm{d}t^2}\varphi(t) - \varphi(t), \; B_v(t) = \tau^2\frac{\mathrm{d}^2}{\mathrm{d}t^2}\varphi(t) - \tau\frac{\mathrm{d}}{\mathrm{d}t}\varphi(t), \quad \alpha \in (0,2), \tag{6.151}$$

where we used $(f * g)^{(n)} = f^{(n)} * g = f * g^{(n)}$.

*Case* $\phi(\alpha) = \tau^\alpha$, $\mu = 0 \leq \alpha \leq \eta = 1$. The function $J$ calculated by [6.132] reads

$$J(s,t,p) = \mathrm{e}^{-p\frac{\tau s - 1}{\ln(\tau s)} + st}, \quad s \in \mathbb{C}_+, \; t > 0, \; p \geq 0.$$

It satisfies condition 6.1, i.e. its branch points are $s = 0$ and $s = \infty$, and it has no singularities in the right complex plane, so the solution kernel $g$ is given by [6.134]. After computing the integrals $\int_0^1 (\tau q)^\alpha \sin(\alpha\pi)\,\mathrm{d}\alpha = \pi\frac{1+\tau q}{\pi^2+\ln^2(\tau q)}$ and $\int_0^1 (\tau q)^\alpha \cos(\alpha\pi)\,\mathrm{d}\alpha = -\frac{(1+\tau q)\ln(\tau q)}{\pi^2+\ln^2(\tau q)}$, required by [6.134], the solution kernel $g$ becomes

$$\begin{aligned} g(x,t) &= \frac{1}{\left(\sqrt{4\pi}\right)^n \pi}\int_0^\infty\int_0^\infty \frac{1}{\left(\sqrt{p}\right)^n} \sin\left(p\pi\frac{1+\tau q}{\pi^2+\ln^2(\tau q)}\right) \\ &\quad \times \mathrm{e}^{p\frac{(1+\tau q)\ln(\tau q)}{\pi^2+\ln^2(\tau q)} - qt - \frac{x^2}{4p}} \mathrm{d}q\mathrm{d}p, \quad x = |\boldsymbol{x}|, \; \boldsymbol{x} \in \mathbb{R}^n, \; t > 0. \end{aligned} \tag{6.152}$$

The solution is given by [6.127], where $B_u$ and $g$ are given by [6.151] and [6.152], respectively.

*Case* $\phi(\alpha) = \tau^\alpha$, $\mu = 0 \leq \alpha \leq \eta = 2$. We treat the one-dimensional case only. Since $\tilde{g}_1$ (see [6.136])

$$\tilde{g}_1(x,s) = \frac{\mathrm{e}^{-|x|\sqrt{\frac{(\tau s)^2-1}{\ln(\tau s)}}}}{2\sqrt{\frac{(\tau s)^2-1}{\ln(\tau s)}}}, \quad x \in \mathbb{R}, \; s \in \mathbb{C}_+, \tag{6.153}$$

satisfies condition 6.2, [6.138] implies that

$$g_1(x,t) = \frac{1}{4\pi i}\int_0^\infty \left( \frac{e^{-|x|\sqrt{\frac{(\tau q e^{-i\pi})^2-1}{\ln(\tau q e^{-i\pi})}}}}{\sqrt{\frac{(\tau q e^{-i\pi})^2-1}{\ln(\tau q e^{-i\pi})}}} - \frac{e^{-|x|\sqrt{\frac{(\tau q e^{i\pi})^2-1}{\ln(\tau q e^{i\pi})}}}}{\sqrt{\frac{(\tau q e^{i\pi})^2-1}{\ln(\tau q e^{i\pi})}}} \right) e^{-qt} dq,$$

$$x \in \mathbb{R},\ t > 0. \qquad [6.154]$$

The solution itself is given by [6.128], where $B_u$, $B_v$ and $g_1$ are given by [6.151] and [6.154], respectively. This generalizes the case treated in [MAI 07b], where $\tau = 1$, $\eta = 1$.

6.3.2.4. *Solution kernel via the post-inversion formula and numerical examples*

The inversion of the Laplace transform can also be obtained by the use of the post-inversion formula. Some recent articles on the subject are [DON 02, FRO 98].

The post-inversion formula yields the original function $f$ of exponential growth from its Laplace transform $\tilde{f}$ via

$$f(t) = \lim_{n\to\infty} \frac{(-1)^n}{n!} s^{n+1} \frac{d^n}{ds^n} \tilde{f}(s)\bigg|_{s=\frac{n}{t}}, \quad t > 0. \qquad [6.155]$$

Note that the first approximation of the original function is obtained from [6.155] for $n = 0$ as $f(t) = s\tilde{f}(s)\big|_{s=\frac{n}{t}}$, $t > 0$. Although the post-inversion formula can be regarded as an analytical result, it can also be used for numerical purposes, where the differentiation can be stopped when required accuracy on values of function $f$ is attained.

We apply [6.155] to the Laplace transform of the solution kernel of the distributed diffusion-wave equation [6.118] in the case when $\phi(\alpha) = \tau^\alpha$, $\alpha \in (0,2)$, for one- and two-dimensional solution kernels $\tilde{g}_1$ and $\tilde{g}_2$, given by [6.153] and [6.140], respectively. For these cases, $\tilde{B}_u$ is given by [6.150]. Therefore, [6.155] yields

$$g_1(x,t) = \frac{1}{2}\lim_{n\to\infty}\frac{(-1)^n}{n!}s^{n+1}\frac{d^n}{ds^n}\left.\frac{e^{-|x|\sqrt{\frac{(\tau s)^2-1}{\ln(\tau s)}}}}{\sqrt{\frac{(\tau s)^2-1}{\ln(\tau s)}}}\right|_{s=\frac{n}{t}}, \quad x\in\mathbb{R},\ t>0, \qquad [6.156]$$

$$g_2(x,t) = \frac{1}{2\pi}\lim_{n\to\infty}\frac{(-1)^n}{n!}s^{n+1}\frac{d^n}{ds^n}K_0\left.\left(x\sqrt{\frac{(\tau s)^2-1}{\ln(\tau s)}}\right)\right|_{s=\frac{n}{t}}, \quad x\in\mathbb{R},\ t>0. \qquad [6.157]$$

Figure 6.20 shows plots of $g_1$ obtained by the use of the analytical expression [6.154] (denoted by dots) and by the use of [6.156] (denoted by the solid line), with $n=15$, for $x=1$, $t\in(0,5)$, $\tau=10^{-3}$.

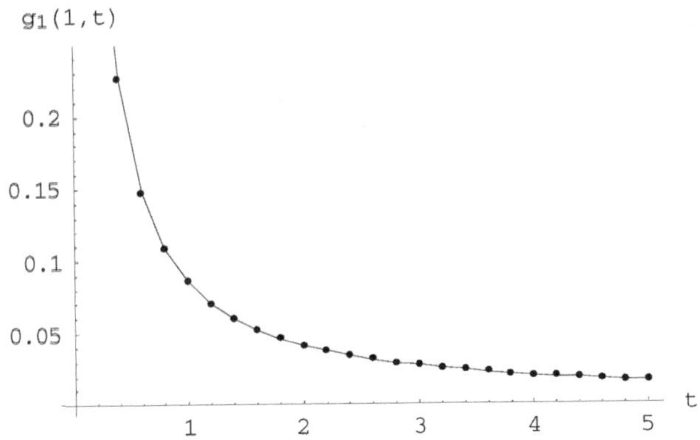


**Figure 6.20.** *Plots of the solution kernel* $g_1(1,t)$

In Figure 6.21, we plot $g_2$ obtained by [6.157] with $n=10$, for the set of values $x=1$, $t\in(0,5)$, $\tau=10^{-3}$.

Derivatives in both [6.156] and [6.157] are calculated symbolically using Mathematica.

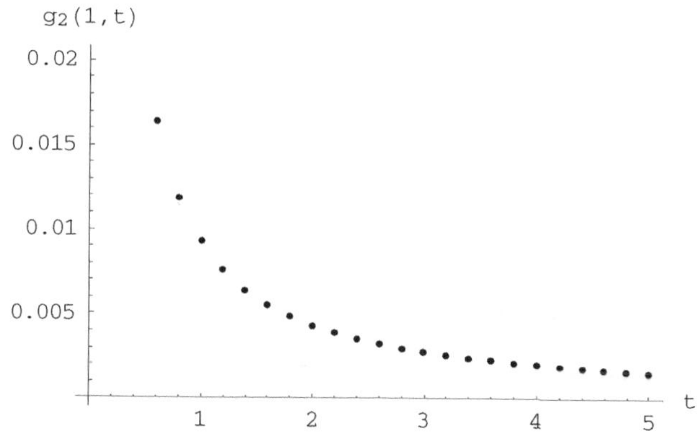


**Figure 6.21.** *Plot of the solution kernel $g_2\,(1,t)$*

## 6.4. Maximum principle for fractional diffusion-wave-type equations

The maximum principle holds true for the classical heat conduction equation

$$\frac{\partial}{\partial t}T\left(\boldsymbol{x},t\right)=\mathcal{D}\,\nabla^2 T\left(\boldsymbol{x},t\right).$$

We will follow [ATA 07b, ATA 09d] and prove the analog of the maximum principle for the fractional telegraph and distributed-order diffusion equation.

### 6.4.1. *Maximum principle for fractional telegraph equation*

We consider the fractional telegraph equation [6.43] when $0\leq\beta\leq\alpha\leq1$. Let $D$ be a rectangle in the $x,t$ plane $D=\{(x,t)\mid 0<x<l,\ 0<t<\hat{t}\}$ with $\hat{t}>0$ given. We denote by $\bar{D}$ the closure of $D$ and by $\gamma$ the part of the boundary of $D$ consisting of the sides $t=0$, $x=0$ and $x=l$. This is a closed set.

We state the following theorem that is an analog of the classical maximum principle. Recall $\tau>0,\ \mathcal{D}>0$.

THEOREM 6.10.– Suppose that $T$ is given and:

– satisfies

$$\tau\,{}_0\mathrm{D}_t^{\alpha}T\left(x,t\right)+{}_0\mathrm{D}_t^{\beta}T\left(x,t\right)=\mathcal{D}\frac{\partial^2}{\partial x^2}T\left(x,t\right),\quad (x,t)\in D, \qquad [6.158]$$

– is continuous in both variables for $(x,t)\in\bar{D}$ and $T\left(x,\cdot\right)$ is Hölder continuous with the exponent $h>\alpha$ for $t\in\left[0,\hat{t}\right]$,

– satisfies the following initial and boundary conditions

$$T(x,0) = g(x), \quad T(0,t) = f_1(t), \quad T(l,t) = f_2(t), \qquad [6.159]$$

– the functions $g$, $f_1$ and $f_2$ are continuous and satisfy

$$g(0) = f_1(0), \quad g(l) = f_2(0). \qquad [6.160]$$

Then the maximum and minimum values of $T$ over a region $\bar{D} = \{(x,t) \mid 0 \leq x \leq l,\ 0 \leq t \leq \hat{t}\}$ are assumed on $\gamma$.

PROOF.– It is enough to prove the theorem for the maximum. Since $T$ is continuous, it reaches its maximum in $\bar{D}$. If $T$ reaches its maximum at the point $(x^*, t^*)$ with $0 < x^* < l, 0 < t^* < \hat{t}$, then at this point we must have

$$\frac{\partial^2}{\partial x^2} T(x^*, t^*) \leq 0. \qquad [6.161]$$

Also, from the generalized Fermat's theorem for stationary points (see [NAH 03, p. 103] and proposition 2.19), at the point $t^*$ where $T(x, \cdot)$ reaches maximum, the inequality

$$\left[{}_0\mathrm{D}_t^{\alpha} T(x,t)\right]_{(x,t)=(x^*,t^*)} \geq 0 \qquad [6.162]$$

holds for each $\alpha \in (0,1)$. If at least in one of [6.161] or [6.162] we have the strict inequality, then [6.43] is not satisfied at $(x^*, t^*)$ since the relaxation time $\tau$ is positive. In the case when $T = T_0 = const.$, we have

$${}_0\mathrm{D}_t^{\alpha} T_0 = \frac{T_0}{\Gamma(1-\alpha)} t^{-\alpha} \quad \text{and} \quad \frac{\partial^2}{\partial x^2} T_0 = 0,$$

again leading to contradiction that [6.158] is satisfied in $D$.

So, suppose that in [6.161] and [6.162], we have equality. Suppose further that the maximum of $T$ on the boundary $\gamma$ is $M$ while the maximum of $T$ in $D$ is attained at the point $(x^*, t^*)$ and that $T(x^*, t^*) = M + \varepsilon$, $\varepsilon > 0$. We will show that this leads to a contradiction. Consider a function

$$U(x,t) = T(x,t) - kt^{\alpha}, \quad (x,t) \in \bar{D}.$$

Choose $k > 0$ so that $k < \frac{\varepsilon}{2(\hat{t})^{\alpha}}$. Then, $\sup_{(x,t)\in D} U(x,t) \geq M + \frac{\varepsilon}{2}$ and $\sup_{(x,t)\in\gamma} U(x,t) \leq M$. Since $U$ is continuous in $\bar{D}$, it reaches its maximum at the point $(x_1, t_1) \in D$ (this follows from the assumption on $(x^*, t^*)$). We will show that $U$ cannot have the maximum in $D$. This will imply that the hypothesis $(x^*, t^*) \in D$ is wrong and the assertion will be proved. As in [6.161] and [6.162], $\frac{\partial^2}{\partial x^2} U(x_1, t_1) \le 0$, $\left[{}_0\mathrm{D}_t^{\alpha} U(x,t)\right]_{(x,t)=(x_1,t_1)} \ge 0$, $\left[{}_0\mathrm{D}_t^{\beta} U(x,t)\right]_{(x,t)=(x_1,t_1)} \ge 0$. We have

$$\begin{aligned} \frac{\partial^2}{\partial x^2} U(x_1, t_1) &= \frac{\partial^2}{\partial x^2} T(x_1, t_1), \\ \left[{}_0\mathrm{D}_t^{\alpha} U(x,t)\right]_{(x,t)=(x_1,t_1)} &= \left[{}_0\mathrm{D}_t^{\alpha} T(x,t)\right]_{(x,t)=(x_1,t_1)} - k\Gamma(1+\alpha), \\ \left[{}_0\mathrm{D}_t^{\beta} U(x,t)\right]_{(x,t)=(x_1,t_1)} &= \left[{}_0\mathrm{D}_t^{\beta} T(x,t)\right]_{(x,t)=(x_1,t_1)} - k\frac{\Gamma(1+\alpha)}{\Gamma(\alpha+1-\beta)} t_1^{\alpha-\beta}. \end{aligned} \quad [6.163]$$

Thus,

$$\tau \left[{}_0\mathrm{D}_t^{\alpha} T(x,t)\right]_{(x,t)=(x_1,t_1)} + \left[{}_0\mathrm{D}_t^{\beta} T(x,t)\right]_{(x,t)=(x_1,t_1)} - \mathcal{D}\frac{\partial^2}{\partial x^2} T(x_1, t_1) \ge 0,$$

since at $(x_1, t_1)$ we have the maximum. By [6.163] and [6.158]

$$\begin{aligned} &\tau \left[{}_0\mathrm{D}_t^{\alpha} T(x,t)\right]_{(x,t)=(x_1,t_1)} + \left[{}_0\mathrm{D}_t^{\beta} T(x,t)\right]_{(x,t)=(x_1,t_1)} - \mathcal{D}\frac{\partial^2}{\partial x^2} T(x_1, t_1) \\ &\quad = -k\Gamma(1+\alpha)\left(1 + \frac{t_1^{\alpha-\beta}}{\Gamma(\alpha+1-\beta)}\right) < 0. \end{aligned}$$

This is a contradiction, $(x_1, t_1) \notin D$ and thus $(x^*, t^*) \in \gamma$. ■

REMARK 6.3.– The condition that $T$ is continuous in $\bar{D}$ can be relaxed by requiring that $T$ is continuous in $D$ and that for $(x,t) \in \gamma$, the limit inferior, $\liminf T(x,t)$, exists (see [WID 75]).

PROPOSITION 6.6.– If $T$ is continuous in $D$ and for $(x,t) \in \gamma$, $\liminf T(x,t)$ exists, then $T$ reaches its maximum and minimum on $\gamma$.

PROOF.– We will assume that $\liminf_{(x,t)\to(x_0,t_0)} T(x,t) \ge 0$ for all $(x_0, t_0) \in \gamma$ and by this, we will show that $T(x,t) \ge 0$, $(x,t) \in D$. This implies the assertion in the case of the non-negative minimum and thus we will have the assertion in the arbitrary case.

Let us assume that $\liminf_{(x,t)\to(x_0,t_0)} T(x,t) \geq 0$, for all $(x_0,t_0) \in \gamma$. By the definition of the limit inferior, this means that for all $\varepsilon > 0$, there exists $\delta_\varepsilon$ such that $(x,t) \in L_{\delta_\varepsilon}((x_0,t_0),\delta_\varepsilon)$ implies $T(x,t) \geq -\varepsilon$ where $L_{\delta_\varepsilon}((x_0,t_0),\delta_\varepsilon)$ denotes a ball with radius $\delta_\varepsilon$ centered at $(x_0,t_0)$. Fix $\varepsilon$ and for every $(x_0,t_0) \in \gamma$ find $L_{\delta_\varepsilon}$. Due to the compactness, there exist a finite number of balls with radii $\delta_{\varepsilon_i}$, $i = 1,...,n$, which cover $\gamma$; find $\delta = \min_{i=1,...,n} \delta_{\varepsilon_i}$. Consider a rectangle $D_0 \subset D$ with the boundary $\gamma_0$ that lies on the distance $\delta_0 < \delta$ from $\gamma$ so that $T(x,t) \geq -\varepsilon$, $(x,t) \in D_0$. Letting $\varepsilon \to 0$, we obtain $T(x,t) \geq 0$ in $D$. ■

Theorem 6.10 is used to prove the uniqueness problem ([6.158] and [6.159]).

PROPOSITION 6.7.– The solution to the initial boundary value problem ([6.158] and [6.159]) for $0 \leq \beta \leq \alpha \leq 1$ is unique.

PROOF.– Let $T_1$ and $T_2$ be two solutions to [6.158] and [6.159]. By applying theorem 6.10 to $T = T_1 - T_2$, we obtain that $T_1 = T_2$. ■

### 6.4.2. *Maximum principle for distributed-order diffusion equation*

In this section, we will extend the results presented in [ATA 07b, LUC 09] and prove the maximum principle for a distributed diffusion equation. It extends the classical maximum principle for parabolic partial differential equations (see [WID 75], for example).

We consider a special case of [6.111]

$$\int_\mu^\eta \phi(\alpha)\, {}_0^C\mathrm{D}_t^\alpha T(\boldsymbol{x},t)\,\mathrm{d}\alpha = \nabla^2 T(\boldsymbol{x},t), \quad (\boldsymbol{x},t) \in D \times \left[0,\hat{t}\right], \qquad [6.164]$$

where $\phi$ is continuous and non-negative for $\alpha \in [\mu,\eta]$, $0 \leq \mu < \eta \leq 1$. We use the notation $\boldsymbol{x} = (x_1,...,x_n) \in D \subset \mathbb{R}^n$, $t \in \left[0,\hat{t}\right]$, and $D$ is a bounded domain in $\mathbb{R}^n$. We denote by $\partial D$ the boundary of $D$ and $\bar{D} = D \cup \partial D$.

THEOREM 6.11.– Suppose that $T$ satisfies [6.164], that it is continuous for $(\boldsymbol{x},t) \in \bar{D} \times \left[0,\hat{t}\right]$ and that $T(\boldsymbol{x},\cdot)$, $\boldsymbol{x} \in \bar{D}$, is Hölder continuous with the exponent $h > \eta$ for $t \in \left[0,\hat{t}\right]$. Moreover, assume that $T$ satisfies initial and boundary conditions

$$T\left(\boldsymbol{x},0^+\right) = g(\boldsymbol{x}), \quad \boldsymbol{x} \in D, \quad T(\boldsymbol{x},t) = f(\boldsymbol{x},t), \quad \boldsymbol{x} \in \partial D,$$

where $g$ and $f$ are continuous.

Then, the maximum and minimum values of $T$ over the region $\bar{D} \times [0, \hat{t}]$ are reached either on $\partial D \times [0, \hat{t}]$ or on $\gamma \equiv \bar{D} \cup \{0\}$.

PROOF.– It is enough to prove the theorem for a positive maximum. Since $T(\boldsymbol{x}, t)$ is continuous, it reaches its maximum in $\bar{D} \times [0, \hat{t}]$. If $T$ reaches its maximum at the point $(\boldsymbol{x}^*, t^*)$, $\boldsymbol{x}^* \in D$, $0 < t^* < \hat{t}$, then at this point we must have

$$\frac{\partial^2}{\partial x_i^2} T(\boldsymbol{x}^*, t^*) \leq 0, \quad (\boldsymbol{x}^*, t^*) \in D \times [0, \hat{t}], \quad i = 1, \ldots, n, \qquad [6.165]$$

so that

$$\nabla^2 T(\boldsymbol{x}^*, t^*) \leq 0, \quad (\boldsymbol{x}^*, t^*) \in D \times [0, \hat{t}]. \qquad [6.166]$$

Also, from proposition 2.19 at the point $t^*$, where $T$ reaches the positive maximum, the inequality

$$\left[ {}_0^C\mathrm{D}_t^{\alpha} T(\boldsymbol{x}, t) \right]_{(\boldsymbol{x},t)=(\boldsymbol{x}^*,t^*)} \geq 0, \quad (\boldsymbol{x}^*, t^*) \in D \times [0, \hat{t}] \qquad [6.167]$$

holds for each $\alpha \in (0, 1)$. Then, since $\phi(\alpha) \geq 0$, $\alpha \in [\mu, \eta]$, we have

$$\int_\mu^\eta \phi(\alpha) \left[ {}_0^C\mathrm{D}_t^{\alpha} T(\boldsymbol{x}, t) \right]_{(\boldsymbol{x},t)=(\boldsymbol{x}^*,t^*)} \mathrm{d}\alpha \geq 0, \quad (\boldsymbol{x}^*, t^*) \in D \times [0, \hat{t}]. \qquad [6.168]$$

If at least in one of [6.166] or [6.168] the strict inequality holds, then [6.164] is not satisfied at $(\boldsymbol{x}^*, t^*)$.

So, suppose that in [6.166] and [6.168], we have equalities. Suppose further that the maximum of $T$ on $\partial D \times [0, \hat{t}]$ or at $\bar{D} \times \{0\}$ is $M$, while the maximum of $T$ in $D \times [0, \hat{t}]$ is attained at the point $(\boldsymbol{x}^*, t^*)$ and that $T(\boldsymbol{x}^*, t^*) = M + \varepsilon$, $\varepsilon > 0$. We will show that this, again, leads to a contradiction. Consider

$$U(\boldsymbol{x}, t) = T(\boldsymbol{x}, t) - k t^{\bar{\alpha}}, \quad 0 < \bar{\alpha} < 1.$$

Clearly

$$U(\boldsymbol{x}^*, t^*) = T(\boldsymbol{x}^*, t^*) - k (t^*)^{\bar{\alpha}}.$$

Choose $k > 0$ so that $k < \frac{\varepsilon}{2(\hat{t})^{\bar{\alpha}}}$. Then

$$\sup_{(\boldsymbol{x},t)\in D\times\left[0,\hat{t}\right]} U(\boldsymbol{x},t) \geq M + \frac{\varepsilon}{2} \quad \text{and} \quad \sup_{(\boldsymbol{x},t)\in \partial D\times\left[0,\hat{t}\right]\cup\gamma} U(\boldsymbol{x},t) \leq M.$$

Since $U$ is continuous, it reaches its maximum at a point $(\boldsymbol{x}_1, t_1)$, so that $U(\boldsymbol{x}_1, t_1) \geq U(\boldsymbol{x}, t)$ and, as in [6.165] or [6.168],

$$\frac{\partial^2}{\partial x_i^2} U(\boldsymbol{x}_1, t_1) \leq 0, \qquad \int_\mu^\eta \phi(\alpha) \left[{}_0^C\mathrm{D}_t^\alpha U(\boldsymbol{x},t)\right]_{(\boldsymbol{x},t)=(\boldsymbol{x}_1,t_1)} \mathrm{d}\alpha \geq 0.$$

We will show that $U$ cannot have a maximum in $D \times \left[0, \hat{t}\right]$ at $(\boldsymbol{x}_1, t_1)$. This will imply that the hypothesis $(\boldsymbol{x}^*, t^*) \in D \times \left[0, \hat{t}\right]$ is wrong and the assertion will be proved. Since

$$\frac{\partial^2}{\partial x_i^2} U(\boldsymbol{x}_1, t_1) = \frac{\partial^2}{\partial x_i^2} T(\boldsymbol{x}_1, t_1),$$

$$\left[{}_0^C\mathrm{D}_t^\alpha U(\boldsymbol{x},t)\right]_{(\boldsymbol{x},t)=(\boldsymbol{x}_1,t_1)} = \left[{}_0^C\mathrm{D}_t^\alpha T(\boldsymbol{x},t)\right]_{(\boldsymbol{x},t)=(\boldsymbol{x}_1,t_1)} - k\frac{\Gamma(1+\bar{\alpha})}{\Gamma(\bar{\alpha}+1-\alpha)} t_1^{\bar{\alpha}-\alpha}, \tag{6.169}$$

it follows that

$$\int_\mu^\eta \phi(\alpha) \left[{}_0^C\mathrm{D}_t^\alpha T(\boldsymbol{x},t)\right]_{(\boldsymbol{x},t)=(\boldsymbol{x}_1,t_1)} \mathrm{d}\alpha$$
$$= \int_\mu^\eta \phi(\alpha) \left[{}_0^C\mathrm{D}_t^\alpha U(\boldsymbol{x},t)\right]_{(\boldsymbol{x},t)=(\boldsymbol{x}_1,t_1)} \mathrm{d}\alpha$$
$$+ \int_\mu^\eta k\phi(\alpha) \frac{\Gamma(1+\bar{\alpha})}{\Gamma(\bar{\alpha}+1-\alpha)} t_1^{\bar{\alpha}-\alpha} \mathrm{d}\alpha. \tag{6.170}$$

Also, since at $(\boldsymbol{x}_1, t_1)$ we have the maximum of $U$, we must have (see [6.165], [6.167] and [6.168])

$$\int_\mu^\eta \phi(\alpha) \left[{}_0^C\mathrm{D}_t^\alpha U(\boldsymbol{x},t)\right]_{(\boldsymbol{x},t)=(\boldsymbol{x}_1,t_1)} \mathrm{d}\alpha \geq 0, \qquad \frac{\partial^2}{\partial x_i^2} U(\boldsymbol{x}_1, t_1) \leq 0,$$

or

$$\int_{\mu}^{\eta} \phi(\alpha) \left[{}_0^C \mathrm{D}_t^{\alpha} U(\boldsymbol{x}, t)\right]_{(\boldsymbol{x},t)=(\boldsymbol{x}_1,t_1)} \mathrm{d}\alpha - \nabla^2 U(\boldsymbol{x}_1, t_1) \geq 0. \qquad [6.171]$$

By putting [6.169] and [6.170] in [6.171], we obtain

$$\int_{\mu}^{\eta} \phi(\alpha) \left[{}_0^C \mathrm{D}_t^{\alpha} T(\boldsymbol{x}, t)\right]_{(\boldsymbol{x},t)=(\boldsymbol{x}_1,t_1)} \mathrm{d}\alpha - \nabla^2 T(\boldsymbol{x}_1, t_1) - \int_{\mu}^{\eta} k\phi(\alpha) \frac{\Gamma(1+\bar{\alpha})}{\Gamma(\bar{\alpha}+1-\alpha)} t_1^{\bar{\alpha}-\alpha} \mathrm{d}\alpha \geq 0,$$

or

$$\int_{\mu}^{\eta} \phi(\alpha) \left[{}_0^C \mathrm{D}_t^{\alpha} T(\boldsymbol{x}, t)\right]_{(\boldsymbol{x},t)=(\boldsymbol{x}_1,t_1)} \mathrm{d}\alpha - \nabla^2 T(\boldsymbol{x}_1, t_1) \geq \int_{\mu}^{\eta} k\phi(\alpha) \frac{\Gamma(1+\bar{\alpha})}{\Gamma(\bar{\alpha}+1-\alpha)} t_1^{\bar{\alpha}-\alpha} \mathrm{d}\alpha > 0.$$

Thus, at $(\boldsymbol{x}_1, t_1)$, equation [6.164] is not satisfied and a maximum of $T$ is not attained in $D \times \left[0, \hat{t}\right]$. ■

The condition that $T$ is continuous in $\bar{D} \times \left[0, \hat{t}\right]$ can be relaxed by requiring that $T$ is continuous in $D \times \left[0, \hat{t}\right]$ and that for $(\boldsymbol{x}_0, t_0) \in \partial D \times \left[0, \hat{t}\right] \cup \gamma$, the limit inferior $\liminf T$ exists, i.e. if $\epsilon > 0$, then for each $(\boldsymbol{x}_0, t_0) \in \partial D \times \left[0, \hat{t}\right] \cup \gamma$, there is a $\delta_0$ such that $T(\boldsymbol{x}, t) \geq -\epsilon$ for all $(\boldsymbol{x}, t) \in D \times \left[0, \hat{t}\right]$ within the distance $\delta_0$ of $(\boldsymbol{x}_0, t_0)$. Applying the same argument as in [ATA 07b] (see also [WID 75]), we can prove the following assertion.

PROPOSITION 6.8.– If $T$ is continuous in $D \times \left[0, \hat{t}\right]$ and for $(\boldsymbol{x}_0, t_0) \in \partial D \times \left[0, \hat{t}\right] \cup \gamma$, $\liminf T$ exists, then $T$ reaches its maximum and minimum on $\partial D \times \left[0, \hat{t}\right] \cup \gamma$.

REMARK 6.4.– Suppose that $g(\boldsymbol{x}) \geq 0$, $\boldsymbol{x} \in D$ and $f(\boldsymbol{x}, t) \geq 0$, $(\boldsymbol{x}, t) \in D \times \left[0, \hat{t}\right]$. Then, by theorem 6.11, $T(\boldsymbol{x}, t) \geq 0$, $(\boldsymbol{x}, t) \in \bar{D} \times \left[0, \hat{t}\right]$.

Chapter 7

# Fractional Heat Conduction Equations

The classical heat conduction equation in one spatial dimension can be written as

$$\frac{\partial}{\partial t}T(x,t) = \mathcal{D}\frac{\partial^2}{\partial x^2}T(x,t), \quad x \in \mathbb{R},\ t > 0, \qquad [7.1]$$

where $T$ is the temperature and $\mathcal{D} = \frac{\lambda}{\rho c}$ is the thermal diffusivity. The heat conduction equation [7.1] is a parabolic partial differential equation. The physical consequence is that the speed of disturbance propagation is infinite.

We refer to [JOS 89] for a review of several generalizations of the heat conduction equation [7.1] using the integer-order derivatives. Among these generalizations, we mention the telegraph equation

$$\begin{aligned} \tau\frac{\partial^2}{\partial t^2}T(x,t) + \frac{\partial}{\partial t}T(x,t) &= \mathcal{D}\frac{\partial^2}{\partial x^2}T(x,t), \\ & x \in \mathbb{R},\ t > 0, \end{aligned} \qquad [7.2]$$

where $\tau$ is the relaxation time. The telegraph equation is a hyperbolic partial differential equation; hence, the speed of disturbance propagation is finite $v = \sqrt{\frac{\mathcal{D}}{\tau}}$. Thus, [7.2] describes the propagation of the heat waves through the media. We also mention the Jeffreys-type heat conduction equation

$$\begin{aligned} \frac{\partial^2}{\partial t^2}T(x,t) + \tau\frac{\partial}{\partial t}T(x,t) &= \mathcal{D}\frac{\partial^2}{\partial x^2}T(x,t) + \mathcal{D}\tau'\frac{\partial}{\partial t}\frac{\partial^2}{\partial x^2}T(x,t), \\ & x \in \mathbb{R},\ t > 0, \end{aligned} \qquad [7.3]$$

where $\tau$ and $\tau'$ are relaxation times. In the classical framework, it generalizes both the heat conduction equation [7.1], which is obtained by setting $\tau = \tau' = 0$ in [7.3], and the telegraph equation [7.2], which is obtained by setting $\tau' = 0$ in [7.3]. The Jeffreys-type heat conduction equation [7.3], although being (like the telegraph equation) the second-order partial differential equation in the time variable, is of the parabolic type, thus having the speed of the disturbance propagation infinite.

Within the framework of fractional calculus, many authors generalize the classical heat conduction [7.1] and telegraph equations [7.2] by substituting the integer-order partial derivative(s) with the derivative(s) of the fractional order (see Chapter 6). In this kind of fractional-type generalizations of [7.1] and [7.2], it is not specified what the corresponding system of balance and constitutive equation is and whether this system is physically acceptable.

Namely, [7.1], [7.2] and [7.3] are obtained from the system consisting of two equations. The first equation is the energy balance equation, which states that the difference between the ingoing and outgoing heat flux $q$ in the infinitesimally small part of the media induces its heating, i.e. the change of its temperature $T$. The energy balance equation for a rigid conductor in a simplified form can be written as

$$\rho c \frac{\partial}{\partial t} T(x,t) = -\frac{\partial}{\partial x} q(x,t), \quad x \in \mathbb{R},\ t > 0, \qquad [7.4]$$

see [SIL 97]. The second equation is the constitutive equation, describing the thermic properties of the rigid conductor, i.e. the material that conducts the heat. We mention three constitutive equations, corresponding to [7.1], [7.2] and [7.3].

– The Fourier heat conduction law

$$q(x,t) = -\lambda \frac{\partial}{\partial x} T(x,t), \quad x \in \mathbb{R},\ t > 0, \qquad [7.5]$$

is the local constitutive equation describing the property of the conductor that the heat flux $q$ vanishes if there is no spatial distribution of temperature (temperature gradient $\frac{\partial}{\partial x}T$) in the conductor, i.e. if, at the same time instant, different parts of the media have the same temperature. Combined with [7.4], the Fourier law leads to the classical heat conduction equation [7.1].

– The Cattaneo heat conduction law

$$\tau \frac{\partial}{\partial t} q(x,t) + q(x,t) = -\lambda \frac{\partial}{\partial x} T(x,t), \quad x \in \mathbb{R},\ t > 0, \qquad [7.6]$$

see [CAT 58, CHE 63]. Equation [7.6] is equivalent to

$$q(x,t) = -\frac{\lambda}{\tau} e^{-\frac{t}{\tau}} *_t \frac{\partial}{\partial x} T(x,t), \quad x \in \mathbb{R},\ t > 0, \qquad [7.7]$$

which is of the form of equation [6.1.20] in [ERI 02, p. 76] so that the Cattaneo law represents a time non-local constitutive equation, i.e. a constitutive equation with memory effects. By combining [7.6] with the energy balance equation [7.4], we obtain the telegraph equation [7.2].

Note that systems [7.4] and [7.7] reduce to the telegraph equation in the form

$$\frac{\partial}{\partial t}T(x,t) = \frac{\mathcal{D}}{\tau}\mathrm{e}^{-\frac{t}{\tau}} *_t \frac{\partial^2}{\partial x^2}T(x,t), \quad x \in \mathbb{R},\ t > 0.$$

– The Jeffreys-type heat conduction law

$$\left(1+\tau\frac{\partial}{\partial t}\right) q(x,t) = -\lambda\left(1+\tau'\frac{\partial}{\partial t}\right)\frac{\partial}{\partial x}T(x,t), \quad x \in \mathbb{R},\ t > 0, \qquad [7.8]$$

see [JOS 89], combined with the energy balance equation [7.4] leads to the Jeffreys-type heat conduction equation [7.3].

In [7.5]–[7.8], we used $\rho > 0$ to denote the density, $c > 0$ is the specific heat capacity of the media, while $\tau, \tau' > 0$ and $\lambda > 0$ are relaxation times and the heat conductivity coefficient.

To avoid ambiguity when introducing the fractional derivatives in heat conduction equations as in Chapter 6, we propose the fractional-type generalizations of the constitutive equations [7.6] and [7.8], while preserving the energy balance equation in the form [7.4]. Similar generalizations of the heat conduction equation are given in [COM 97, QI 11]. We proceed with the fractional generalizations of the constitutive equations.

The space–time fractional Cattaneo heat conduction law reads

$$\tau\, {}_0^C\mathrm{D}_t^{\alpha} q(x,t) + q(x,t) = -\lambda\, \mathcal{E}_x^{\beta} T(x,t), \quad x \in \mathbb{R},\ t > 0, \qquad [7.9]$$

with $\tau > 0$ and $\lambda > 0$ being generalized times of relaxation and the generalized heat conductivity coefficient, as well as $\alpha, \beta \in (0,1)$. Here, $\mathcal{E}_x^{\beta}$ is the symmetrized Caputo fractional derivative given by [2.65]. Constitutive equation [7.9] solved with respect to heat flux $q$ reads

$$q(x,t) = \lambda \frac{\mathrm{d}}{\mathrm{d}t} e_{\alpha}(t) *_t \mathcal{E}_x^{\beta} T(x,t), \quad x \in \mathbb{R},\ t > 0, \qquad [7.10]$$

where $e_{\alpha}(t) = E_{\alpha}(-\frac{t^{\alpha}}{\tau}, t > 0$, with $E_{\alpha}$ being the one-parameter Mittag-Leffler function. Contrary to the case of the Cattaneo heat conduction law [7.7], which implies time non-locality, the constitutive equation [7.10] implies both time and space non-locality. Note that the space–time fractional Cattaneo heat conduction law [7.9] generalizes both Fourier [7.5] and Cattaneo [7.6] heat conduction laws. Indeed, let $\beta \to 1$. Then $\mathcal{E}_x^\beta \to \frac{\partial}{\partial x}$, and if in addition $\alpha \to 1$, then [7.9] reduces to [7.6]. Let again $\beta \to 1$. If $\tau = 0$ or $\alpha = 0$, then [7.9] reduces to [7.5] with $\lambda \to \frac{\lambda}{1+\tau}$.

The fractional Jeffreys-type constitutive equation reads

$$\left(1+\tau\,{}_0\mathrm{D}_t^\alpha\right) q\left(x,t\right) = -\lambda\left(1+\tau'\,{}_0\mathrm{D}_t^\alpha\right) \frac{\partial}{\partial x} T\left(x,t\right), \quad x \in \mathbb{R},\ t > 0, \qquad [7.11]$$

where $\tau, \tau' > 0$ and $\lambda$ are generalized times of relaxation and generalized heat conductivity coefficient and $\alpha \in (0,1)$. Note that the fractional Jeffreys-type constitutive equation [7.11] reduces to the Fourier law in two cases: if $\alpha = 0$, and $\tau = \tau'$. In the case when $\alpha = 1$, the constitutive equation [7.11] becomes the classical Jeffreys-type constitutive equation.

Proposed constitutive laws: [7.9] and [7.11] have to satisfy the entropy inequality, i.e. the second law of thermodynamics. For the Cattaneo-type equation, such analysis was presented in [RUG 94] via the method of Lagrange multipliers and in [SIL 97, p. 130] by the analysis of cyclic processes. Since space–time fractional Cattaneo heat conduction equation [7.9] is non-local, both in time and space, a direct application of the methods of rational extended thermodynamics, see [MÜL 93], is not possible. In [ATA 04b], an expansion formula was proposed for fractional derivatives in terms of a (finite) number of moments of a function. These moments were then treated as internal variables. This procedure, in principle, leads to the possibility of applying the method of [MÜL 93] to equation [7.9]. However, we will not be concerned with this analysis. We simply assume that in [7.9] we have $\tau, \tau' \geq 0$ and $\lambda > 0$. We also note that the entropy production in problems of heat conduction (diffusion) is treated in [ESS 00, HOF 98, LI 03]. In these works, time and/or space derivatives were fractionalized. The space derivative is defined as $\frac{\mathrm{d}^\alpha}{\mathrm{d}x^\alpha} u(x) = \mathcal{F}^{-1}[(\mathrm{i}\omega)^\alpha \mathcal{F}u](x)$, $x \in \mathbb{R}$.

We refer to [ATA 10a, ATA 12b] for the results presented in this chapter.

## 7.1. Cattaneo-type space–time fractional heat conduction equation

Following [ATA 12b], we combine the energy balance equation [7.4] and the space–time fractional Cattaneo heat conduction law [7.9] to obtain

$$\begin{aligned} \frac{\partial}{\partial t} T\left(x,t\right) &= -\frac{\partial}{\partial x} \bar{q}\left(x,t\right), \quad x \in \mathbb{R},\ t > 0, \\ \tau\,{}_0^C\mathrm{D}_t^\alpha \bar{q}\left(x,t\right) + \bar{q}\left(x,t\right) &= -\mathcal{D}\,\mathcal{E}_x^\beta T\left(x,t\right), \quad x \in \mathbb{R},\ t > 0, \end{aligned}$$

where $\bar{q} = \frac{q}{\rho c}$, $\mathcal{D} = \frac{\lambda}{\rho c}$, which, after the change $\bar{q} \to q$ and $\mathcal{D} \to \lambda$, becomes

$$\frac{\partial}{\partial t}T(x,t) = -\frac{\partial}{\partial x}q(x,t), \quad x \in \mathbb{R}, \quad t > 0, \qquad [7.12]$$

$$\tau\, {}_0^C\mathrm{D}_t^{\alpha} q(x,t) + q(x,t) = -\lambda\, \mathcal{E}_x^{\beta} T(x,t), \quad x \in \mathbb{R},\ t > 0, \qquad [7.13]$$

We subject to systems [7.12] and [7.13] the initial and boundary conditions

$$T(x,0) = T_0(x), \quad q(x,0) = 0, \quad x \in \mathbb{R}, \qquad [7.14]$$

$$\lim_{x \to \pm\infty} T(x,t) = 0, \quad \lim_{x \to \pm\infty} q(x,t) = 0, \quad t > 0, \qquad [7.15]$$

where $T_0$ denotes the initial temperature distribution, while constants $\alpha$ and $\beta$ satisfy $\alpha, \beta \in (0,1)$.

Note that systems [7.12] and [7.13] are equivalent to

$$\frac{\partial}{\partial t}T(x,t) = -\lambda \frac{\mathrm{d}}{\mathrm{d}t} e_{\alpha}(t) *_t \frac{\partial}{\partial x} \mathcal{E}_x^{\beta} T(x,t), \quad x \in \mathbb{R},\ t > 0, \qquad [7.16]$$

which is obtained by substituting [7.10] into [7.12]. We call [7.16] the Cattaneo-type space–time fractional heat conduction equation.

The advantage of the use of the operator $\mathcal{E}_x^{\beta}$, also called the fractional symmetrized gradient, in constitutive equation [7.13], lies in the fact that the order of the fractional gradient describes the ability of a material to conduct heat. If in systems [7.12]–[7.15] we put $\beta = 1$ and $\alpha = 0$, then it models the classical heat conduction or diffusion. By the decrease of $\beta$, the resistance to heat conduction increases, i.e. the diffusion decreases. In the limiting case $\beta = 0$, the material becomes an ideal isolator and systems [7.12]– [7.15] describe a perfect thermal isolator with the property $T(x,t) = T_0(x)$. The parameter $\alpha$ measures the influence of the history of the process on the heat flux $q$.

If $T(\cdot,t) = const.$, $t \geq 0$, then $\mathcal{E}_x^{\beta} T = 0$, which further implies $\tau\, {}_0^C\mathrm{D}_t^{\alpha} q + q = 0$. The unique solution of this equation is given by $q(x,t) = \frac{1}{\tau} q(x,0) \cdot e_{\alpha}(t)$, which with [7.14] gives $q = 0$. Therefore, we have that $T(\cdot,t) = const.$ with [7.14] implies $q = 0$; hence, the space–time fractional Cattaneo heat conduction law [7.9] describes the expected physical fact that if the temperature is constant in space, and there are no sources of heat, there is no heat flux. Classical heat conduction laws, as well as those with fractional time generalizations, also exhibit such physical properties.

The use of the fractional gradient in space brings a new quality: even though there is a spatial distribution of temperature, if the fractional gradient of temperature vanishes, there is no heat flux. Indeed, if $T(\cdot,t) \neq const.$, $t \geq 0$, and $\beta \to 0$, then $\mathcal{E}_x^\beta T(\cdot,t) \to 0$, and again $q = 0$. Thus, when $\beta$ tends to zero, equation [7.9] with [7.4] describes an ideal heat isolator. In such a body, an initial space distribution of the temperature would not change in time. Indeed, since [7.4] with $q = 0$ leads to $\frac{\partial}{\partial t}T = 0$, we have that $T(x,t) = T(x) = T_0(x)$. We conclude that materials which are good isolators could be modeled with small $\beta$. Similar properties of the operator $\mathcal{E}_x^\beta$ are used in [ATA 09f] in the case of the space-fractional wave equation.

### 7.1.1. *Existence and uniqueness of a solution*

We study the existence and uniqueness of a generalized solution to the initial-boundary value problem [7.12]–[7.15]. We work with distributions depending on two variables: $T = T(x,t)$ and $q = q(x,t)$.

Let us explain the idea of the proof that is to follow. We will make formal calculations within the distribution theory, which will be justified at the end by showing that the convolution with respect to $x$ can be written as the linear combination of derivatives of convolution of functions in $L^1(\mathbb{R})$ and $L^\infty(\mathbb{R})$. Recall, if $f \in L^1(\mathbb{R})$ and $g \in L^\infty(\mathbb{R})$, then $f * g \in L^1(\mathbb{R})$. We explain this more precisely. Since we will apply the Fourier transform with respect to $x$ and the Laplace transform with respect to $t$ to distributions $T$ and $q$, we consider these distributions within the space $\mathcal{K}' \otimes \mathcal{K}'_+$, which is the subspace of $\mathcal{K}'(\mathbb{R}^2)$, consisting of distributions having support in $\mathbb{R} \times [0,\infty)$. Actually, our solutions $T$ and $q$ will be elements of $\mathcal{D}'_{L^1}$ for fixed $\varphi \in \mathcal{K}$, i.e. $\langle T(x,t), \varphi(t)\rangle \in \mathcal{D}'_{L^1}$, and elements of $\mathcal{K}'_+$ for fixed $\psi \in \mathcal{K}$, i.e. $\langle T(x,t), \psi(x)\rangle \in \mathcal{K}'_+$. Note that $\mathcal{S}' \subset \mathcal{K}'$ ($\mathcal{S}$ is dense in $\mathcal{K}$) and $\mathcal{S}'_+ \subset \mathcal{K}'_+$. We refer to section 1.3 for definitions of all the spaces mentioned. We use the usual notation in which we have the dual paring written in the form of an integral. A divergent integral has sense in the space of distributions. For example, if $f \in \mathcal{D}'_{L^1}$, $f = \sum_{\alpha<k} \mathrm{D}^\alpha F_\alpha$, where $F_\alpha \in L^1(\mathbb{R})$ and $\mathrm{D}^\alpha$ is the derivative in the distributional sense, and $g \in \mathcal{D}'_{L^\infty}$, $g = \sum_{\beta<l} \mathrm{D}^\beta G_\beta$, with $G_\beta \in L^\infty(\mathbb{R})$, then

$$f * g = \sum_{\substack{\alpha<k \\ \beta<l}} \mathrm{D}^{\alpha+\beta}(F_\alpha * G_\beta),$$

where the convolutions $F_\alpha * G_\beta$ exist in the sense of the Lebesgue integral. We refer to section 1.3 for the definition of $\mathcal{D}'_{L^p}$.

DEFINITION 7.1.– *Generalized solutions to the initial problems* [7.12]–[7.14] *are* $T, q \in \mathcal{K}'(\mathbb{R}^2)$, *with support in* $\mathbb{R} \times [0, \infty)$, *if*

$$\begin{aligned} \frac{\partial}{\partial t} T(x,t) &= -\frac{\partial}{\partial x} q(x,t) + T_0(x)\delta(t), && [7.17] \\ \tau\, {}_0^C\mathrm{D}_t^\alpha q(x,t) + q(x,t) &= -\lambda\, \mathcal{E}_x^\beta T(x,t), && [7.18] \end{aligned}$$

*hold in* $\mathcal{K}'(\mathbb{R}^2)$.

We state the theorem on the existence of a unique solution to the initial-boundary value problem [7.12]–[7.15].

THEOREM 7.1.– Let $T_0 \in \mathcal{D}'_{L^1}$. Then there exist unique generalized solutions $T, q \in \mathcal{K}'(\mathbb{R}^2)$, with support in $\mathbb{R} \times [0, \infty)$, to the initial value problem [7.12]–[7.14].

PROOF.– Applying the Laplace transform to [7.17] and [7.18] with respect to $t$, we obtain

$$\begin{aligned} s\tilde{T}(x,s) &= -\frac{\partial}{\partial x}\tilde{q}(x,s) + T_0(x), \quad x \in \mathbb{R}, \ \operatorname{Re} s > 0, \\ (1+\tau s^\alpha)\tilde{q}(x,s) &= -\lambda\, \mathcal{E}_x^\beta \tilde{T}(x,s), \quad x \in \mathbb{R}, \ \operatorname{Re} s > 0. \end{aligned}$$

Substituting the second equation into the first one, we obtain

$$\frac{\partial}{\partial x}\mathcal{E}_x^\beta \tilde{T}(x,s) - \frac{s(1+\tau s^\alpha)}{\lambda}\tilde{T}(x,s) = -\frac{1+\tau s^\alpha}{\lambda} T_0(x), \quad x \in \mathbb{R}, \ \operatorname{Re} s > 0. \quad [7.19]$$

Equation [7.19] is of the type

$$\frac{\partial}{\partial x}\mathcal{E}_x^\beta u(x) - \omega\, u(x) = -\nu\, u_0(x), \quad x \in \mathbb{R}, \qquad [7.20]$$

where

$$\omega = \omega(s) = \frac{s(1+\tau s^\alpha)}{\lambda} \quad \text{and} \quad \nu = \nu(s) = \frac{1+\tau s^\alpha}{\lambda}, \quad \operatorname{Re} s > 0.$$

The unique solution $u \in \mathcal{D}'_{L^1}$ to [7.20]

$$u(x) = \frac{\nu}{\pi} u_0(x) *_x \int_0^\infty \frac{1}{\rho^{1+\beta}\sin\frac{\beta\pi}{2} + \omega}\cos(\rho x)\mathrm{d}\rho, \quad x \in \mathbb{R}. \qquad [7.21]$$

is obtained as the inverse Fourier transform of

$$\hat{u}(\xi) = \frac{\nu\, \hat{u}_0(\xi)}{|\xi|^{1+\beta} \sin \frac{\beta\pi}{2} + \omega}, \quad \xi \in \mathbb{R},$$

if $u_0 \in \mathcal{D}'_{L^1}$ and $\omega, \nu \in \mathbb{C}$, $\omega \in \mathbb{C} \setminus (-\infty, 0]$, since the integral belongs to $\mathcal{D}'_{L^\infty}$. In order to see that $\omega(s) \in \mathbb{C} \setminus (-\infty, 0]$ for $\operatorname{Re} s > 0$, we prove that $\operatorname{Im} \omega(s) = 0$ does not imply $\operatorname{Re} \omega(s) < 0$. Indeed, if we put $s = r\mathrm{e}^{\mathrm{i}\varphi}$, $\phi \in (-\frac{\pi}{2}, \frac{\pi}{2})$, then $\operatorname{Im} \omega(s) = 0$ is equivalent to $r^\alpha = -\frac{\sin\varphi}{\tau \sin((\alpha+1)\varphi)}$. We use it to calculate $\operatorname{Re} \omega(s)$ and obtain $\operatorname{Re} \omega(s) = \frac{r}{\lambda} \frac{\sin(\alpha\varphi)}{\sin((\alpha+1)\varphi)} > 0$. Therefore, by [7.21], we have the solution to [7.19] in the form

$$\begin{aligned} \tilde{T}(x, s) &= \frac{1 + \tau s^\alpha}{\lambda\pi} T_0(x) *_x \int_0^\infty \frac{1}{\rho^{1+\beta} \sin \frac{\beta\pi}{2} + \omega(s)} \cos(\rho x) \mathrm{d}\rho \\ &= \frac{1}{\pi} T_0(x) *_x \int_0^\infty \frac{\tau s^\alpha + 1}{\tau s^{\alpha+1} + s + \lambda\rho^{1+\beta} \sin \frac{\beta\pi}{2}} \cos(\rho x) \mathrm{d}\rho, \quad x \in \mathbb{R}, \\ &\qquad \operatorname{Re} s > 0. \end{aligned} \qquad [7.22]$$

We apply the inverse Laplace transform to [7.22] and obtain

$$T(x, t) = \frac{1}{2\pi^2 \mathrm{i}} T_0(x) \left( \tau \frac{t^{-\alpha-1}}{\Gamma(-\alpha)} + \delta(t) \right) *_{x,t} Q(x, t), \quad x \in \mathbb{R},\ t > 0, \ [7.23]$$

where

$$\begin{aligned} Q(x, t) &= 2\pi \mathrm{i} \mathcal{L}^{-1} \left[ \int_0^\infty \frac{\cos(\rho x)}{\tau s^{\alpha+1} + s + \lambda\rho^{1+\beta} \sin \frac{\beta\pi}{2}} \mathrm{d}\rho \right](t) \\ &= \int_{s_0 - \mathrm{i}\infty}^{s_0 + \mathrm{i}\infty} \int_0^\infty \frac{\cos(\rho x) \mathrm{e}^{st}}{\tau s^{\alpha+1} + s + \lambda\rho^{1+\beta} \sin \frac{\beta\pi}{2}} \mathrm{d}\rho \mathrm{d}s, \quad x \in \mathbb{R},\ t > 0. \ [7.24] \end{aligned}$$

We will show that $Q \in \mathcal{K}'(\mathbb{R}^2)$, with support in $\mathbb{R} \times [0, \infty)$. Consider the divergent integral [7.24]. We introduce the parameterization $s = s_0 + \mathrm{i}p$, $p \in \mathbb{R}$ in [7.24], so that

$$\begin{aligned} Q(x, t) &= \int_{-\infty}^\infty \int_0^\infty \frac{\cos(\rho x) \mathrm{e}^{s_0 t} \mathrm{e}^{\mathrm{i}pt}}{\left[\tau s^{\alpha+1} + s\right]_{s = s_0 + \mathrm{i}p} + \lambda\rho^{1+\beta} \sin \frac{\beta\pi}{2}} \mathrm{d}\rho\, \mathrm{d}p \\ &= I(x, t) + I^+(x, t) + I^-(x, t), \quad x \in \mathbb{R},\ t > 0, \end{aligned}$$

where $I$, $I^{+}$ and $I^{-}$ are given below. We see that the integral

$$I\left(x,t\right)=\int_{-p_0}^{p_0}\int_0^{\infty}\frac{\cos(\rho x)\mathrm{e}^{s_0t}\mathrm{e}^{ipt}}{\left[\tau s^{\alpha+1}+s\right]_{s=s_0+ip}+\lambda\rho^{1+\beta}\sin\frac{\beta\pi}{2}}\mathrm{d}\rho\,\mathrm{d}p,\quad x\in\mathbb{R},\ t>0,$$

is absolutely convergent, since for $s_0$ large enough

$$\left|I\left(x,t\right)\right|\ \leq\ \mathrm{e}^{s_0t}\int_{-p_0}^{p_0}\int_0^{\infty}\frac{1}{\mathrm{Re}\left(\left[\tau s^{\alpha+1}+s\right]_{s=s_0+ip}\right)+\lambda\rho^{1+\beta}\sin\frac{\beta\pi}{2}}\mathrm{d}\rho\,\mathrm{d}p,\quad x\in\mathbb{R},\ t>0. \qquad [7.25]$$

In [7.25], $p_0$ is chosen so that $\mathrm{Re}\left(\left[\tau s^{\alpha+1}+s\right]_{s=s_0+ip}\right)=\tau r^{\alpha+1}\cos\left(\left(\alpha+1\right)\varphi\right)+r\cos\varphi>0$, with $r=\sqrt{s_0^2+p^2}$, $\tan\varphi=\frac{p}{s_0}$. Note that for $p=0$ and $\rho=0$ the integrand is well defined, due to the existence of $s_0>0$. Thus, the integral $I$ exists and belongs to $L^{\infty}\left(\mathbb{R}\right)$ with respect to $x$ and it is exponentially bounded with respect to $t$. Next, we consider

$$\begin{aligned}I^{+}\left(x,t\right)&=\int_{p_0}^{\infty}\int_0^{\infty}\frac{\cos(\rho x)\mathrm{e}^{s_0t}\mathrm{e}^{ipt}}{\left[\tau s^{\alpha+1}+s\right]_{s=s_0+ip}+\lambda\rho^{1+\beta}\sin\frac{\beta\pi}{2}}\mathrm{d}\rho\,\mathrm{d}p\\&=J_1^{+}\left(x,t\right)+\frac{\partial^2}{\partial x^2}J_2^{+}\left(x,t\right),\quad x\in\mathbb{R},\ t>0,\end{aligned} \qquad [7.26]$$

where $J_1^{+}$ and $J_2^{+}$ are defined below. Note that $J_2^{+}$ is the essential part of this decomposition. Set

$$J_1^{+}\left(x,t\right)=\int_{p_0}^{\infty}\int_0^{1}\frac{\cos(\rho x)\mathrm{e}^{s_0t}\mathrm{e}^{ipt}}{\left[\tau s^{\alpha+1}+s\right]_{s=s_0+ip}+\lambda\rho^{1+\beta}\sin\frac{\beta\pi}{2}}\mathrm{d}\rho\,\mathrm{d}p,\quad x\in\mathbb{R},\ t>0,$$

which yields

$$\begin{aligned}\left|J_1^{+}\left(x,t\right)\right|&\leq\mathrm{e}^{s_0t}\int_{p_0}^{\infty}\int_0^{1}\frac{\left|\cos(\rho x)\right|}{\left|\left[\tau s^{\alpha+1}+s\right]_{s=s_0+ip}+\lambda\rho^{1+\beta}\sin\frac{\beta\pi}{2}\right|}\mathrm{d}\rho\,\mathrm{d}p\\&\leq\mathrm{e}^{s_0t}\int_{p_0}^{\infty}\int_0^{1}\frac{1}{\mathrm{Im}\left(\left[\tau s^{\alpha+1}+s\right]_{s=s_0+ip}\right)}\mathrm{d}\rho\,\mathrm{d}p,\quad x\in\mathbb{R},\ t>0.\end{aligned}$$

The integral $J_1^+$ exists, since $\operatorname{Im}\left(\left[\tau s^{\alpha+1}+s\right]_{s=s_0+ip}\right) = \tau r^{\alpha+1}\sin\left((\alpha+1)\varphi\right)+r\sin\varphi>0$ and $\operatorname{Im}\left(\left[\tau s^{\alpha+1}+s\right]_{s=s_0+ip}\right)\sim p^{\alpha+1}$ for $p$ large enough and belongs to $L^\infty(\mathbb{R})$ with respect to $x$, while it is exponentially bounded with respect to $t$. We put

$$\begin{aligned} J_2^+(x,t) &= \int_{p_0}^\infty\int_1^\infty \frac{(1-\cos(\rho x))\,\mathrm{e}^{s_0 t}\mathrm{e}^{ipt}}{\rho^2\left(\left[\tau s^{\alpha+1}+s\right]_{s=s_0+ip}+\lambda\rho^{1+\beta}\sin\frac{\beta\pi}{2}\right)}\mathrm{d}\rho\,\mathrm{d}p, \\ & \qquad x\in\mathbb{R},\ t>0, \end{aligned}$$

from which we have

$$\left|J_2^+(x,t)\right|\le \mathrm{e}^{s_0 t}\int_{p_0}^\infty\int_1^\infty \frac{1}{\rho^2\operatorname{Im}\left(\left[\tau s^{\alpha+1}+s\right]_{s=s_0+ip}\right)}\mathrm{d}\rho\,\mathrm{d}p,\quad x\in\mathbb{R},\ t>0.$$

By the same arguments as for $J_1^+$, we find that the integral $J_2^+$ exists and belongs to $L^\infty(\mathbb{R})$ with respect to $x$, while it is exponentially bounded with respect to $t$. Thus, for $I^+$ given by [7.26], we have $I^+\in\mathcal{K}'\left(\mathbb{R}^2\right)$, with support in $\mathbb{R}\times[0,\infty)$. Similarly, as for $I^+$, we prove the existence for

$$\begin{aligned} I^-(x,t) &= \int_{-\infty}^{-p_0}\int_0^\infty \frac{\cos(\rho x)\mathrm{e}^{s_0 t}\mathrm{e}^{ipt}}{\left[\tau s^{\alpha+1}+s\right]_{s=s_0+ip}+\lambda\rho^{1+\beta}\sin\frac{\beta\pi}{2}}\mathrm{d}\rho\,\mathrm{d}p \\ &= \int_{p_0}^\infty\int_0^\infty \frac{\cos(\rho x)\mathrm{e}^{s_0 t}\mathrm{e}^{-iqt}}{\left[\tau s^{\alpha+1}+s\right]_{s=s_0-iq}+\lambda\rho^{1+\beta}\sin\frac{\beta\pi}{2}}\mathrm{d}\rho\,\mathrm{d}q \\ &= J_1^-(x,t)+\frac{\partial^2}{\partial x^2}J_2^-(x,t),\quad x\in\mathbb{R},\ t>0. \end{aligned} \qquad [7.27]$$

We have

$$J_1^-(x,t)=\int_{p_0}^\infty\int_0^1 \frac{\cos(\rho x)\mathrm{e}^{s_0 t}\mathrm{e}^{-iqt}}{\left[\tau s^{\alpha+1}+s\right]_{s=s_0-iq}+\lambda\rho^{1+\beta}\sin\frac{\beta\pi}{2}}\mathrm{d}\rho\,\mathrm{d}q,\quad x\in\mathbb{R},\ t>0,$$

so that

$$\begin{aligned}\left|J_1^-(x,t)\right| &\le \mathrm{e}^{s_0 t}\int_{p_0}^{\infty}\int_0^1 \frac{|\cos(\rho x)|}{\left|\left[\tau s^{\alpha+1}+s\right]_{s=s_0-\mathrm{i}q}+\lambda\rho^{1+\beta}\sin\frac{\beta\pi}{2}\right|}\mathrm{d}\rho\,\mathrm{d}q \\ &\le \mathrm{e}^{s_0 t}\int_{p_0}^{\infty}\int_0^1 \frac{1}{-\operatorname{Im}\left(\left[\tau s^{\alpha+1}+s\right]_{s=s_0-\mathrm{i}q}\right)}\mathrm{d}\rho\,\mathrm{d}q,\quad x\in\mathbb{R},\ t>0.\end{aligned}$$

The integral $J_1^-$ exists and belongs to $L^\infty(\mathbb{R})$ with respect to $x$ and to $\mathcal{K}'_+$ with respect to $t$, since $\operatorname{Im}\left(\left[\tau s^{\alpha+1}+s\right]_{s=s_0-\mathrm{i}q}\right)=\tau r^{\alpha+1}\sin((\alpha+1)\varphi)+r\sin\varphi<0$ and $\operatorname{Im}\left(\left[\tau s^{\alpha+1}+s\right]_{s=s_0+\mathrm{i}p}\right)\sim p^{\alpha+1}$ for $p$. Consider

$$\begin{aligned}J_2^-(x,t) &= \int_{p_0}^{\infty}\int_1^{\infty}\frac{(1-\cos(\rho x))\,\mathrm{e}^{s_0 t}\mathrm{e}^{-\mathrm{i}qt}}{\rho^2\left(\left[\tau s^{\alpha+1}+s\right]_{s=s_0-\mathrm{i}q}+\lambda\rho^{1+\beta}\sin\frac{\beta\pi}{2}\right)}\mathrm{d}\rho\,\mathrm{d}q, \\ &\qquad x\in\mathbb{R},\ t>0.\end{aligned}$$

We have

$$\begin{aligned}\left|J_2^-(x,t)\right| &\le \mathrm{e}^{s_0 t}\int_{p_0}^{\infty}\int_1^{\infty}\frac{1}{-\rho^2\operatorname{Im}\left(\left[\tau s^{\alpha+1}+s\right]_{s=s_0-\mathrm{i}q}\right)}\mathrm{d}\rho\,\mathrm{d}q, \\ &\qquad x\in\mathbb{R},\ t>0,\ \ x\in\mathbb{R},\ t>0.\end{aligned}$$

The integral $J_2^-$ exists and belongs to $L^\infty(\mathbb{R})$ with respect to $x$ and to $\mathcal{K}'_+$ with respect to $t$. Thus, for $I^-$ given by [7.27], we have $I^-\in\mathcal{K}'(\mathbb{R}^2)$, with support in $\mathbb{R}\times[0,\infty)$.

Therefore, for $Q$ given in the form

$$Q=I+J_1^++\frac{\partial^2}{\partial x^2}J_2^++J_1^-+\frac{\partial^2}{\partial x^2}J_2^-$$

we proved that $Q\in\mathcal{K}'(\mathbb{R}^2)$, with support in $\mathbb{R}\times[0,\infty)$, since all the addenda belong to this space. The assumption that $T_0\in\mathcal{D}'_{L^1}$ implies that the convolution with respect to $x$ in [7.23] is well defined since $Q$ belongs to $\mathcal{D}'_{L^\infty}$ with respect to $x$. Note that the convolution with respect to the variable $t$ is also well defined since we have elements of $\mathcal{K}'_+$.

By [7.18], we have

$$q(x,t) = -\lambda\, \mathcal{L}^{-1}\left(\frac{1}{1+\tau s^{\alpha}}\right) *_t \mathcal{E}_x^{\beta}\, T(x,t), \quad x \in \mathbb{R},\ t > 0,$$

which is unique in $\mathcal{K}'(\mathbb{R}^2)$, with support in $\mathbb{R} \times [0, \infty)$. ■

### 7.1.2. *Explicit form of the solution*

In this section, we give an explicit form of the generalized solution $T$ to [7.16] and use the obtained formula in several numerical examples.

THEOREM 7.2.– Let $T, q \in \mathcal{K}'(\mathbb{R}^2)$, with support in $\mathbb{R} \times [0, \infty)$, be the generalized solution to the initial problem [7.12], –[7.14] whose existence and uniqueness is guaranteed by theorem 7.1. Then

$$T(x,t) = \frac{1}{\pi} T_0(x) *_x \int_0^{\infty} S(\rho,t)\cos(\rho x)\mathrm{d}\rho, \quad x \in \mathbb{R},\ t > 0, \qquad [7.28]$$

$$S(\rho,t) = I(\rho,t) + R(\rho,t), \quad \rho > 0,\ t > 0, \qquad [7.29]$$

where

$$I(\rho,t) = -\frac{1}{\pi}\int_0^{\infty} \frac{\tau q^{\alpha}\theta(\rho)\sin(\alpha\pi)\mathrm{e}^{-qt}\,\mathrm{d}q}{(\tau q^{\alpha+1})^2 + 2\tau q^{\alpha+1}(q-\theta(\rho))\cos(\alpha\pi) + (q-\theta(\rho))^2},$$
$$\rho > 0,\ t > 0, \qquad [7.30]$$

$$R(\rho,t) = 2\mathrm{Re}\left(\left.\frac{(\tau s^{\alpha}+1)\mathrm{e}^{st}}{\tau(\alpha+1)s^{\alpha}+1}\right|_{s=s_0(\rho)}\right), \quad \rho > 0,\ t > 0, \qquad [7.31]$$

with $\theta(\rho) = \lambda\rho^{1+\beta}\sin\frac{\beta\pi}{2}$ and $s_0 = r_0\mathrm{e}^{\mathrm{i}\varphi_0}$ is from lemma 7.1 below.

REMARK 7.1.– Note that the generalized solution to [7.16] is given by [7.28] and that, in fact, in the proof of this theorem we calculate the fundamental solution to [7.16] (generalized solution with $T_0(x) = \delta(x)$) as

$$F(x,t) = \frac{1}{\pi}\int_0^{\infty} S(\rho,t)\cos(\rho x)\mathrm{d}\rho, \quad x \in \mathbb{R},\ t > 0. \qquad [7.32]$$

REMARK 7.2 (Limiting cases).– In the analysis that follows, we will calculate solution [7.28] for some fixed parameters and compare it with some limiting cases for which there are explicit formulas.

(i) When $\beta = 1$, i.e. when in [7.13] the symmetrized fractional gradient is replaced by the classical gradient, the solution is calculated as

$$\begin{aligned} T(x,t) &= \frac{1}{4\pi} T_0(x) *_x \int_0^\infty \left( \sqrt{1+\tau q^\alpha \mathrm{e}^{-\mathrm{i}\alpha\pi}} \mathrm{e}^{\mathrm{i}|x|\sqrt{q(1+\tau q^\alpha \mathrm{e}^{-\mathrm{i}\alpha\pi})}} \right. \\ &\quad \left. + \sqrt{1+\tau q^\alpha \mathrm{e}^{\mathrm{i}\alpha\pi}} \mathrm{e}^{-\mathrm{i}|x|\sqrt{q(1+\tau q^\alpha \mathrm{e}^{\mathrm{i}\alpha\pi})}} \right) \frac{\mathrm{e}^{-qt}}{\sqrt{q}} \, \mathrm{d}q, \quad x \in \mathbb{R},\ t > 0. \end{aligned} \qquad [7.33]$$

For similar formulas, see [ATA 07b].

(ii) For $\beta = 0$ the solution is obviously of the form

$$T(x,t) = T_0(x), \quad x \in \mathbb{R},\ t > 0.$$

(iii) For $\alpha = 1$ in heat conduction law [7.13], we have a classical time partial derivative instead of a fractional one, and [7.13] represents the space-fractional Cattaneo heat conduction law. The solution is given by [7.28] and $S$ takes the form ($\rho > 0,\ t > 0$)

$$S(\rho,t) = \begin{cases} \left( \dfrac{\sinh\left(\sqrt{1-4\tau\theta(\rho)}\frac{t}{2\tau}\right)}{\sqrt{1-4\tau\theta(\rho)}} + \cosh\left(\sqrt{1-4\tau\theta(\rho)}\frac{t}{2\tau}\right)\right) \mathrm{e}^{-\frac{t}{2\tau}}, \\ \qquad \rho \le \left(\dfrac{1}{4\tau\lambda \sin\frac{\beta\pi}{2}}\right)^{\frac{1}{1+\beta}}, \\ \left( \dfrac{\sin\left(\sqrt{4\tau\theta(\rho)-1}\frac{t}{2\tau}\right)}{\sqrt{4\tau\theta(\rho)-1}} + \cos\left(\sqrt{4\tau\theta(\rho)-1}\frac{t}{2\tau}\right)\right) \mathrm{e}^{-\frac{t}{2\tau}}, \\ \qquad \rho > \left(\dfrac{1}{4\tau\lambda \sin\frac{\beta\pi}{2}}\right)^{\frac{1}{1+\beta}}, \end{cases} \qquad [7.34]$$

with $\theta(\rho) = \lambda\rho^{1+\beta}\sin\frac{\beta\pi}{2}$. A similar calculation is done in [ORS 03].

(iv) Limiting case $\alpha = 0$ is also of particular interest, since then on the left-hand side of [7.13] we have two terms containing the heat flux $q$ and no terms with its derivatives. Therefore, [7.13] represents the space-fractional Fourier heat conduction law. The solution is given by [7.28] and $S$ takes the form

$$S(\rho,t) = \mathrm{e}^{-\frac{\theta(\rho)}{1+\tau}t}, \quad \rho > 0,\ t > 0, \qquad [7.35]$$

with $\theta(\rho) = \lambda\rho^{1+\beta}\sin\frac{\beta\pi}{2}$. A similar calculation is done in [HAN 01].

For the proof of theorem 7.2, we need the following lemma.

LEMMA 7.1.– Let $\tau, \theta > 0$ and $0 < \alpha < 1$. Then there exist unique $r_0 > 0$ and $\varphi_0 \in (\frac{\pi}{\alpha+1}, \pi)$ (depending on $\tau, \theta$ and $\alpha$) such that $s_0 = r_0 e^{i\varphi_0}$ and $\overline{s}_0$, the complex conjugated number of $s_0$, are zeros of the multiplicity one of the function

$$f(s) = \tau s^{\alpha+1} + s + \theta, \quad f : \mathbb{C} \to \mathbb{C}.$$

We point out an interesting consequence of the previous lemma.

COROLLARY 7.1.– Let $\tau, \theta > 0$, $0 < \alpha < 1$ and $v \in \mathcal{S}'_+$. Then the fractional differential equation

$$\tau {}_0^C D_t^\alpha \frac{d}{dt} u(t) + \frac{d}{dt} u(t) + \theta u(t) = v(t), \quad t > 0, \tag{7.36}$$

has a unique solution in $\mathcal{S}'_+$.

PROOF.– According to [ATA 10c, Theorem 3.2], equation [7.36] has a unique solution in $\mathcal{S}'_+$ if and only if $\tau s^{\alpha+1} + s + \theta \neq 0$, for all $\operatorname{Re} s > 0$. We have shown in lemma 7.1 that the equation $\tau s^{\alpha+1} + s + \theta = 0$ has two complex conjugated zeros with negative real parts, i.e. $s_0^{(\pm)} = r_0 e^{\pm i\varphi_0}$, with $\varphi_0 \in (\frac{\pi}{\alpha+1}, \pi)$. Thus, the claim follows from lemma 7.1. ■

REMARK 7.3.–

(i) The solution to [7.36] is given by

$$u = K *_t v, \quad K(t) = \mathcal{L}^{-1}\left(\frac{1}{\tau s^{\alpha+1} + s + \theta}\right)(t), \quad t > 0, \tag{7.37}$$

where the existence of the inverse Laplace transform above actually follows from [ATA 10c, Theorem 3.2] and lemma 7.1. The locally integrable function $K$ can be calculated, in a similar way as we will calculate $S$ in the proof of theorem 7.2 below, as

$$\begin{aligned} K(t) &= \frac{1}{\pi}\int_0^\infty \frac{\tau q^{\alpha+1}\sin(\alpha\pi)}{(\tau q^{\alpha+1})^2 + 2\tau q^{\alpha+1}(q-\theta)\cos(\alpha\pi) + (q-\theta)^2} e^{-qt} dq \\ &\quad + 2\operatorname{Re}\left(\left.\frac{e^{st}}{\tau(\alpha+1)s^\alpha + 1}\right|_{s=s_0}\right), \quad t > 0, \end{aligned}$$

where $s_0$ is the solution to systems [7.39] and [7.40].

(ii) For $\alpha = 0$, equation [7.36] is an ordinary differential equation of the first order and $K$ is given by

$$K(t) = \frac{1}{\tau + 1} \mathrm{e}^{-\frac{\theta}{\tau+1} t}, \quad t > 0.$$

For $\alpha = 1$, equation [7.36] is an ordinary differential equation of the second order with solution [7.37] where $K$ takes the form ($t > 0$)

$$K(t) = \begin{cases} \frac{2}{\sqrt{1-4\tau\theta}} \mathrm{e}^{-\frac{t}{2\tau}} \sinh(\frac{\sqrt{1-4\tau\theta}\, t}{2\tau}), & 4\tau\theta < 1, \\ \frac{2}{\sqrt{4\tau\theta - 1}} \mathrm{e}^{-\frac{t}{2\tau}} \sin(\frac{\sqrt{4\tau\theta-1}\, t}{2\tau}), & 4\tau\theta > 1. \end{cases} \tag{7.38}$$

(iii) In Figure 7.1 we present the plots of solution $u$ for fixed parameters $\tau = 0.1$, $\theta = 10$, right-hand side $v = \delta$ and various values of parameter $\alpha$. We compare plots of solution $u$ calculated for several values of $\alpha \in (0, 1)$ by [7.37] (solid lines) with the plots of solution $u$ calculated in the limiting case $\alpha = 1$ by [7.38] (dashed line). We note that for $\alpha = 0.99$, the solid line almost overlaps the dashed line.

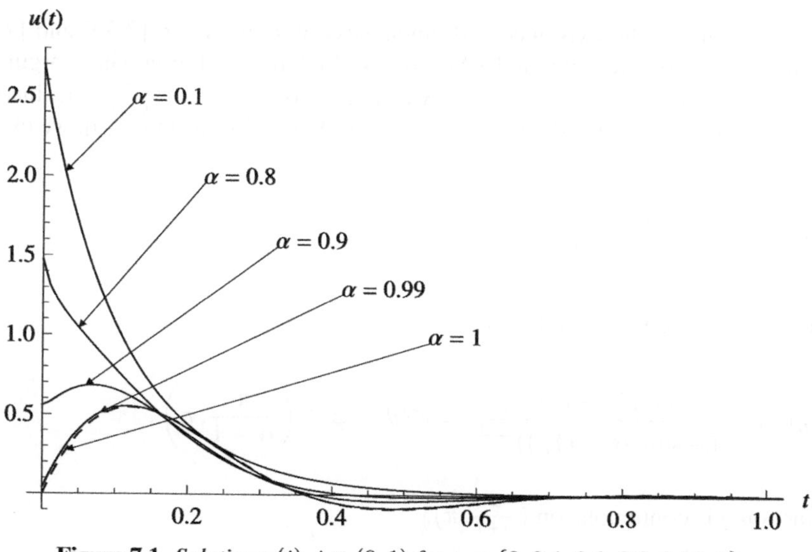


**Figure 7.1.** *Solution* $u(t)$, $t \in (0, 1)$ *for* $\alpha \in \{0, 0.1, 0.8, 0.9, 0.99, 1\}$.

We conclude this section by giving the proofs of lemma 7.1 and theorem 7.2.

PROOF OF LEMMA 7.1.– Let $s \in \mathbb{C}$. Set $s = r\mathrm{e}^{\mathrm{i}\varphi}$. After separating the real and imaginary parts in $f(s) = 0$, we obtain systems

$$\tau r^{\alpha+1}\cos((\alpha+1)\varphi) + r\cos\varphi + \theta = 0, \quad [7.39]$$

$$\tau r^{\alpha+1}\sin((\alpha+1)\varphi) + r\sin\varphi = 0. \quad [7.40]$$

Let $s_0 = r_0\mathrm{e}^{\mathrm{i}\varphi_0}$ be a solution to systems [7.39] and [7.40]. Then $\bar{s}_0 = r_0\mathrm{e}^{-\mathrm{i}\varphi_0}$ is also a solution to [7.39] and [7.40] since both equations are invariant under change $\varphi \to -\varphi$. Therefore, if the system has solutions, then they are complex conjugated. This allows us to consider the upper complex plane only. Suppose first $\varphi \in (0, \frac{\pi}{2}]$. Then, $(\alpha+1)\varphi \in (0,\pi)$, since $\alpha \in (0,1)$. Hence, $\sin\varphi > 0$ and $\sin((\alpha+1)\varphi) > 0$; thus, [7.40] cannot be satisfied for $\varphi \in (0, \frac{\pi}{2}]$. For $\varphi = 0$, equation [7.39] cannot be satisfied since $\tau, \theta > 0$, while for $\varphi = \pi$ equation [7.40] cannot be satisfied. Therefore, if there are solutions to [7.39] and [7.40], their real parts must be negative, i.e. $\varphi \in (\frac{\pi}{2}, \pi)$.

We will examine the existence and uniqueness of solutions to [7.39] and [7.40] following the method presented in [NAB 10]. We look for a solution whose argument $\varphi \in (\frac{\pi}{2}, \pi)$. In order to have [7.40] valid, we have to consider $\varphi \in (\frac{\pi}{\alpha+1}, \pi)$, since for such $\varphi$ we have $\sin\varphi > 0$ and $\sin((\alpha+1)\varphi) < 0$. For $r \neq 0$, from [7.40], we obtain

$$r = \left(\frac{\sin\varphi}{-\tau\sin((\alpha+1)\varphi)}\right)^{\frac{1}{\alpha}},$$

which we insert into [7.39] and obtain

$$g(\varphi) = \frac{(\sin\varphi)^{\frac{1}{\alpha}}\sin(\alpha\varphi)}{(-\sin((\alpha+1)\varphi))^{\frac{\alpha+1}{\alpha}}} = \tau^{\frac{1}{\alpha}}\theta, \quad \varphi \in \left(\frac{\pi}{\alpha+1}, \pi\right).$$

Function $g$ is continuous on $(\frac{\pi}{\alpha+1}, \pi)$,

$$\lim_{\varphi\to\frac{\pi}{\alpha+1}} g(\varphi) = \infty \quad \text{and} \quad \lim_{\varphi\to\pi} g(\varphi) = 0,$$

hence it follows from the mean value theorem that there exists $\varphi_0 \in (\frac{\pi}{\alpha+1}, \pi)$ such that $g(\varphi_0) = \tau^{\frac{1}{\alpha}}\theta$. Therefore, a solution to [7.39] and [7.40] exists and is given by $s_0 = r_0\mathrm{e}^{\mathrm{i}\varphi_0}$, where $r_0 = \left(\frac{\sin\varphi_0}{-\tau\sin((\alpha+1)\varphi_0)}\right)^{\frac{1}{\alpha}}$.

Next, we will prove that $g$ is a strictly monotonically decreasing function on $(\frac{\pi}{\alpha+1}, \pi)$. We calculated

$$\frac{\mathrm{d}}{\mathrm{d}\varphi} g(\varphi) = -\frac{(\sin\varphi)^{\frac{1-\alpha}{\alpha}}}{\alpha(-\sin((\alpha+1)\varphi))^{\frac{2\alpha+1}{\alpha}}} A(\varphi)$$

with

$$A(\varphi) = \alpha(\alpha+1)\sin^2\varphi + (\alpha+1)\sin^2(\alpha\varphi) - \alpha\sin^2((\alpha+1)\varphi).$$

The sign of $\frac{\mathrm{d}g}{\mathrm{d}\varphi}$ depends on the sign of $A$, so we transform $A$ as

$$\begin{aligned} A(\varphi) &= \alpha^2\sin^2\varphi + (\alpha+1)\sin^2(\alpha\varphi) - \alpha(\sin^2((\alpha+1)\varphi) - \sin^2\varphi) \\ &= \alpha^2\sin^2\varphi + (\alpha+1)\sin^2(\alpha\varphi) - \alpha\sin(\alpha\varphi)\sin((\alpha+2)\varphi) \\ &= \alpha^2\sin^2\varphi + \sin^2(\alpha\varphi) - \alpha\sin(\alpha\varphi)(\sin((\alpha+2)\varphi) - \sin(\alpha\varphi)) \\ &= \alpha^2\sin^2\varphi + \sin^2(\alpha\varphi) - 2\alpha\sin\varphi\sin(\alpha\varphi)\cos((\alpha+1)\varphi) \\ &= (\alpha\sin\varphi - \sin(\alpha\varphi))^2 + 2\alpha\sin\varphi\sin(\alpha\varphi)(1 - \cos((\alpha+1)\varphi)). \end{aligned}$$

Since for $\varphi \in (\frac{\pi}{\alpha+1}, \pi)$ we have $\sin\varphi, \sin(\alpha\varphi) > 0$ and $\cos((\alpha+1)\varphi) < 1$, we conclude that $A > 0$. Therefore, $\frac{\mathrm{d}g}{\mathrm{d}\varphi} < 0$ and consequently $g$ is a strictly monotonically decreasing function on $(\frac{\pi}{\alpha+1}, \pi)$.

This now implies that $\varphi_0$ is the unique solution to $g(\varphi) = \tau^{\frac{1}{\alpha}}\theta$ and therefore $s_0$ is the unique solution to [7.39] and [7.40], in the upper half plane. Since $f'(s) = \tau(\alpha+1)s^\alpha + 1 = 0$ has no solution for $\operatorname{Re} s > 0$ and $\alpha \in (0,1)$, multiplicity of the zero $s_0$ is 1. ■

PROOF OF THEOREM 7.2.– Starting from the Laplace transform of $T$ given by expression [7.22] and setting

$$\tilde{S}(\rho, s) = \frac{\tau s^\alpha + 1}{\tau s^{\alpha+1} + s + \lambda\rho^{1+\beta}\sin\frac{\beta\pi}{2}}, \quad \rho > 0, \ \operatorname{Re} s > 0, \qquad [7.41]$$

we come to

$$\tilde{T}(x, s) = \frac{1}{\pi} T_0(x) *_x \int_0^\infty \tilde{S}(s, x)\cos(\rho x)\mathrm{d}\rho, \quad x \in \mathbb{R}, \ \operatorname{Re} s > 0,$$

which after the inversion of the Laplace transform gives [7.28]. The inverse Laplace transform of [7.41] is guaranteed by lemma 7.1 and [ATA 10c, theorem 3.2].

Further, we look for the form of $S$ using the inversion formula for the Laplace transform

$$S(\rho, t) = \frac{1}{2\pi \mathrm{i}} \int_{s_0 - \mathrm{i}\infty}^{s_0 + \mathrm{i}\infty} \tilde{S}(\rho, s) \mathrm{e}^{st} \mathrm{d}s, \quad \rho > 0, \; t > 0,$$

and the Cauchy residues theorem

$$\frac{1}{2\pi \mathrm{i}} \int_{\Gamma} \tilde{S}(\rho, s) \mathrm{e}^{st} \mathrm{d}s = \sum_{n} \mathrm{Res}(\tilde{S}(\rho, s) \mathrm{e}^{st}, s_n),$$

where we chose the integration contour $\Gamma$ to be the Hankel contour (for similar calculations, see, e.g., [KON 10]). Lemma 7.1 implies that $\tilde{S}(\rho, s)\, \mathrm{e}^{st}$ has two complex conjugated poles of the first order, $s_0$ and $\bar{s}_0$, both depending on $\theta = \theta(\rho)$, and therefore

$$\begin{aligned} \sum_{n=1}^{2} \mathrm{Res}(\tilde{S}(\rho, s) \mathrm{e}^{st}, s_n) &= \left. \frac{(\tau s^{\alpha} + 1) \mathrm{e}^{st}}{\tau(\alpha + 1) s^{\alpha} + 1} \right|_{s = s_0(\rho)} \\ &\quad + \left. \frac{(\tau s^{\alpha} + 1) \mathrm{e}^{st}}{\tau(\alpha + 1) s^{\alpha} + 1} \right|_{s = \bar{s}_0(\rho)} = R(\rho, t). \end{aligned}$$

A direct calculation shows that the above sum is a real number, i.e. $R$ takes the form as in [7.31]. On the other hand, integration along the Hankel contour gives

$$\frac{1}{2\pi \mathrm{i}} \int_{\Gamma} \tilde{S}(\rho, s) \mathrm{e}^{st} \mathrm{d}s = \frac{1}{2\pi \mathrm{i}} \int_{s_0 - \mathrm{i}\infty}^{s_0 + \mathrm{i}\infty} \tilde{S}(\rho, s) \mathrm{e}^{st} \mathrm{d}s - I(\rho, t),$$

with $I$ given by [7.30]. Putting it all together ends the proof. ■

### 7.1.3. *Numerical examples*

Now, for particular values of parameters $\tau$, $\lambda$, $\alpha$ and $\beta$, we will calculate the solution $T = T(x,t)$ using formula [7.28] and investigate the initial disturbance propagation. In all examples, we fix arbitrarily chosen parameters $\tau = 0.1$ and $\lambda = 1$, and assume that the initial temperature distribution is the delta distribution, i.e.

$$T_0(x) = \delta(x), \quad x \in \mathbb{R},$$

so that [7.28] reads

$$T(x,t) = \frac{1}{\pi}\int_0^\infty S(\rho,t)\cos(\rho x)\mathrm{d}\rho, \quad x \in \mathbb{R}, \; t > 0, \qquad [7.42]$$

with $S$ given by [7.29]. In the limiting case when $\beta = 1$, solution [7.33] with initial disturbance $\delta$ becomes ($x \in \mathbb{R}, \; t > 0$)

$$\begin{aligned} T(x,t) &= \frac{1}{4\pi}\int_0^\infty \Big(\sqrt{1+\tau q^\alpha \mathrm{e}^{-\mathrm{i}\alpha\pi}}\mathrm{e}^{\mathrm{i}|x|\sqrt{q(1+\tau q^\alpha \mathrm{e}^{-\mathrm{i}\alpha\pi})}} \\ &\quad + \sqrt{1+\tau q^\alpha \mathrm{e}^{\mathrm{i}\alpha\pi}}\mathrm{e}^{-\mathrm{i}|x|\sqrt{q(1+\tau q^\alpha \mathrm{e}^{\mathrm{i}\alpha\pi})}}\Big)\frac{\mathrm{e}^{-qt}}{\sqrt{q}}\mathrm{d}q. \end{aligned} \qquad [7.43]$$

In the limiting case when $\alpha = 1$, solution [7.28] with initial disturbance $\delta$ again becomes [7.42], but now with $S$ given by [7.34], while in the limiting case when $\alpha = 0$ solution [7.28] with initial disturbance $\delta$ again takes the form [7.42]; however, now $S$ is given by [7.35].

Plots of $T$ in all examples are given only for $x \geq 0$, since for $x < 0$ all graphs are symmetric with respect to the $T$ axis, thus having two maxima for $x \in (-\infty, \infty)$. We use the following formulas: [7.42] with $S$ given by [7.29] if $\beta < 1$ and $\alpha < 1$; [7.43] if $\beta = 1$ and $\alpha < 1$; [7.42] with $S$ given by [7.34] if $\beta < 1$ and $\alpha = 1$; [7.42] with $S$ given by [7.35] if $\beta < 1$ and $\alpha = 0$.

Fix $\alpha = 0.25$, $\beta = 0.45$ and consider the behavior of $T$, versus spatial coordinate $x$ for different time instances $t$. Figure 7.2 presents the plot for small times, while in Figure 7.3 the plots are given for larger times.

In both figures, we compare our solution with the solution obtained for $\beta = 1$, i.e. $T$ calculated using [7.43]. Figures 7.2 and 7.3 show the heat conduction phenomena, or the diffusive character of $T$. We see that for each time instant, $T$ has a higher and narrower peak for $\beta < 1$, than for $\beta = 1$. Hence, material conducts heat more slowly in the case when the media is described by the fractional gradient $\beta = 0.45$ than in the case when the gradient is the classical one, i.e. $\beta = 1$.

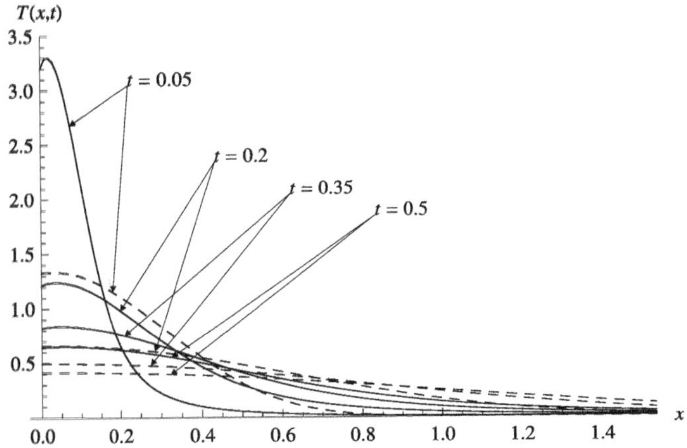


**Figure 7.2.** *Temperature $T$ as a function of $x$ at $t \in \{0.05, 0.2, 0.35, 0.5\}$. Solid lines represent $T$ for $\beta = 0,45$ and dashed lines stand for $\beta = 1$*

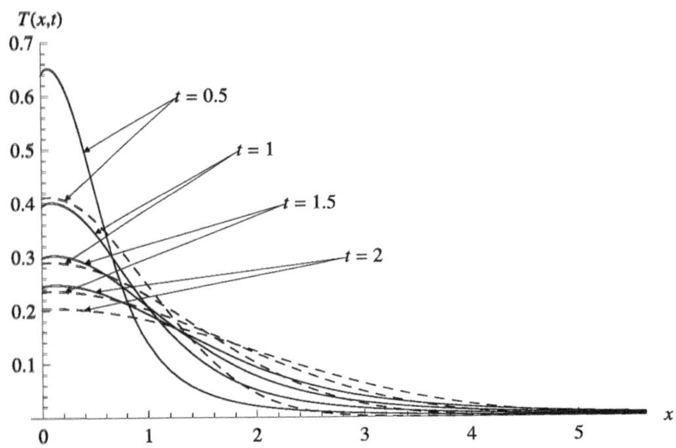


**Figure 7.3.** *Temperature $T$ as a function of $x$ at $t \in \{0.5, 1, 1.5, 2\}$. Solid lines represent $T$ for $\beta = 0,45$ and dashed lines stand for $\beta = 1$*

Next, we fix $\alpha = 0.25$ and the time instant $t = 1$. We consider the behavior of $T$ for various values of $\beta < 1$ versus spatial coordinate $x$. In Figure 7.4, we present the plot for the small values of $\beta$, while in Figure 7.5 the plots are given for larger values of $\beta$. Both figures show that when $\beta$ decreases, the resistance of a material to heat conduction increases. As $\beta$ decreases, the position of the peak tends to the $T$ axis ($T$ tends to $\delta$), and presumably in the limiting case $\beta = 0$ it would be on the $T$ axis, i.e. the initial temperature distribution would not change in time. This indicates that the parameter $\beta$ in the fractional gradient characterizes the ability of media to conduct heat. The limiting case $\beta = 0$ characterizes the media which we call the ideal heat isolator.

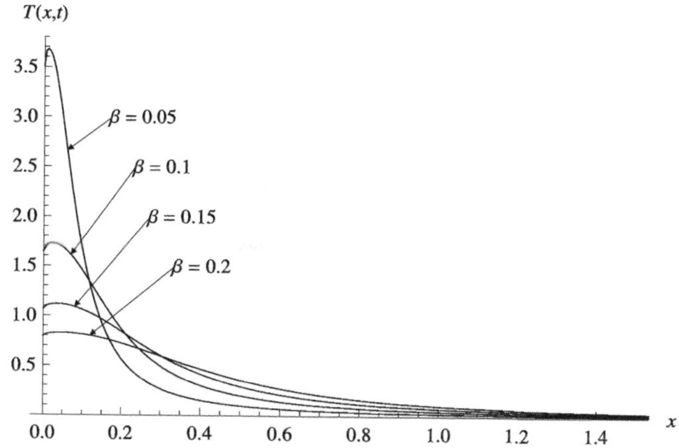


**Figure 7.4.** *Temperature $T$ as a function of $x$ at $t = 1$ for $\beta \in \{0.05, 0.1, 0.15, 0.2\}$*

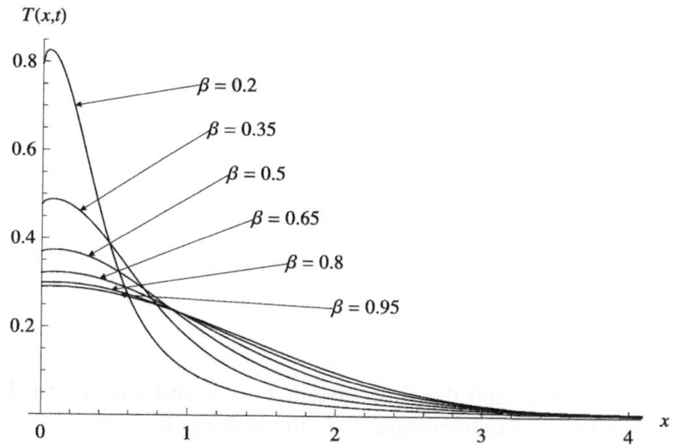


**Figure 7.5.** *Temperature $T$ as a function of $x$ at $t = 1$ for $\beta \in \{0.2, 0.35, 0.5, 0.65, 0.8, 0.95\}$*

Figures 7.6–7.11 present the comparison between the plots of solution for various $\beta < 1$ (solid lines) and solution for $\beta = 1$ (dashed lines). One can notice that, as $\beta \to 1$, the curve of $T$ for $\beta < 1$ tends to the curve of $T$ for $\beta = 1$. In the case $\beta = 0.95$, the graphs of two curves almost overlap.

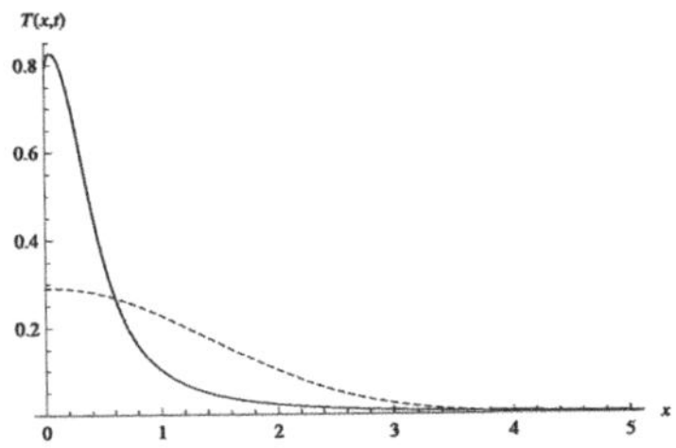


**Figure 7.6.** *T for* $\beta = 0.2$ *and* $\beta = 1$

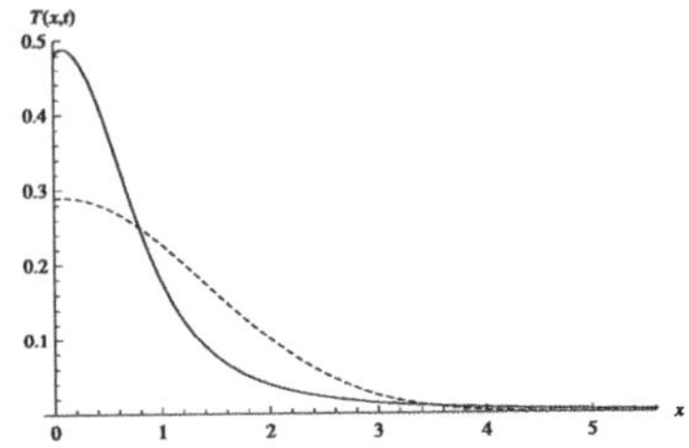


**Figure 7.7.** *T for* $\beta = 0.35$ *and* $\beta = 1$

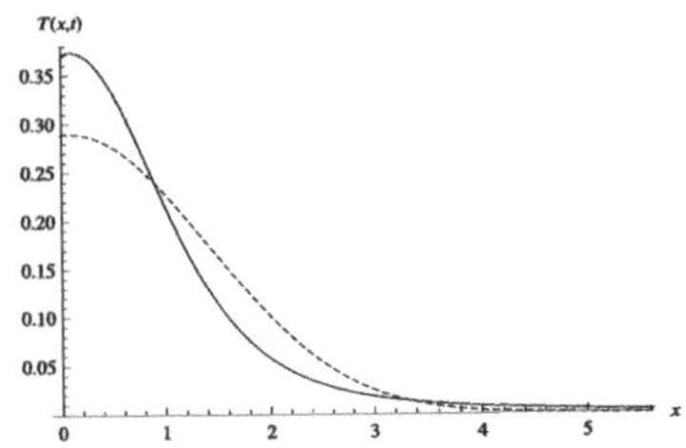


**Figure 7.8.** *T for* $\beta = 0.5$ *and* $\beta = 1$

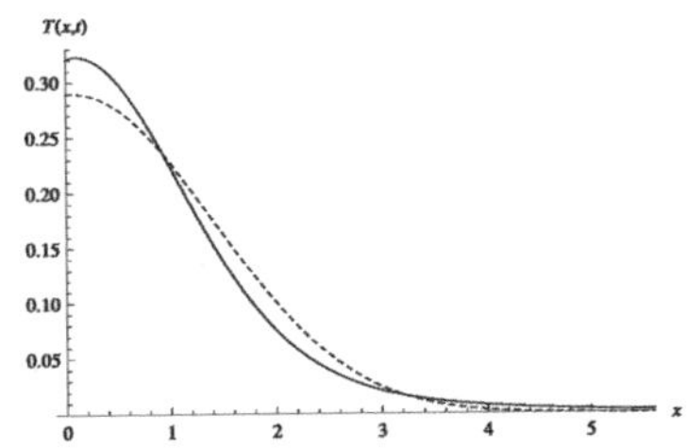


**Figure 7.9.** *T for* $\beta = 0.65$ *and* $\beta = 1$

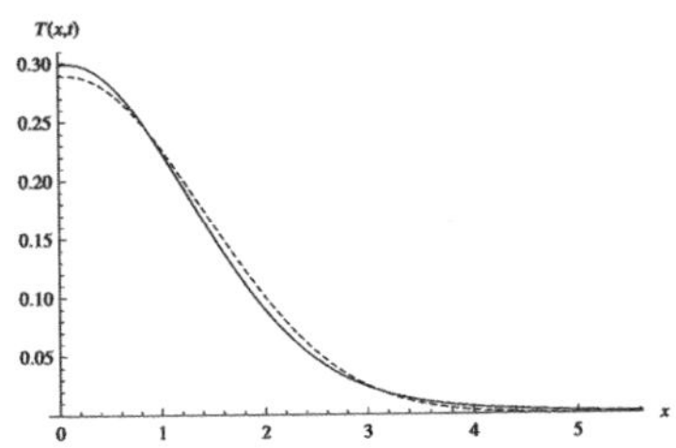


**Figure 7.10.** *T for* $\beta = 0.8$ *and* $\beta = 1$

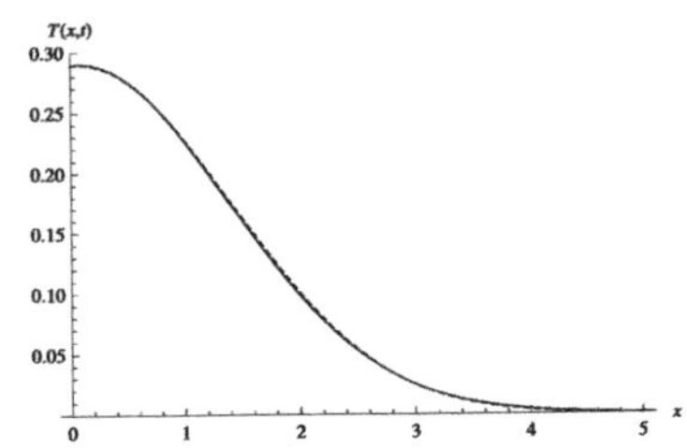


**Figure 7.11.** *T for* $\beta = 0.95$ *and* $\beta = 1$

Now we fix $\beta = 0.45$ and the time instant $t = 2$ and vary $\alpha$. In this case, we examine the dependence of temperature $T$ on the history (described by the parameter $\alpha$) for the same type of spatial non-locality. Figure 7.12 presents the behavior of $T$ versus spatial coordinate $x$ for different values of parameter $\alpha < 1$ (solid lines), compared to the behavior of $T$ in the limiting cases $\alpha = 1$ (dashed line) and $\alpha = 0$ (dotted-dashed line). From Figure 7.12, we see that in the case of the space–time fractional Cattaneo heat conduction law, the plots of temperature $T$ lie between the plots of temperature in the two limiting cases: space-fractional Cattaneo and Fourier heat conduction laws.

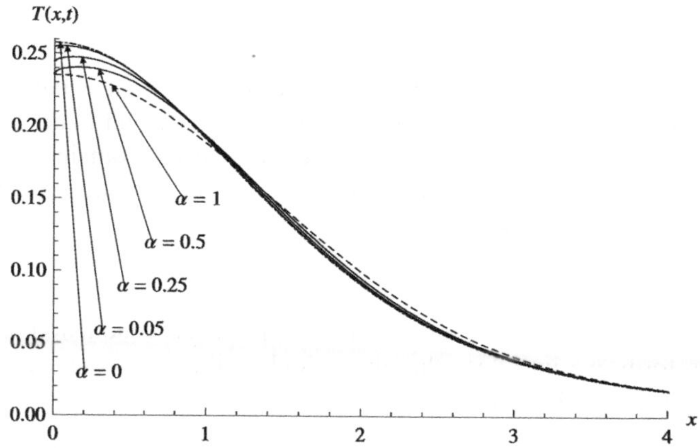


**Figure 7.12.** *Temperature $T$ as a function of $x \in (0,4)$ at $t = 2$ for $\alpha \in \{0, 0.05, 0.25, 0.5, 1\}$.*

## 7.2. Fractional Jeffreys-type heat conduction equation

Following [ATA 10a], we write the model of the fractional Jeffreys-type heat conduction as a system consisting of the energy balance equation [7.4] and the fractional Jeffreys-type heat conduction law [7.11]. This system reads:

$$\rho c \frac{\partial}{\partial t} T(x,t) = -\frac{\partial}{\partial x} q(x,t), \quad x \in \mathbb{R},\ t > 0, \qquad [7.44]$$

$$\left(1 + \tau\, {}_0\mathrm{D}_t^{\alpha}\right) q(x,t) = -\lambda \left(1 + \tau'\, {}_0\mathrm{D}_t^{\alpha}\right) \frac{\partial}{\partial x} T(x,t), \quad x \in \mathbb{R},\ t > 0, \qquad [7.45]$$

with the initial

$$T(x,0) = T_0(x), \quad q(x,0) = 0, \quad \frac{\partial}{\partial x} T(x,0) = 0, \quad x \in \mathbb{R},$$

and boundary data

$$\lim_{x \to \pm\infty} q(x,t) = 0, \quad \lim_{x \to \pm\infty} T(x,t) = 0, \quad t > 0.$$

Note that [7.44] and [7.45] can be written as a single equation as

$$\left(1 + \tau\, {}_0\mathrm{D}_t^{\alpha}\right) \frac{\partial}{\partial t} T(x,t) = \mathcal{D} \left(1 + \tau'\, {}_0\mathrm{D}_t^{\alpha}\right) \frac{\partial^2}{\partial x^2} T(x,t), \quad x \in \mathbb{R},\ t > 0. \qquad [7.46]$$

### 7.2.1. *Solution to the Cauchy problem*

In this section, the fractional Jeffreys-type heat conduction equation, given by system [7.44] and [7.45], is solved by means of the Laplace and Fourier transforms, applied to system [7.48]–[7.50] in the time and space domains, respectively. Inversions of transforms are calculated and a solution is obtained in a convolution form.

#### 7.2.1.1. *Solution of system [7.44], [7.45]*

In order to write the fractional Jeffreys-type heat conduction equation in dimensionless form, we introduce the dimensionless time, coordinate, temperature, heat flux and relaxation times as

$$\bar{t} = \frac{t}{t^*}, \quad \bar{x} = \frac{x}{x^*}, \quad \bar{T} = \frac{T}{T^*}, \quad \bar{q} = \frac{q}{q^*}, \quad \bar{\tau} = \frac{\tau}{(t^*)^{\alpha}}, \quad \bar{\tau}' = \frac{\tau'}{(t^*)^{\alpha}}, \qquad [7.47]$$

where the bar "–" denotes the dimensionless quantity and the asterisk "*" denotes the constant having the dimension of the quantity. Letting

$$t^* = (\tau')^{\frac{1}{\alpha}}, \quad x^* = \sqrt{\frac{\lambda}{\rho c}(\tau')^{\frac{1}{\alpha}}}, \quad \frac{q^*}{T^*} = \sqrt{\frac{\rho c \lambda}{(\tau')^{\frac{1}{\alpha}}}}$$

(and therefore $\bar{\tau}' = 1$) in [7.47], and plugging [7.47] into [7.44] and [7.45], the fractional Jeffreys-type heat conduction equation in dimensionless form, after omitting the bar, reads

$$\frac{\partial}{\partial t} T(x,t) = -\frac{\partial}{\partial x} q(x,t), \quad x \in \mathbb{R}, \ t > 0,$$

$$(1 + \tau\, {}_0\mathrm{D}_t^{\alpha})\, q(x,t) = -(1 + {}_0\mathrm{D}_t^{\alpha}) \frac{\partial}{\partial x} T(x,t), \quad x \in \mathbb{R}, \ t > 0. \qquad [7.48]$$

The initial and boundary conditions are

$$T(x,0) = T_0(x), \quad q(x,0) = 0, \quad \frac{\partial}{\partial x} T(x,0) = 0, \quad x \in \mathbb{R}, \qquad [7.49]$$

$$\lim_{x \to \pm\infty} q(x,t) = 0, \quad \lim_{x \to \pm\infty} T(x,t) = 0, \quad t > 0. \qquad [7.50]$$

Applying the Laplace transform

$$\tilde{f}(\cdot, s) \equiv \mathcal{L}[f(\cdot,t)](\cdot,s) = \int_0^{\infty} f(\cdot,t)\, \mathrm{e}^{-st} \mathrm{d}t, \quad \mathrm{Re}\, s > 0,$$

to [7.48], with initial conditions [7.49] taken into account, we get

$$s\tilde{T}(x,s) - T_0(x) = -\frac{\partial}{\partial x}\tilde{q}(x,s), \quad x \in \mathbb{R}, \ \mathrm{Re}\, s > 0,$$
$$(1+\tau s^{\alpha})\,\tilde{q}(x,s) = -(1+s^{\alpha})\frac{\partial}{\partial x}\tilde{T}(x,s), \quad x \in \mathbb{R}, \ \mathrm{Re}\, s > 0. \qquad [7.51]$$

The Fourier transform

$$\hat{g}(\xi,\cdot) \equiv \mathcal{F}[g(x,\cdot)](\xi,\cdot) = \int_{-\infty}^{\infty} g(x,\cdot)\,\mathrm{e}^{-\mathrm{i}\xi x}\mathrm{d}x, \quad \xi \in \mathbb{R}.$$

applied to [7.51], with the boundary conditions [7.50] taken into account, yields

$$s\hat{\tilde{T}}(\xi,s) - \hat{T}_0(\xi) = -\mathrm{i}\xi\hat{\tilde{q}}(\xi,s), \quad \xi \in \mathbb{R}, \ \mathrm{Re}\, s > 0,$$
$$(1+\tau s^{\alpha})\,\hat{\tilde{q}}(x,s) = -(1+s^{\alpha})\,\mathrm{i}\xi\hat{\tilde{T}}(\xi,s), \quad \xi \in \mathbb{R}, \ \mathrm{Re}\, s > 0. \qquad [7.52]$$

Solving [7.52] with respect to $\hat{\tilde{T}}$, we obtain

$$\hat{\tilde{T}}(\xi,s) = \frac{1+\tau s^{\alpha}}{1+s^{\alpha}}\frac{1}{s\frac{1+\tau s^{\alpha}}{1+s^{\alpha}}+\xi^2}\hat{T}_0(\xi), \quad \xi \in \mathbb{R}, \ \mathrm{Re}\, s > 0.$$

Introducing the solution kernel

$$S = \mathcal{L}^{-1}\left[\mathcal{F}^{-1}\left[\hat{\tilde{S}}\right]\right],$$

where

$$\hat{\tilde{S}}(\xi,s) = \frac{1+\tau s^{\alpha}}{1+s^{\alpha}}\frac{1}{s\frac{1+\tau s^{\alpha}}{1+s^{\alpha}}+\xi^2}, \quad \xi \in \mathbb{R}, \ \mathrm{Re}\, s > 0, \qquad [7.53]$$

we obtain the solution to the fractional Jeffreys-type heat conduction equation [7.48]–[7.50] as

$$T(x,t) = T_0(x) *_x S(x,t), \quad x \in \mathbb{R}, \ t > 0, \qquad [7.54]$$

where $*_x$ denotes the convolution, defined as $(f *_x g)(x) = \int_{-\infty}^{\infty} f(\xi)\,g(x-\xi)d\xi$, $x \in \mathbb{R}$.

7.2.1.2. *Inversions of* $\widehat{\tilde{S}}$

In order to obtain the solution kernel, the first step is to find the inverse Fourier transform of $\widehat{\tilde{S}}$, given by [7.53]. After using a well-known inversion

$$\mathcal{F}^{-1}\left[\frac{1}{\lambda+\xi^2}\right](x) = \frac{\mathrm{e}^{-|x|\sqrt{\lambda}}}{2\sqrt{\lambda}}, \quad \lambda \in \mathbb{C}/(-\infty,0], \ x \in \mathbb{R}, \qquad [7.55]$$

for example, see [KON 10], in [7.53], we obtain

$$\tilde{S}(x,s) = \frac{1}{2}\sqrt{\frac{1}{s}\frac{1+\tau s^\alpha}{1+s^\alpha}}\mathrm{e}^{-|x|\sqrt{s\frac{1+\tau s^\alpha}{1+s^\alpha}}}, \quad x \in \mathbb{R}, \ \operatorname{Re} s > 0. \qquad [7.56]$$

Let us explain that one can apply [7.55] to [7.53] if

$$\lambda(s) \in \mathbb{C}/(-\infty,0] \quad \text{for} \quad \operatorname{Re} s > 0, \qquad [7.57]$$

where

$$\lambda(s) = s\frac{1+\tau s^\alpha}{1+s^\alpha}, \quad \operatorname{Re} s > 0. \qquad [7.58]$$

Take $s = R\mathrm{e}^{\mathrm{i}\phi}$ in [7.58] and for $R > 0, \ \phi \in \left(-\frac{\pi}{2}, \frac{\pi}{2}\right)$, obtain

$$\lambda\left(R\mathrm{e}^{\mathrm{i}\phi}\right) = R(\cos\phi + \mathrm{i}\sin\phi)\frac{1+\tau R^\alpha\cos(\alpha\phi) + \mathrm{i}\tau R^\alpha\sin(\alpha\phi)}{1+R^\alpha\cos(\alpha\phi)+\mathrm{i}R^\alpha\sin(\alpha\phi)}.$$

Real and imaginary parts of $\lambda$ are

$$\begin{aligned}\operatorname{Re}\lambda\left(R\mathrm{e}^{\mathrm{i}\phi}\right) &= \frac{R}{C^2+D^2}\left([AC+BD]\cos\phi + [AD-BC]\sin\phi\right),\\ \operatorname{Im}\lambda\left(R\mathrm{e}^{\mathrm{i}\phi}\right) &= \frac{R}{C^2+D^2}\left([AC+BD]\sin\phi - [AD-BC]\cos\phi\right)\end{aligned}$$

where $A$, $B$, $C$ and $D$ are introduced by

$$\begin{aligned}&A = 1+\tau R^\alpha\cos(\alpha\phi), \quad B = \tau R^\alpha\sin(\alpha\phi),\\ &C = 1+R^\alpha\cos(\alpha\phi), \quad D = R^\alpha\sin(\alpha\phi).\end{aligned}$$

Note that

$$AD - BC = (1 - \tau) R^{\alpha} \sin(\alpha\phi).$$

Let $\phi \in \left[0, \frac{\pi}{2}\right)$. Then $\sin\phi, \sin(\alpha\phi) \geq 0$, and $\cos\phi, \cos(\alpha\phi) > 0$, since $\alpha \in (0,1)$. Therefore, we have $B, D \geq 0$. Additionally, let $\tau \in (0,1]$, then $AD - BC \geq 0$. Hence, $\operatorname{Re}\lambda\left(Re^{\mathrm{i}\phi}\right) > 0$, if $\phi \in \left[0, \frac{\pi}{2}\right)$, and $\tau \in (0,1]$. Now, let $\tau > 1$, then $AD - BC \leq 0$. Hence, $\operatorname{Im}\lambda\left(Re^{\mathrm{i}\phi}\right) \geq 0$, if $\phi \in \left[0, \frac{\pi}{2}\right)$, and $\tau > 1$. Note that $\operatorname{Im}\lambda\left(Re^{\mathrm{i}\phi}\right) = 0$, if $\phi = 0$, but then $\operatorname{Re}\lambda\left(Re^{\mathrm{i}\phi}\right)\big|_{\phi=0} > 0$.

Let $\phi \in \left(-\frac{\pi}{2}, 0\right)$. Then $\sin\phi, \sin(\alpha\phi) < 0$, while $\cos\phi, \cos(\alpha\phi) > 0$. Therefore, we have $A, C > 0$, $B, D < 0$. Additionally, let $\tau \in (0,1]$, then $AD - BC < 0$. Hence, $\operatorname{Re}\lambda\left(Re^{\mathrm{i}\phi}\right) > 0$, if $\phi \in \left(-\frac{\pi}{2}, 0\right)$, and $\tau \in (0,1]$. Now, let $\tau > 1$, then $AD - BC > 0$. Hence, $\operatorname{Im}\lambda\left(Re^{\mathrm{i}\phi}\right) < 0$, if $\phi \in \left(-\frac{\pi}{2}, 0\right)$, and $\tau > 1$. The conclusion is that condition [7.57] is satisfied and one can apply [7.55] to [7.53].

The second step is the inversion of the Laplace transform in [7.56]. We have

$$S(\cdot, t) = \mathcal{L}^{-1}\left[\tilde{S}(\cdot, s)\right](t) = \frac{1}{2\pi\mathrm{i}} \int_{\gamma} \tilde{S}(\cdot, s)\, \mathrm{e}^{st} \mathrm{d}s, \quad t > 0,$$

where $\gamma = \{s \in \mathbb{C} \mid \operatorname{Re} s = s_0,\ \operatorname{Im} s \in (-\infty, \infty)\}$ is the Bromwich contour. Function $\tilde{S}$, given by [7.56], has $s = 0$ and $s = \infty$ as only branch points and it has no singularities in the complex plane. Hence, it can be evaluated using the Cauchy theorem

$$\oint_{\Gamma} \tilde{S}(x, s)\, \mathrm{e}^{st} \mathrm{d}s = 0, \quad x \in \mathbb{R},\ t > 0, \qquad [7.59]$$

where $\Gamma = \Gamma_1 \cup \Gamma_2 \cup \Gamma_\varepsilon \cup \Gamma_3 \cup \Gamma_4 \cup \gamma_0$ is the contour given in Figure 7.13. In the limit when $R \to \infty$, the integral along contour $\Gamma_1$ reads ($x \in \mathbb{R}$, $t > 0$)

$$\begin{aligned} &\lim_{R\to\infty} \int_{\Gamma_1} \tilde{S}(x, s)\, \mathrm{e}^{st} \mathrm{d}s \\ &= \frac{1}{2} \lim_{R\to\infty} \int_{\varphi_0}^{\pi} \sqrt{\frac{1}{R\mathrm{e}^{\mathrm{i}\varphi}} \frac{1 + \tau R^{\alpha} \mathrm{e}^{\mathrm{i}\alpha\varphi}}{1 + R^{\alpha} \mathrm{e}^{\mathrm{i}\alpha\varphi}}}\, \mathrm{e}^{-|x|\sqrt{R\mathrm{e}^{\mathrm{i}\varphi} \frac{1 + \tau R^{\alpha} \mathrm{e}^{\mathrm{i}\alpha\varphi}}{1 + R^{\alpha} \mathrm{e}^{\mathrm{i}\alpha\varphi}}} + Rt\mathrm{e}^{\mathrm{i}\varphi}}\, \mathrm{i}R\mathrm{e}^{\mathrm{i}\varphi} \mathrm{d}\varphi, \end{aligned}$$

where $\lim_{R\to\infty} \varphi_0 = \frac{\pi}{2}$. Evaluating the absolute value of $\int_{\Gamma_1} \tilde{S}(x,s)e^{st}ds$, we obtain $(x \in \mathbb{R}, t > 0)$

$$\lim_{R\to\infty} \left| \int_{\Gamma_1} \tilde{S}(x,s)\, \mathrm{e}^{st} \mathrm{d}s \right|$$
$$\leq \frac{1}{2} \lim_{R\to\infty} \int_{\varphi_0}^{\pi} \sqrt{R} \left| \sqrt{\frac{1+\tau R^{\alpha} \mathrm{e}^{\mathrm{i}\alpha\varphi}}{1+R^{\alpha}\mathrm{e}^{\mathrm{i}\alpha\varphi}}} \right| \left| \mathrm{e}^{-|x|\sqrt{R\mathrm{e}^{\mathrm{i}\varphi} \frac{1+\tau R^{\alpha}\mathrm{e}^{\mathrm{i}\alpha\varphi}}{1+R^{\alpha}\mathrm{e}^{\mathrm{i}\alpha\varphi}}}} \right| \mathrm{e}^{Rt\cos\varphi} \mathrm{d}\varphi. \quad [7.60]$$

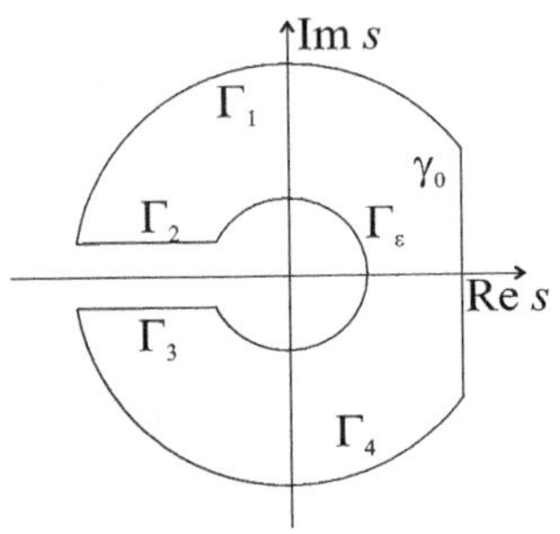


**Figure 7.13.** *Integration contour* $\Gamma$

Since $\lim_{R\to\infty} \sqrt{\frac{1+\tau R^{\alpha}\mathrm{e}^{\mathrm{i}\alpha\varphi}}{1+R^{\alpha}\mathrm{e}^{\mathrm{i}\alpha\varphi}}} = \sqrt{\tau}$, $\cos\varphi < 0$, and $\cos\frac{\varphi}{2} > 0$ for $\varphi \in \left(\frac{\pi}{2}, \pi\right)$, [7.60] takes the form

$$\lim_{R\to\infty} \left| \int_{\Gamma_1} \tilde{S}(x,s)\, \mathrm{e}^{st} \mathrm{d}s \right| \leq \frac{1}{2}\sqrt{\tau} \lim_{R\to\infty} \int_{\varphi_0}^{\pi} \sqrt{R} \mathrm{e}^{R\left(t\cos\varphi - \frac{1}{\sqrt{R}}|x|\sqrt{\tau}\cos\frac{\varphi}{2}\right)} \mathrm{d}\varphi = 0.$$

Similar arguments are valid for the integral along $\Gamma_4$. In the limit when $\varepsilon \to 0$, the absolute value of integral along $\Gamma_\varepsilon$ is given by the formula similar to [7.60] and it is estimated as $(x \in \mathbb{R}, t > 0)$

$$\lim_{\varepsilon\to 0} \left| \int_{\Gamma_\varepsilon} \tilde{S}(x,s)\, \mathrm{e}^{st} \mathrm{d}s \right|$$
$$\leq \frac{1}{2} \lim_{\varepsilon\to 0} \int_{\pi}^{-\pi} \sqrt{\varepsilon} \left| \sqrt{\frac{1+\tau\varepsilon^{\alpha}\mathrm{e}^{\mathrm{i}\alpha\varphi}}{1+\varepsilon^{\alpha}\mathrm{e}^{\mathrm{i}\alpha\varphi}}} \right| \left| \mathrm{e}^{-|x|\sqrt{\varepsilon\mathrm{e}^{\mathrm{i}\varphi} \frac{1+\tau\varepsilon^{\alpha}\mathrm{e}^{\mathrm{i}\alpha\varphi}}{1+\varepsilon^{\alpha}\mathrm{e}^{\mathrm{i}\alpha\varphi}}}} \right| \mathrm{e}^{\varepsilon t\cos\varphi} \mathrm{d}\varphi = 0.$$

In the limit when $R \to \infty$, $\varepsilon \to 0$, integrals along contours $\Gamma_2$, $\Gamma_3$ and $\gamma_0$ give $(x \in \mathbb{R},\ t > 0)$

$$\lim_{\substack{R\to\infty\\ \varepsilon\to 0}} \int_{\Gamma_2} \tilde{S}(x,s)\,\mathrm{e}^{st}\mathrm{d}s = \frac{1}{2\mathrm{i}} \int_0^\infty \sqrt{\frac{1+\tau q^\alpha \mathrm{e}^{\mathrm{i}\alpha\pi}}{1+q^\alpha \mathrm{e}^{\mathrm{i}\alpha\pi}}}\,\mathrm{e}^{-qt-\mathrm{i}|x|\sqrt{q}\sqrt{\frac{1+\tau q^\alpha \mathrm{e}^{\mathrm{i}\alpha\pi}}{1+q^\alpha \mathrm{e}^{\mathrm{i}\alpha\pi}}}} \frac{\mathrm{d}q}{\sqrt{q}},$$

$$\lim_{\substack{R\to\infty\\ \varepsilon\to 0}} \int_{\Gamma_3} \tilde{S}(x,s)\,\mathrm{e}^{st}\mathrm{d}s = \frac{1}{2\mathrm{i}} \int_0^\infty \sqrt{\frac{1+\tau q^\alpha \mathrm{e}^{-\mathrm{i}\alpha\pi}}{1+q^\alpha \mathrm{e}^{-\mathrm{i}\alpha\pi}}}\,\mathrm{e}^{-qt+\mathrm{i}|x|\sqrt{q}\sqrt{\frac{1+\tau q^\alpha \mathrm{e}^{-\mathrm{i}\alpha\pi}}{1+q^\alpha \mathrm{e}^{-\mathrm{i}\alpha\pi}}}} \frac{\mathrm{d}q}{\sqrt{q}},$$

$$\lim_{\substack{R\to\infty\\ \varepsilon\to 0}} \int_{\gamma_0} \tilde{S}(x,s)\,\mathrm{e}^{st}\mathrm{d}s = 2\pi\mathrm{i}S(x,t).$$

Now, by the Cauchy theorem [7.59], $S$ is calculated as $(x \in \mathbb{R},\ t > 0)$

$$\begin{aligned} S(x,t) &= \frac{1}{4\pi}\int_0^\infty \left( \sqrt{\frac{1+\tau q^\alpha \mathrm{e}^{-\mathrm{i}\alpha\pi}}{1+q^\alpha \mathrm{e}^{-\mathrm{i}\alpha\pi}}}\,\mathrm{e}^{\mathrm{i}|x|\sqrt{q}\sqrt{\frac{1+\tau q^\alpha \mathrm{e}^{-\mathrm{i}\alpha\pi}}{1+q^\alpha \mathrm{e}^{-\mathrm{i}\alpha\pi}}}} \right. \\ &\quad \left. + \sqrt{\frac{1+\tau q^\alpha \mathrm{e}^{\mathrm{i}\alpha\pi}}{1+q^\alpha \mathrm{e}^{\mathrm{i}\alpha\pi}}}\,\mathrm{e}^{-\mathrm{i}|x|\sqrt{q}\sqrt{\frac{1+\tau q^\alpha \mathrm{e}^{\mathrm{i}\alpha\pi}}{1+q^\alpha \mathrm{e}^{\mathrm{i}\alpha\pi}}}} \right) \frac{\mathrm{e}^{-qt}}{\sqrt{q}}\mathrm{d}q. \end{aligned} \qquad [7.61]$$

Having the resolvent calculated by [7.61], the solution itself is obtained from [7.54].

Note that the proposed equation [7.46] is not supposed to exhibit the property of the finite speed of the disturbance propagation. This property can be obtained as a condition for the convergence of integrals in the contour $\Gamma$, like in [KON 10], for example.

### 7.2.2. *Numerical examples*

In several numerical examples that are to follow, the behavior of the solution to systems [7.48]–[7.50] is investigated for the initial condition $T_0(x) = \delta(x)$, $x \in \mathbb{R}$, where $\delta$ is the Dirac distribution. Then, the solution $T$, given by [7.54], is obtained as $T = S$, where $S$ is given by [7.61]. Figures 7.14 and 7.15 present plots of $T$, as described above, for the set of parameters $\alpha = 0.25$ and $\tau = 0.7$. Figure 7.14 clearly displays the diffusive character of $T$. Figures 7.16 and 7.17 present the solution $T$ when $\alpha$ changes as $\alpha \in \{0.05, 0.25, 0.5, 0.75, 0.95\}$, while $\tau = 0.005$ is fixed.

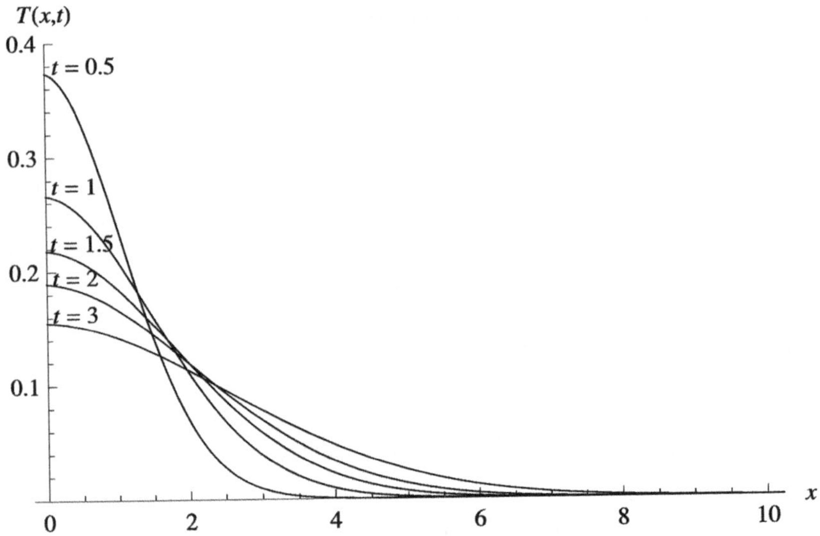


**Figure 7.14.** *Plots of* $T(x,t)$ *for* $T_0 = \delta, t \in \{0.5, 1, 1.5, 2, 3\}, x \in (0, 10)$.

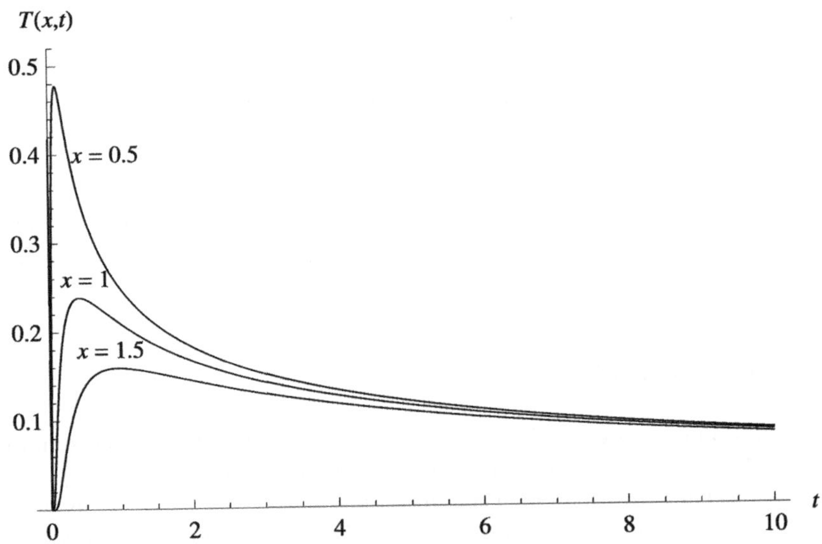


**Figure 7.15.** *Plots of* $T(x,t)$ *for* $T_0 = \delta, t \in (0, 10), x \in \{0.5, 1, 1.5\}$.

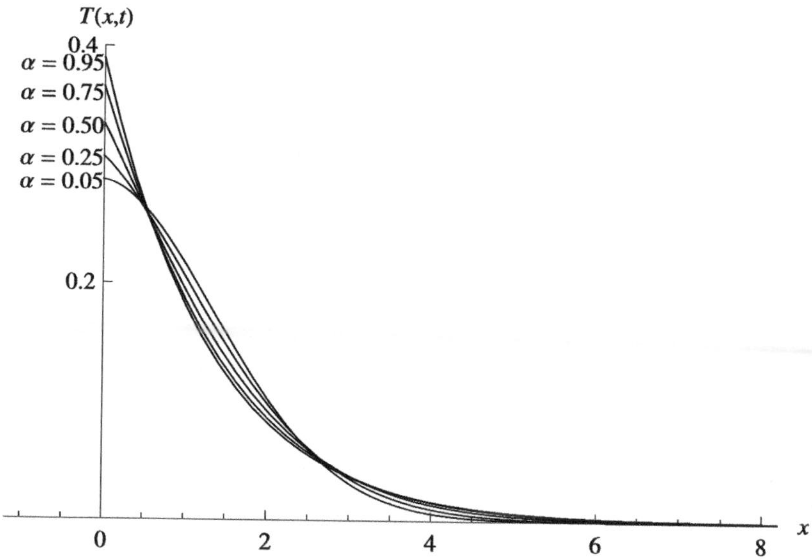


**Figure 7.16.** *Plots of* $T(x, 0.5)$ *for* $T_0 = \delta$, $x \in (0, 8)$ .

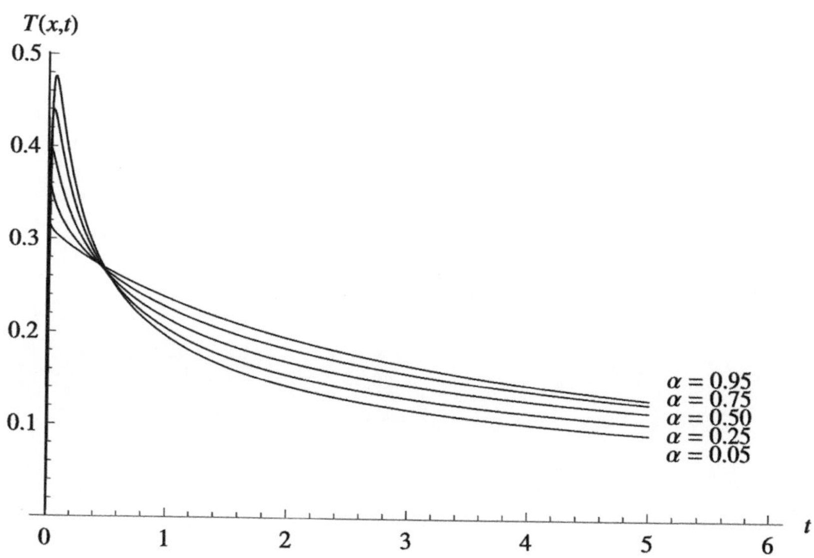


**Figure 7.17.** *Plots of* $T(0.5, t)$ *for* $T_0 = \delta$, $t \in (0, 5)$ .

# Bibliography

[ABD 99] ABDULLAH S., "On solvability of convolution equations in spaces of generalized distributions with restricted growth", *Journal of Mathematical Analysis and Applications*, vol. 233, pp. 301–313, 1999.

[ABR 64] ABRAMOWITZ M., STEGUN I.A., *Handbook of Mathematical Functions*, National Bureau of Standards, 1964.

[AGR 01] AGARWAL R.P., MEEHAN M., O'REGAN D., *Nonlinear Integral Equations and Inclusions*, Nova Science, Huntington, NY, 2001.

[AGR 02a] AGRAWAL O.P., "Formulation of Euler-Lagrange equations for fractional variational problems", *Journal of Mathematical Analysis and Applications*, vol. 272, pp. 368–379, 2002.

[AGR 02b] AGRAWAL O.P., "Solution for a fractional diffusion-wave equation defined in a bounded domain", *Nonlinear Dynamics*, vol. 29, pp. 145–155, 2002.

[AGR 04] AGRAWAL O.P., "A general formulation and solution scheme for fractional optimal control problems", *Nonlinear Dynamics*, vol. 38, pp. 323–337, 2004.

[AGR 06] AGRAWAL O.P., "Fractional variational calculus and the transversality conditions", *Journal of Physics A: Mathematical and General*, vol. 39, pp. 10375–10384, 2006.

[AGR 07] AGRAWAL O.P., "Fractional variational calculus in terms of Riesz fractional derivatives", *Journal of Physics A: Mathematical and Theoretical*, vol. 40, pp. 6287–6303, 2007.

[AGR 08] AGRAWAL O.P., "A general finite element formulation for fractional variational problems", *Journal of Mathematical Analysis and Applications*, vol. 337, pp. 1–12, 2008.

[ALK 05] AL-KHALED K., MOMANI S., "An approximate solution for a fractional diffusion-wave equation using the decomposition method", *Applied Mathematics and Computation*, vol. 165, pp. 473–483, 2005.

[ALM 09] ALMEIDA R., TORRES D.F.M., "Calculus of variations with fractional derivatives and fractional integrals", *Applied Mathematics Letters*, vol. 22, pp. 1816–1820, 2009.

[ALM 11] ALMEIDA R., TORRES D.F.M., "Necessary and sufficient conditions for the fractional calculus of variations with Caputo derivatives", *Communications in Nonlinear Science and Numerical Simulation*, vol. 16, pp. 1490–1500, 2011.

[ALM 12] ALMEIDA R., MALINOWSKA A.B., TORRES D.F.M., "Fractional Euler-Lagrange differential equations via Caputo derivatives", in BALEANU D., TENREIRO MACHADO J.A., LUO A.C.J. (eds), *Fractional Dynamics and Control*, Springer, New York, pp. 109–118, 2012.

[ANA 09] ANASTASSIOU G.A., *Fractional Differentiation Inequalities*, Springer Science+Business Media LLC, New York, 2009.

[ANT 86] ANTHONY K.H., "A new approach to thermodynamics of irreversible processes by means of Lagrange formalism", in KILMISTER C.W. (ed.), *Disequilibrium and Self-Organisation*, D. Reidel, Dordrecht, the Netherlands, pp. 75–92, 1986.

[ARE 01] ARENDT W., BATTY C.J.K., HIEBER M., NEUBRANDER F., *Vector-Valued Laplace Transforms and Cauchy Problems*, Birkhäuser Verlag, Basel, 2001.

[ART 73] ARTHURS A.M. "Dual extremum principles and error bounds for a class of boundary value problems", *Journal of Mathematical Analysis and Applications*, vol. 41, pp. 781–795, 1973.

[ART 80] ARTHURS A.M., *Complementary Variational Principles*, Clarendon Press, Oxford, 1980.

[ATA 83] ATANACKOVIĆ T.M., DJUKIC D.S., "An extremum variational principle for a class of boundary value problems", *Journal of Mathematical Analysis and Applications*, vol. 93, pp. 344–362, 1983.

[ATA 97] ATANACKOVIĆ T.M., *Stability Theory of Elastic Rods*, World Scientific, New Jersey, 1997.

[ATA 00] ATANACKOVIĆ T.M., GURAN A., *Theory of Elasticity for Scientists and Engineers*, Birkhäuser, Boston, MA, 2000.

[ATA 02a] ATANACKOVIĆ T.M., "A generalized model for the uniaxial isothermal deformation of a viscoelastic body", *Acta Mechanica*, vol. 159, pp. 77–86, 2002.

[ATA 02b] ATANACKOVIĆ T.M., "A modified Zener model of a viscoelastic body", *Continuum Mechanics and Thermodynamics*, vol. 14, pp. 137–148, 2002.

[ATA 02c] ATANACKOVIĆ T.M., STANKOVIĆ B., "Dynamics of a viscoelastic rod of fractional derivative type", *Zeitschrift für Angewandte Mathematik und Mechanik*, vol. 82, pp. 377–386, 2002.

[ATA 03] ATANACKOVIĆ T.M. "On a distributed derivative model of a viscoelastic body", *Comptes Rendus Mécanique*, vol. 331, pp. 687–692, 2003.

[ATA 04a] ATANACKOVIĆ T.M., SPASIC D.T., "On viscoelastic compliant contact-impact models", *Journal of Applied Mechanics – Transactions of ASME*, vol. 71, pp. 134–138, 2004.

[ATA 04b] ATANACKOVIĆ T.M., STANKOVIĆ B., "An expansion formula for fractional derivatives and its applications", *Fractional Calculus and Applied Analysis*, vol. 7, pp. 365–378, 2004.

[ATA 04c] ATANACKOVIĆ T.M., STANKOVIĆ B., "On a system of differential equations with fractional derivatives arising in rod theory", *Journal of Physics A: Mathematical and General*, vol. 37, pp. 1241–1250, 2004.

[ATA 04d] ATANACKOVIĆ T.M., STANKOVIĆ B., "Stability of an elastic rod on a fractional derivative type of foundation", *Journal of Sound and Vibration*, vol. 227, pp. 149–161, 2004.

[ATA 05a] ATANACKOVIĆ T.M., BUDINCEVIC M., PILIPOVIĆ S., "On a fractional distributed-order oscillator", *Journal of Physics A: Mathematical and General*, vol. 38, pp. 6703–6713, 2005.

[ATA 05b] ATANACKOVIĆ T.M., PILIPOVIĆ S., "On a class of equations arising in linear viscoelasticity theory", *Zeitschrift für Angewandte Mathematik und Mechanik*, vol. 85, pp. 748–754, 2005.

[ATA 06] ATANACKOVIĆ T.M., OPARNICA L., PILIPOVIĆ S., "On a model of viscoelastic rod in unilateral contact with a rigid wall", *IMA Journal of Applied Mathematics*, vol. 71, pp. 1–13, 2006.

[ATA 07a] ATANACKOVIĆ T.M., OPARNICA L., PILIPOVIĆ S., "On a nonlinear distributed order fractional differential equation", *Journal of Mathematical Analysis and Applications*, vol. 328, pp. 590–608, 2007.

[ATA 07b] ATANACKOVIĆ T.M., PILIPOVIĆ S., ZORICA D., "Diffusion wave equation with two fractional derivatives of different order", *Journal of Physics A: Mathematical and Theoretical*, vol. 40, pp. 5319–5333, 2007.

[ATA 08a] ATANACKOVIĆ T.M., KONJIK S., PILIPOVIĆ S., "Variational problems with fractional derivatives: Euler-Lagrange equations", *Journal of Physics A: Mathematical and Theoretical*, vol. 41, pp. 095201–095213, 2008.

[ATA 08b] ATANACKOVIĆ T.M., STANKOVIĆ B., "On a numerical scheme for solving differential equations of fractional order", *Mechanics Research Communications*, vol. 35, pp. 429–438, 2008.

[ATA 09a] ATANACKOVIĆ T.M., KONJIK S., PILIPOVIĆ S., *et al.*, "Variational problems with fractional derivatives: invariance conditions and Noether theorem", *Nonlinear Analysis*, vol. 71, pp. 1504–1517, 2009.

[ATA 09b] ATANACKOVIĆ T.M., OPARNICA L., PILIPOVIĆ S., "Distributional framework for solving fractional differential equation", *Integral Transforms and Special Functions*, vol. 20, pp. 215–222, 2009.

[ATA 09c] ATANACKOVIĆ T.M., PILIPOVIĆ S., ZORICA D., "Existence and calculation of the solution to the time distributed order diffusion equation", *Physica Scripta*, vol. T136, 014012, pp. 1–6, 2009.

[ATA 09d] ATANACKOVIĆ T.M., PILIPOVIĆ S., ZORICA D., "Time distributed-order diffusion-wave equation. I. Volterra type equation", *Proceedings of the Royal Society A: Mathematical, Physical and Engineering Sciences*, vol. 465, pp. 1869–1891, 2009.

[ATA 09e] ATANACKOVIĆ T.M., PILIPOVIĆ S., ZORICA D., "Time distributed-order diffusion-wave equation. II. Applications of the Laplace and Fourier transformations", *Proceedings of the Royal Society A: Mathematical, Physical and Engineering Sciences*, vol. 465, pp. 1893–1917, 2009.

[ATA 09f] ATANACKOVIĆ T.M., STANKOVIĆ B., "Generalized wave equation in nonlocal elasticity", *Acta Mechanica*, vol. 208, pp. 1–10, 2009.

[ATA 10a] ATANACKOVIĆ T.M., GRILLO A., WITTUM G., *et al.*, "Fractional Jeffreys-type diffusion equation", *Proceedings of the 4th IFAC Workshop on Fractional Differentiation and its Applications*, University of Extremadura, Badajoz, Spain, 18–20 October, 2010.

[ATA 10b] ATANACKOVIĆ T.M., KONJIK S., OPARNICA L., *et al.*, "Generalized Hamilton's principle with fractional derivatives", *Journal of Physics A: Mathematical and Theoretical*, vol. 43, 255203, pp. 1–12, 2010.

[ATA 10c] ATANACKOVIĆ T.M., OPARNICA L., PILIPOVIĆ S., "Semilinear ordinary differential equation coupled with distributed order fractional differential equation", *Nonlinear Analysis: Theory, Methods and Applications*, vol. 72, pp. 4101–4114, 2010.

[ATA 11a] ATANACKOVIĆ T.M., DOLICANIN D., KONJIK S., *et al.*, "Dissipativity and stability for a nonlinear differential equation with distributed order symmetrized fractional derivative", *Applied Mathematics Letters*, vol. 24, pp. 1020–1025, 2011.

[ATA 11b] ATANACKOVIĆ T.M., KONJIK S., OPARNICA L., *et al.*, "Thermodynamical restrictions and wave propagation for a class of fractional order viscoelastic rods", *Abstract and Applied Analysis*, vol. 2011, Article ID 975694, pp. 1–32, 2011.

[ATA 11c] ATANACKOVIĆ T.M., PILIPOVIĆ S., "Hamilton's principle with variable order fractional derivatives", *Fractional Calculus and Applied Analysis*, vol. 14, pp. 94–109, 2011.

[ATA 11d] ATANACKOVIĆ T.M., PILIPOVIĆ S., ZORICA D., "Distributed-order fractional wave equation on a finite domain: creep and forced oscillations of a rod", *Continuum Mechanics and Thermodynamics*, vol. 23, pp. 305–318, 2011.

[ATA 11e] ATANACKOVIĆ T.M., PILIPOVIĆ S., ZORICA D., "Distributed-order fractional wave equation on a finite domain. Stress relaxation in a rod", *International Journal of Engineering Science*, vol. 49, pp. 175–190, 2011.

[ATA 12a] ATANACKOVIĆ T.M., JANEV M., PILIPOVIĆ S., *et al.*, "Complementary variational principles with fractional derivatives", *Acta Mechanica*, vol. 223, pp. 685–704, 2012.

[ATA 12b] ATANACKOVIĆ T.M., KONJIK S., OPARNICA L., *et al.*, "The Cattaneo type space-time fractional heat conduction equation", *Continuum Mechanics and Thermodynamics*, vol. 24, pp. 293–311, 2012.

[ATA 12c] ATANACKOVIĆ T.M., KONJIK S., PILIPOVIĆ S. "Fractionalization of constitutive equations in viscoelasticity", *Proceedings of the 5th Symposium on Fractional Differentiation and Its Applications*, Hohai University, Nanjing, China, 14–17 May, 2012.

[ATA 12d] ATANACKOVIĆ T.M., PILIPOVIĆ S., STANKOVIĆ B., "A theory of linear differential equations with fractional derivatives", *Bulletin de l'Académie Serbe des Sciences et des Arts, Sciences Mathématiques*, vol. 37, pp. 73–97, 2012.

[ATA 12e] ATANACKOVIĆ T.M., PILIPOVIĆ S., ZORICA D., "On a system of equations arising in viscoelasticity theory of fractional type", arXiv:1205.5343:22 pp, 1–22,2012.

[ATA 13a] ATANACKOVIĆ T.M., JANEV M., PILIPOVIĆ S., *et al.*, "An expansion formula for fractional derivatives of variable order", *Central European Journal of Physics*, pp. 1–11, 2013.

[ATA 13b] ATANACKOVIĆ T.M., PILIPOVIĆ S., STANKOVIĆ B., "On a class of fractional order differential equations", 2013, forthcoming.

[ATA 13c] ATANACKOVIĆ T.M., PILIPOVIĆ S., ZORICA D., "Forced oscillations of a body attached to a viscoelastic rod of fractional derivative type", *International Journal of Engineering Science*, vol. 64, pp. 54–65, 2013.

[ATA 13d] ATANACKOVIĆ T.M., PILIPOVIĆ S., ZORICA D., "An initial value problem arising in mechanics", *Archive of Applied Mechanics*, 2013.

[ATA 13e] ATANACKOVIĆ T.M., ZORICA D., "On the Bagley-Torvik equation", *Journal of Applied Mechanics – Transactions of the ASME*, vol. 80, 041013, pp 1–4, 2013.

[ATA 14a] ATANACKOVIĆ T.M., JANEV M., KONJIK S., *et al.*, "Expansion formula for fractional derivatives in variational problems", *Journal of Mathematical Analysis and Applications*, vol. 409, pp. 911–924, 2014.

[ATA 14b] ATANACKOVIĆ T.M., PILIPOVIĆ S., STANKOVIĆ B., ZORICA D., *Fractional Calculus with Applications in Mechanics: Wave Propagation, Impact and Variational Principles*, ISTE, London, John Wiley & Sons, New York, 2014.

[AUB 06] AUBERT G., KORNPROBST P., *Mathematical Problems in Image Processing, Partial Differential Equations and the Calculus of Variation*, Springer Science+Business media LLC, New York, 2006.

[BAC 00] BACLIC B.S., ATANACKOVIĆ T.M., "Stability and creep of a fractional derivative order viscoelastic rod", *Bulletin de l'Académie Serbe des Sciences et des Arts, Sciences Mathématiques*, vol. 121, pp. 115–131, 2000.

[BAG 86] BAGLEY R.L., TORVIK P.J., "On the fractional calculus model of viscoelastic behavior", *Journal of Rheology*, vol. 30, pp. 133–155, 1986.

[BAG 00a] BAGLEY R.L., TORVIK P.J., "On the existence of the order domain and the solution of distributed order equations. I", *International Journal of Applied Mathematics*, vol. 2, pp. 865–882, 2000.

[BAG 00b] BAGLEY R.L., TORVIK P.J., "On the existence of the order domain and the solution of distributed order equations. II", *International Journal of Applied Mathematics*, vol. 2, pp. 965–987, 2000.

[BAI 07] BAI J., FENG X.C., "Fractional order anisotropic diffusion for image denoising", *IEEE Transactions on Image Processing*, vol. 16, pp. 2492–2502, 2007.

[BAL 04] BALEANU D., AVKAR T., "Lagrangians with linear velocities within Riemann-Liouville fractional derivatives", *Il Nuovo Cimento B*, vol. 119, pp. 73–79, 2004.

[BAL 08] BALEANU D., TRUJILLO J.J., "On exact solutions of a class of fractional Euler-Lagrange equations", *Nonlinear Dynamics*, vol. 52, pp. 331–335, 2008.

[BAL 12a] BALEANU D., DIETHELM K., SCALAS E., *et al.*, *Fractional Calculus: Models and Numerical Methods*, World Scientific, Singapore, 2012.

[BAL 12b] BALEANU D., TENREIRO MACHADO J.A., LUO A.C.J. (eds), *Fractional Dynamics and Control*, Springer, New York, 2012.

[BER 68] BERG L., *Asymptotische Darstellungen und Entwicklungen*, VEB Deutscher Verlag der Wissenschaften, Berlin, 1968.

[BEY 95] BEYER H., KEMPFLE S., "Definition of physically consistent damping laws with fractional derivatives", *Zeitschrift für Angewandte Mathematik und Mechanik*, vol. 75, pp. 623–635, 1995.

[BLU 89] BLUMAN G.W., KUMEI S., *Symmetries and Differential Equations*, vol. 5, Springer, Berlin, 1989.

[BOR 12] BORN M., VON KÁRMÁN T., "On fluctuations in spatial grids", *Physikalishe Zeitschrift*, vol. 13, pp. 297–309, 1912.

[BRO 99] BROGLIATO B., *Nonsmooth Mechanics: Models, Dynamics and Control*, Springer, New York, 1999.

[BUC 98] BUCKWAR E., LUCHKO Y., "Invariance of a partial differential equation of fractional order under the Lie group of scaling transformation", *Journal of Mathematical Analysis and Applications*, vol. 227, pp. 81–97, 1998.

[BUT 00] BUTZER P.L., WESTPHAL U., "An introduction to fractional calculus", in HILFER R. (ed.), *Applications of Fractional Calculus in Physics*, World Scientific, Singapore, pp. 1–85, 2000.

[CAM 08] CAMARGO R.F., CHIACCHIO A.O., CAPELAS DE OLIVEIRA E., "Differentiation to fractional orders and the fractional telegraph equation", *Journal of Mathematical Physics*, vol. 49, 033505 pp. 1–12, 2008.

[CAM 09] CAMARGO R.F., CHARNET R., CAPELAS DE OLIVEIRA E., "On some fractional Green's functions", *Journal of Mathematical Physics*, vol. 50, 043514 PP. 1–12, 2009.

[CAN 87] CANAVATI J.A., "The Riemann-Liouville integral", *Nieuw Archief Voor Wiskunde*, vol. 5, pp. 53–75, 1987.

[CAN 02] CANTWELL B.J., *Introduction to Symmetry Analysis*, vol. 5, Cambridge University Press, Cambridge, 2002.

[CAP 67] CAPUTO M., "Linear models of dissipation whose *Q* is almost frequency independent – II", *Geophysical Journal of the Royal Astronomical Society*, vol. 13, pp. 529–539, 1967.

[CAP 71a] CAPUTO M., MAINARDI F., "Linear models of dissipation in an elastic solids", *Rivista del Nuovo Cimento (Ser. II)*, vol. 1, pp. 161–198, 1971.

[CAP 71b] CAPUTO M., MAINARDI F., "A new dissipation model based on memory mechanism", *Pure and Applied Geophysics*, vol. 91, pp. 134–147, 1971.

[CAP 01] CAPUTO M. "Distributed order differential equation modeling dielectric induction and diffusion", *Fractional Calculus and Applied Analysis*, vol. 4, pp. 421–442, 2001.

[CAR 11] CARPINTERI A., CORNETTI P., SAPORA A., "A fractional calculus approach to nonlocal elasticity", *European Physical Journal – Special Topics*, vol. 193, pp. 193–204, 2011.

[CAT 48] CATTANEO C., "Sulla conductione del calore", *Atti del Seminario Matematico e Fisico dell' Università di Modena e Reggio Emilia*, vol. 3, pp. 83–101, 1948.

[CAT 58] CATTANEO M.C., "Sur une forme de l'équation de la chaleur éliminant le paradoxe d'une propagation instantanée", *Comptes Rendus de l'Acadmie des Sciences – Series I – Mathematics*, vol. 247, pp. 431–433, 1958.

[CHA 09] CHALLAMEL N., RAKOTOMANANA L., LE MARREC L., "A dispersive wave equation using nonlocal elasticity", *Comptes Rendus Mécanique*, vol. 337, pp. 591–595, 2009.

[CHA 13] CHALLAMEL N., ZORICA D., ATANACKOVIĆ T.M., *et al.*, "On the fractional generalization of Eringen's nonlocal elasticity for wave propagation", *Comptes Rendus Mécanique*, vol. 341, pp. 298–303, 2013.

[CHE 63] CHESTER M., "Second sound in solids", *Physical Review*, vol. 131, pp. 2013–2015, 1963.

[CHE 76] CHEN W.F., ATSUTA T., *Theory of Beam-Columns: In-Plane Behavior and Design*, vol. 1, McGraw-Hill, New York, 1976.

[CHE 02] CHECHKIN A.V., GORENFLO R., SOKOLOV I.M., "Retarding subdiffusion and accelerating superdiffusion governed by distributed-order fractional diffusion equations", *Physical Review E*, vol. 66, 046129 pp. 1–7, 2002.

[COI 03] COIMBRA C.F.M., "Mechanics with variable-order differential operators", *Annalen der Physik (Leipzig)*, vol. 12, pp. 692–703, 2003.

[COM 97] COMPTE A., METZLER R., "The generalized Cattaneo equation for the description of anomalous transport processes", *Journal of Physics A: Mathematical and General*, vol. 30, pp. 7277–7289, 1997.

[COT 09] COTTONE G., DI PAOLA M., ZINGALES M., "Elastic waves propagation in 1D fractional non-local continuum", *Physica E: Low-dimensional Systems and Nanostructures*, vol. 42, pp. 95–103, 2009.

[DAC 08] DACOROGNA B., *Direct Methods in the Calculus of Variations*, Springer, New York, 2008.

[DAS 08] DAS S., *Functional Fractional Calculus*, Springer, Berlin, 2008.

[DAU 00] DAUTRY R., LIONS J.L., *Mathematical Analysis and Numerical Methods for Science and Technology*, vol. 5, Springer Verlag, Berlin, 2000.

[DEB 07] DEBNATH L., BHATTA D., *Integral Transforms and Their Applications*, Chapman & Hall/CRC, Boca Raton, FL, 2007.

[DEL 96] DELBOSCO D., RODINO L., "Existence and uniqueness for a nonlinear fractional differential equation", *Journal of Mathematical Analysis and Applications*, vol. 204, pp. 609–625, 1996.

[DEM 67] DEMIDOVIC B.P., *Lectures on Mathematical Theory of Stability*, Nauka, Moscow, 1967. [In Russian]

[DI 09] DI PAOLA M., FAILLA G., ZINGALES M., "Physically-based approach to the mechanics of strong non-local linear elasticity theory", *Journal of Elasticity*, vol. 97, pp. 103–130, 2009.

[DI 08] DI PAOLA M., ZINGALES M., "Long-range cohesive interactions of non-local continuum faced by fractional calculus", *International Journal of Solids and Structures*, vol. 45, pp. 5642–5659, 2008.

[DIA 09] DIAZ G., COIMBRA C.F.M., "Nonlinear dynamics and control of a variable order oscillator with application to the van der Pol equation", *Nonlinear Dynamics*, vol. 56, pp. 145–157, 2009.

[DIE 02a] DIETHELM K., FORD N.J., "Analysis of fractional differential equations", *Journal of Mathematical Analysis and Applications*, vol. 265, pp. 229–248, 2002.

[DIE 02b] DIETHELM K., FORD N.J., "Numerical solution of the Bagley-Torvik equation", *BIT Numerical Mathematics*, vol. 42, pp. 490–507, 2002.

[DIE 04] DIETHELM K., FORD N.J., "Multi-order fractional differential equations their numerical solution", *Applied Mathematics and Computation*, vol. 154, pp. 621–640, 2004.

[DIE 10] DIETHELM K., "The analysis of fractional differential equations: an application-oriented exposition using differential operators of Caputo type", *Lecture Notes in Mathematics*, vol. 2004, Springer Verlag, Berlin, Heidelberg, 2010.

[DJO 08] DJORDJEVIC V.D., ATANACKOVIĆ T.M., "Similarity solutions to nonlinear heat conduction and Burgers/Korteweg-de Vries fractional equations", *Journal of Computational and Applied Mathematics*, vol. 222, pp. 701–714, 2008.

[DOE 55] DOETSCH G., *Handbuch der Laplace-Transformationen*, vol. I and II, Birkhäuser, Basel, 1950, 1955.

[DOE 67] DOETSCH G., *Anleitung zum praktischen Gebrauch der Laplace-Transformation und der ZTransformation*, R. Oldenburg, München, 1967.

[DOK 10] DOKOUMETZIDIS A., MAGIN R., MACHERAS P., "Fractional kinetics in multi-compartmental systems", *Journal of Pharmacokinetics and Pharmacodynamics*, vol. 37, pp. 507–524, 2010.

[DON 02] DONALTO C., "Analytical and numerical inversion of the Laplace-Carson transform by a differential method", *Computer Physics Communications*, vol. 145, pp. 298–309, 2002.

[DRE 03a] DREISIGMEYER D.W., YOUNG P.M., "Extending Bauer's corollary to fractional derivatives", *Journal of Physics A: Mathematical and General*, vol. 37, pp. 117–121, 2003.

[DRE 03b] DREISIGMEYER D.W., YOUNG P.M., "Nonconservative Lagrangian mechanics: a generalized function approach", *Journal of Physics A: Mathematical and General*, vol. 36, pp. 8297–8310, 2003.

[DRO 87] DROZDOV A.D., KOLMANOVSKII V.B., "Stability of viscoelastic rod on compliant basis", *Izvestiia Akademii nauk SSSR. Mekhanika tverdogo tela.*, vol. 22, pp. 162–172, 1987.

[DRO 98] DROZDOV A.D., *Viscoelastic Structures: Mechanics of Growth and Aging*, Academic Press, London, 1998.

[DUI 10] DUISTERMAAT J.J., KOLK J.A.C., *Distributions: Theory and Applications*, Birkhäuser, New York, 2010.

[EID 04] EIDELMAN S.D., KOCHUBEI A.N., "Cauchy problem for fractional diffusion equations", *Journal of Differential Equations*, vol. 199, pp. 211–255, 2004.

[EL 95] EL-SAYED A.M.A., IBRAHIM A.G., "Multivalued fractional differential equations", *Applied Mathematics and Computaton*, vol. 68, pp. 15–25, 1995.

[EL 01] EL-SAYED A.M.A., IBRAHIM A.G., "Set-valued integral equations of fractional-orders", *Applied Mathematics and Computaton*, vol. 118, pp. 113–121, 2001.

[ELI 12] ELISHAKOFF I., PENTARAS D., DUJAT K., *et al.*, *Carbon Nanotubes and Nanosensors: Vibrations, Buckling and Ballistic Impact*, ISTE, London, John Wiley & Sons, New York, 2012.

[ENE 99] ENELUND M., MÄHLER L., RUNESSON K., *et al.*, "Formulation and integration of the standard linear viscoelastic solid with fractional order rate laws", *International Journal of Solids and Structures*, vol. 36, pp. 2417–2442, 1999.

[ENE 10] ENESIZ Y., KESKIN Y., KURNAZ A., "The solution of the Bagley-Torvik equation with the generalized Taylor collocation method", *Journal of the Franklin Institute*, vol. 347, pp. 452–466, 2010.

[ENG 10] ENGLER H. "On the speed of spread for fractional reaction-diffusion equations", *International Journal of Differential Equations*, vol. 2010, Article ID, 315421, pp. 1–16, 2010.

[ERD 53] ERDÉLY A., MAGNUS W., OBERHETTINGER F., *et al.*, *Higher Transcendental Functions*, McGraw-Hill, NewYork, 1953.

[ERD 54] ERDÉLY A., MAGNUS W., OBERHETTINGER F., *et al.*, *Tables of Integral Transforms*, vol. I, McGraw-Hill, New York, 1954.

[ERI 83] ERINGEN A.C., "On differential equations of nonlocal elasticity and solutions of screw dislocations and surface waves", *Journal of Applied Physics*, vol. 54, pp. 4703–4710, 1983.

[ERI 87] ERINGEN A.C., "Theory of nonlocal elasticity and some applications", *Res Mechanica*, vol. 21, pp. 313–342, 1987.

[ERI 02] ERINGEN A.C., *Nonlocal Continuum Field Theories*, Springer Verlag, New York, 2002.

[ESS 00] ESSEX C., SCHULZKY C., FRANZ A., *et al.*, "Tsallis and Rényi entropies in fractional diffusion and entropy production", *Physica A*, vol. 284, pp. 299–308, 2000.

[FAB 92] FABRIZIO M., MORRO A., *Mathematical Problems in Linear Viscoelasticity*, vol. 12, SIAM Studies in Applied Mathematics, Philadelphia, PA, 1992.

[FEN 98] FENANDERÅ., "A fractional derivative railpad model included in a railway track model", *Journal of Sound and Vibration*, vol. 212, pp. 889–903, 1998.

[FIC 92] FICHERA G., "Is the Fourier theory of heat propagation paradoxical", *Rendiconti del Circolo Matematico di Palermo*, vol. 41, pp. 5–28, 1992.

[FIL 89] FILIPPOV V.M., *Variational Principles for Nonpotential Operators (Translations of Mathematical Monographs vol. 77)*, American Mathematical Society, Providence, RI, 1989.

[FR 02] FRÉMOND M., *Nonsmooth Thermomechanics*, Springer, New York, 2002.

[FRE 07] FREDERICO G.S.F., TORRES D.F.M., "A formulation of Nöther's theorem for fractional problems of the calculus of variations", *Journal of Mathematical Analysis and Applications*, vol. 334, pp. 834–846, 2007.

[FRE 10] FREDERICO G.S.F., TORRES D.F.M., "Fractional Noether's theorem in the Riesz-Caputo sense", *Applied Mathematics and Computation*, vol. 217, pp. 1023–1033, 2010.

[FRO 98] FROLOV G.A., KITAEV M.Y., "Improvement of accuracy in numerical methods for inverting Laplace transforms based on Post-Widder formula", *Computers & Mathematics with Applications*, vol. 36, pp. 23–34, 1998.

[GAH 77] GAHOV F.D., *Boundary Value Problems*, Nauka, Moscow, 1977. [In Russian]

[GEL 63] GELFAND I.M., FOMIN S.V., *Calculus of Variations*, Prentice Hall, Englewood Cliffs, NJ, 1963.

[GLO 99] GLOCKER C., "Formulation of spatial contact situations in rigid multibody systems", *Computer Methods in Applied Mechanics and Engineering*, vol. 177, pp. 199–214, 1999.

[GLO 01] GLOCKER C., *Set-Valued Force Laws, Dynamics of Non-Smooth Systems*, Springer Verlag, Berlin, 2001.

[GON 73] GONSOVSKI V.L., ROSSIKHIN Y.A., "Stress waves in a viscoelastic medium with a singular hereditary kernel", *Journal of Applied Mechanics and Technical Physics*, vol. 14, pp. 595–597, 1973.

[GOR 95] GORENFLO R., RUTMAN R., "On ultraslow and intermediate processes", in KIRYAKOVA V., RUSEV P., DIMOVSKI I. (eds), *Transform Methods & Special Functions, Proceedings of the 1st International Workshop, Sofia '94*, pp. 61–81, SCTP, Singapore, 1995.

[GOR 97a] GORENFLO R., LUCHKO Y., ROGOSIN S., Mittag-Leffler type Functions: Notes on Growth Properties and Distribution of Zeros, Department of Mathematics and Informatics, Free University of Berlin, 1997.

[GOR 97b] GORENFLO R., MAINARDI F., *Fractional Calculus: Integral and Differential Equations of Fractional Order. In: Fractals and Fractional Calculus in Continuum Mechanics (eds A. Carpinteri, F. Mainardi)*, vol. 378 of CISM Courses and Lecture Notes. Springer Verlag, Wien and New York, 1997.

[GOR 00] GORENFLO R., LUCHKO Y., MAINARDI F., "Wright functions as scale-invariant solutions of the diffusion-wave equation", *Journal of Computational and Applied Mathematics*, vol. 118, pp. 175–191, 2000.

[GRA 10] GRAHOVAC N.M., ZIGIC M.M., "Modelling of the hamstring muscle group by use of fractional derivatives", *Computers and Mathematics with Applications*, vol. 59, pp. 1695–1700, 2010.

[GRA 12a] GRAHOVAC N., ZIGIC M., SPASIC D., "On impact scripts with both fractional and dry friction type of dissipation", *International Journal of Bifurcation and Chaos*, vol. 22, 1250076 pp. 1–10, 2012.

[GRA 12b] GRAHOVAC N.M., SPASIC D.T., "Multivalued fractional differential equations as a model for an impact of two bodies", *Journal of Vibration and Control*, 2012.

[HAD 96] HADID S.B., LUCHKO Y.F., "An operational method for solving fractional differential equations of an arbitrary real order", *Pan-American Mathematical Journal*, vol. 6, pp. 56–73, 1996.

[HAL 93] HALE J.K., VERDUYN LUNEL S.M., *Introduction to Functional Differential Equations*, Springer, New York, 1993.

[HAN 01] HANYGA A., "Multidimensional solutions of space-fractional diffusion equations", *Proceedings of the Royal Society A: Mathematical, Physical and Engineering Sciences*, vol. 457, pp. 2993–3005, 2001.

[HAN 02a] HANYGA A., "Multi-dimensional solutions of space-time-fractional diffusion equations", *Proceedings of the Royal Society A: Mathematical, Physical and Engineering Sciences*, vol. 458, pp. 429–450, 2002.

[HAN 02b] HANYGA A. "Multidimensional solutions of time-fractional diffusion-wave equations", *Proceedings of the Royal Society A: Mathematical, Physical and Engineering Sciences*, vol. 458, pp. 933–957, 2002.

[HAN 07] HANYGA A., "Anomalous diffusion without scale invariance", *Journal of Physics A: Mathematical and Theoretical*, vol. 40, pp. 5551–5563, 2007.

[HAR 03] HARTLEY T.T., LORENZO C.F., "Fractional-order system identification based on continuous order-distributions", *Signal Processing*, vol. 83, pp. 2287–2300, 2003.

[HAS 61] HASUMI M., "Note on the *n*-dimensional tempered ultra-distributions", *Tohoku Mathematical Journal*, vol. 13, pp. 94–104, 1961.

[HEN 92] HENNEAUX M., TEITELBOIM C., *Quantization of Gauge Systems*, Princeton University Press, Princeton, NJ, 1992.

[HER 11] HERRMANN R., *Fractional Calculus. An Introduction for Physicists*, World Scientific, New Jersey, 2011.

[HEY 03] HEYMANS N., "Constitutive equations for polymer viscoelasticity derived from hierarchical models in cases of failure of time-temperature superposition", *Signal Processing*, vol. 83, pp. 2345–2357, 2003.

[HIL 00] HILFER R. (ed.), *Applications of Fractional Calculus in Physics*, World Scientific, Singapore, 2000.

[HOF 98] HOFFMANN K.H., ESSEX C., SCHULZKY C., "Fractional diffusion and entropy production", *Journal of Non-Equilibrium Thermodynamics*, vol. 23, pp. 166–175, 1998.

[HOL 84] HOLMES P., SPENCE D., "On a Painlevé-type boundary-value problem", *Quarterly Journal of Mechanics and Applied Mathematics*, vol. 37, pp. 525–538, 1984.

[IBR 94] IBRAGIMOV N.H. (ed.), *CRC Handbook of Lie Group Analysis of Differential Equations*, vol. IIII, CRC Press, Boca Raton, FL, 1994.

[IGN 10] IGNACZAK J., OSTOJA-STARZEWSKI M., *Thermoelasticity with Finite Wave Speeds*, Oxford University Press, Oxford, 2010.

[ING 04] INGMAN D., SUZDALNITSKY J., "Control of damping oscillations by fractional differential operator with time-dependent order", *Computer Methods in Applied Mechanics and Engineering*, vol. 193, pp. 5585–5595, 2004.

[ITO 08] ITO K., KUNISCH K., *Lagrange Multiplier Approach to Variational Problems and Applications*, SIAM, Philadelphia, PA, 2008.

[JAN 11] JANEV M., ATANACKOVIĆ T., PILIPOVIĆ S., *et al.*, "Image denoising by a direct variational minimization", *EURASIP Journal on Advances in Signal Processing*, vol. 2011, 8, pp. 1–16, 2011.

[JOS 89] JOSEPH D.D., PREZIOSI L., "Heat waves", *Reviews of Modern Physics*, vol. 61, pp. 41–73, 1989.

[JOU 01] JOU D., CASAS-VÁZQUEZ J., LEBON G., *Extended Irreversible Thermodynamics*, Springer, Berlin, 2001.

[KEM 98a] KEMPFLE S., "Causality criteria for solutions of linear fractional differential equations", *Fractional Calculus and Applied Analysis*, vol. 1, pp. 351–364, 1998.

[KEM 98b] KEMPFLE S., BEYER H., "Global and causal solutions of fractional linear differential equations", in KIRYAKOVA V. RUSEV P., DIMOVSKI I. (eds), *Transform Methods & Special Functions, Proceedings of the 2nd International Workshop, Varna '96*, IMI-BAS, Sofia, pp. 210–226, 1998.

[KIL 76] KILCHEVSKI N.A., *Dynamic Contact of Solid Bodies*, Naukova Dumka, Kiev, 1976.

[KIL 04] KILBAS A.A., SAIGO M., SAXENA R.K., "Generalized Mittag–Leffler function and generalized fractional calculus operators", *Integral Transforms and Special Functions*, vol. 15, pp. 31–49, 2004.

[KIL 06] KILBAS A.A., SRIVASTAVA H.M., TRUJILLO J.J., *Theory and Applications of Fractional Differential Equations*, Elsevier B.V., Amsterdam, 2006.

[KIR 94] KIRYAKOVA V., *Generalized Fractional Calculus and Applications*, Longman Scientific & Technical, Harlow, Essex, UK, 1994.

[KIR 02] KIRILLOV O.N., SEYRANIAN A.P., "Solution to the Hermann-Smith problem", *Doklady Physics*, vol. 47, pp. 767–771, 2002.

[KLI 01] KLIMEK M., "Fractional sequential mechanics – models with symmetric fractional derivative", *Czechoslovak Journal of Physics*, vol. 51, pp. 1348–1354, 2001.

[KLI 09] KLIMEK M., *On Solutions of Linear Fractional Differential Equations of a Variational Type*, The Publishing Office of Czestochowa University of Technology, Czestochowa, 2009.

[KOC 08] KOCHUBEI A.N., "Distributed order calculus and equations of ultraslow diffusion", *Journal of Mathematical Analysis and Applications*, vol. 340, pp. 252–281, 2008.

[KOM 87] KOMATSU H., "Laplace transforms of hyperfunctions – a new foundation of the Heviside calculus", *Journal of the Faculty of Science, the University of Tokyo, Section IA, Mathematics*, vol. 34, pp. 805–820, 1987.

[KOM 96] KOMATSU H., *Solutions of Differential Equations by Means of Laplace Hyperfunctions. Structures of Solutions of Differential Equations (Katata/Kyoto, 1995)*, World Scientific, River Edge, NJ, 1996.

[KON 10] KONJIK S., OPARNICA L., ZORICA D., "Waves in fractional Zener type viscoelastic media", *Journal of Mathematical Analysis and Applications*, vol. 365, pp. 259–268, 2010.

[KON 11] KONJIK S., OPARNICA L., ZORICA D., "Waves in viscoelastic media described by a linear fractional model", *Integral Transforms and Special Functions*, vol. 22, pp. 283–291, 2011.

[KWO 05] KWOK S.F., "A falling body problem through the air in view of the fractional derivative approach", *Physica A*, vol. 350, pp. 199–206, 2005.

[LAN 00] LANGTHJEM M.A., SUGIYAMA Y., "Dynamic stability of columns subjected to follower loads: a survey", *Journal of Sound and Vibration*, vol. 238, pp. 809–851, 2000.

[LAN 06] LANGLANDS T.A.M., "Solution of a modified fractional diffusion equation", *Physica A*, vol. 367, pp. 136–144, 2006.

[LAZ 06] LAZOPULOS K.A., "Non-local continuum mechanics and fractional calculus", *Mechanics Research Communications*, vol. 33, pp. 753–757, 2006.

[LEI 04] LEINE R.I., NIJMEIJER H., *Dynamics and Bifurcations of Non-Smooth Mechanical Systems*, Springer Verlag, Berlin, 2004.

[LI 01] LI G.G., ZHU Z.Y., CHENG C.J., "Dynamical stability of viscoelastic column with fractional derivative constitutive relation", *Applied Mathematics and Mechanics*, vol. 22, pp. 294–303, 2001.

[LI 03] LI X., ESSEX C., DAVISON M., *et al.*, "Fractional diffusion, irreversibility and entropy", *Journal of Non-Equilibrium Thermodynamics*, vol. 28, pp. 279–291, 2003.

[LI 10] LI Y., ZHAO W., "Haar wavelet operational matrix of fractional order integration and its applications in solving the fractional order differential equations", *Applied Mathematics and Computation*, vol. 216, pp. 2276–2285, 2010.

[LIU 06] LIU J.G., XU Y.M., "Higher-order fractional constitutive equations of viscoelastic materials involving three different parameters and their relaxation and creep functions", *Mechanics of Time-dependent Materials*, vol. 10, pp. 263–279, 2006.

[LOR 02] LORENZO C.F., HARTLEY T.T., "Variable order and distributed order fractional operators", *Nonlinear Dynamics*, vol. 29, pp. 57–98, 2002.

[LOR 13] LORENZO C.F., HARTLEY T.T., ADAMS J.L., "Time-varying initialization and corrected Laplace transform of the Caputo derivative", *Proceedings of the 6th Workshop on Fractional Differentiation and Its Applications, Part of 2013 IFAC Joint Conference SSSC*, Grenoble, France, 4–6 February, 2013.

[LUC 99] LUCHKO Y., GORENFLO R., "An operational method for solving fractional differential equations with the Caputo derivatives", *Acta Mathematica Vietnamica*, vol. 2, pp. 207–233, 1999.

[LUC 09] LUCHKO Y., "Maximum principle for the generalized time-fractional diffusion equation", *Journal of Mathematical Analysis and Applications*, vol. 351, pp. 218–223, 2009.

[MAC 04] MACEWEN K.W., "The follower load problem and bifurcation to periodic motions", *International Journal of Non-Linear Mechanics*, vol. 39, pp. 555–568, 2004.

[MAG 48] MAGNUS W., OBERHETTINGER F., *Formeln und Sätze für die Speciellen Funktionen der Mathematischen Physik*, Springer Verlag, Berlin, 1948.

[MAG 06] MAGIN R.L., *Fractional Calculus in Bioengineering*, Begell House Publishers, Redding, CA, 2006.

[MAI 96] MAINARDI F., "Fractional relaxation-oscillation and fractional diffusion-wave phenomena", *Chaos, Solitons & Fractals*, vol. 7, pp. 1461–1477, 1996.

[MAI 97a] MAINARDI F., "Fractional calculus: some basic problems in continuum and statistical mechanics", in CARPINTERI A., MAINARDI F. (eds), *Fractals and Fractional Calculus in Continuum Mechanics*, vol. 378 of *CISM Courses and Lecture Notes*, Springer Verlag, Wien/New York, 1997.

[MAI 97b] MAINARDI F., TOMIROTTI M., "Seismic pulse propagation with constant $Q$ and stable probability distribution", *Annali di Geofisica*, vol. 40, pp. 1311–1328, 1997.

[MAI 00] MAINARDI F., GORENFLO R., "On Mittag-Leffler-type functions in fractional evolution processes", *Journal of Computational and Applied Mathematics*, vol. 118, pp. 283–299, 2000.

[MAI 01] MAINARDI F., LUCHKO Y., PAGNINI G., "The fundamental solution of the space-time fractional diffusion equation", *Fractional Calculus and Applied Analysis*, vol. 4, pp. 153–192, 2001.

[MAI 07a] MAINARDI F., MURA A., GORENFLO R., *et al.*, "The two forms of fractional relaxation of distributed order", *Journal of Vibration and Control*, vol. 13, pp. 1249–1268, 2007.

[MAI 07b] MAINARDI F., PAGNINI G., GORENFLO R., "Some aspects of fractional diffusion equations of single and distributed order", *Applied Mathematics and Computation*, vol. 187, pp. 295–305, 2007.

[MAI 08] MAINARDI F., PAGNINI G., MURA A., *et al.*, "Time-fractional diffusion of distributed order", *Journal of Vibration and Control*, vol. 14, pp. 1267–1290, 2008.

[MAI 10] MAINARDI F., *Fractional Calculus and Waves in Linear Viscoelasticity*, Imperial College Press, London, 2010.

[MAK 91] MAKRIS N., CONSTANTINOU M.C., "Fractional-derivative Maxwell model for viscous dampers", *Journal of Structural Engineering*, vol. 117, pp. 2708–2724, 1991.

[MAK 92] MAKRIS N., CONSTANTINOU M.C., "Spring-viscous damper systems for combined seismic and vibration isolation", *Earthquake Engineering and Structural Dynamics*, vol. 21, pp. 649–664, 1992.

[MAK 93a] MAKRIS N., CONSTANTINOU M.C., "Models of viscoelasticity with complex-order derivatives", *Journal of Engineering Mechanics*, vol. 119, pp. 1453–1464, 1993.

[MAK 93b] MAKRIS N., DARGUSH G.F., CONSTANTINOU M.C., "Dynamic analysis of generalized viscoelastic fluids", *Journal of Engineering Mechanics*, vol. 119, pp. 1663–1679, 1993.

[MAL 12] MALINOWSKA A.B., TORRES D.F.M., *Introduction to the Fractional Calculus of Variations*, Imperial College Press, London, 2012.

[MAR 49] MARDEN M., *The Geometry of the Zeros of a Polynomial in a Complex Variable*, American Mathematical Society, New York, 1949.

[MAU 99] MAUGIN G.A., *Nonlinear Waves in Elastic Crystals*, Oxford University Press, Oxford, 1999.

[MET 04] METZLER R., KLAFTER J., "The restaurant at the end of the random walk: recent developments in the description of anomalous transport by fractional dynamics", *Journal of Physics A: Mathematical and General*, vol. 37, pp. R161–R208, 2004.

[MIC 11] MICHELITSCH T.M. "The self-similar field and its application to a diffusion problem", *Journal of Physics A: Mathematical and Theoretical*, vol. 44, 465206, pp. 1–13, 2011.

[MIC 12] MICHELITSCH T.M., MAUGIN G.A., RAHMAN M., *et al.*, "An approach to generalized one-dimensional self-similar elasticity", *International Journal of Engineering Science*, vol. 61, pp. 103–111, 2012.

[MIK 87] MIKUSIŃSKI J., *Operational Calculus*, Pergamon Press, Oxford, 1987.

[MIL 93] MILLER K.S., ROSS B., *An Introduction to the Fractional Calculus and Fractional Differential Equations*, John Wiley and Sons, New York, 1993.

[MIS 08] MISHURA Y.S., *Stochastic Calculus for Fractional Brownian Motion and Related Processes*, Springer, Berlin, 2008.

[MÜL 93] MÜLLER I., RUGGERI T., *Rational Extended Thermodynamics*, Springer, Berlin, 1993.

[MUS 05] MUSLIH S.I., BALEANU D., "Hamiltonian formulation of systems with linear velocities within Riemann-Liouville fractional derivatives", *Journal of Mathematical Analysis and Applications*, vol. 304, pp. 599–606, 2005.

[NAB 04] NABER M., "Distributed order fractional sub-diffusion", *Fractals*, vol. 12, pp. 23–32, 2004.

[NAB 10] NABER M., "Linear fractionally damped oscillator", *International Journal of Differential Equations*,'vol. 2010, Article ID 197020 pp. 1–12, 2010.

[NAH 03] NAHUSHEV A.M., *Fractional Calculus and its Applications*, FIZMATLIT, Moscow, 2003. [In Russian]

[NAR 01] NARAHARI ACHAR B.N., HANNEKEN J.W., ENCK T., *et al.*, "Dynamics of the fractional oscillator", *Physica A*, vol. 297, pp. 361–367, 2001.

[NAR 02] NARAHARI ACHAR B.N., HANNEKEN J. W., CLARKE T., "Response characteristics of a fractional oscillator", *Physica A*, vol. 309, pp. 275–288, 2002.

[NÄS 13] NÄSHOLM S.P., HOLM S., "On a fractional Zener elastic wave equation", *Fractional Calculus and Applied Analysis*, vol. 16, pp. 26–50, 2013.

[NOW 63] NOWACKI W., *Dynamics of Elastic Systems*, Chapman & Hall, London, 1963.

[ODI 07] ODIBAT Z.M., Shawagfeh N.T., "Generalized Taylor's formula", *Applied Mathematics and Computation*, vol. 24, pp. 286–293, 2007.

[ODZ 13] ODZIJEWICZ T., MALINOWSKA A.B., TORRES D.F.M., "Fractional variational calculus of variable order", in ALMEIDA A., CASTRO L., SPECK F.O. (eds), *Advances in Harmonic Analysis and Operator Theory: Advances and Applications*, vol. 229, Springer, Basel, pp. 291–301, 2013.

[OLD 74] OLDHAM K.B., SPANIER J., *The Fractional Calculus: Theory and Applications of Differentiation and Integration to Arbitrary Order*, Academic Press, London, 1974.

[OLV 86] OLVER P.J., *Application of Lie Groups to Differential Equations*, Springer, Berlin, 1986.

[OPA 02] OPARNICA L., "Generalized fractional calculus with applications in mechanics", *Matematicki vesnik*, vol. 53, pp. 151–158, 2002.

[ORS 03] ORSINGHER E., ZHAO X., "The space-fractional telegraph equation and the related fractional telegraph process", *Chinese Annals of Mathematics*, vol. 24B, pp. 45–56, 2003.

[ORT 11] ORTIGUEIRA M.D., *Fractional Calculus for Scientists and Engineers, Lecture Notes in Electrical Engineering*, vol. 84, Springer, Dordrecht, the Netherlands, 2011.

[OUS 95] OUSTALOUP A., *La dérivation non entière. Théorie, synthèse et applications*, Hermes, Paris, 1995.

[OZE 04] OZER T., "On symmetry group properties and general similarity forms of the Benney equations in the Lagrangian variables", *Journal of Computational and Applied Mathematics*, vol. 169, pp. 297–313, 2004.

[OZE 05] OZER T., "Symmetry group analysis of Benney system and an application for shallow-water equations", *Mechanics Research Communications*, vol. 32, pp. 241–254, 2005.

[PAP 97] PAPOULIA K.D., KELLY J.M., "Visco-hyperelastic model for filled rubbers used in vibration isolation", *ASME Journal of Engineering Materials and Technology*, vol. 119, pp. 292–297, 1997.

[PER 90] PERONA P., MALIK J., "Scale-space and edge detection using anisotropic diffusion", *IEEE Transactions on Pattern Analysis and Machine Intelligence*, vol. 12, pp. 629–639, 1990.

[PET 05] PETROVIC L.M., SPASIC D.T., ATANACKOVIĆ T.M., "On a mathematical model of a human root dentin", *Dental Materials*, vol. 21, pp. 125–128, 2005.

[PET 08] PETROVIC L.M., ATANACKOVIĆ T.M., "A model for shrinkage strain in photo polymerization of dental composites", *Dental Materials*, vol. 24, pp. 556–560, 2008.

[PET 11] PETRÁS I., *Fractional-Order Nonlinear Systems: Modeling, Analysis and Simulation*, Nonlinear Physical Science, Springer, Berlin, Heidelberg, 2011.

[PFE 96] PFEIFFER F., GLOCKER C., *Multibody Dynamics with Unilateral Contacts*, John Wiley and Sons, New York, 1996.

[POD 99] PODLUBNY I., *Fractional Differential Equations*, Academic Press, San Diego, CA, 1999.

[POO 12a] POOSEH S., ALMEIDA R., TORRES D.F.M., "Approximation of fractional integrals by means of derivatives", *Computers and Mathematics with Applications*, vol. 64, pp. 3090–3100, 2012.

[POO 12b] POOSEH S., ALMEIDA R., TORRES D.F.M., "Expansion formulas in terms of integer order derivatives for the Hadamard fractional integral and derivative", *Numerical Functional Analysis and Optimization*, vol. 33, pp. 301–319, 2012.

[POO 13] POOSEH S., ALMEIDA R., TORRES D.F.M., "Numerical approximations of fractional derivatives with applications", *Asian Journal of Control*, vol. 15, pp. 698–712, 2013.

[POP 13] POPOVIC J.K., PILIPOVIĆ S., ATANACKOVIĆ T.M., "Two compartmental fractional derivative model with fractional derivatives of different order", *Communications in Nonlinear Science and Numerical Simulation*, vol. 18, pp. 2507–2514, 2013.

[PRI 96] PRITZ T., "Analysis of four-parameter fractional derivative model of real solid materials", *Journal of Sound and Vibration*, vol. 195, pp. 103–115, 1996.

[PRI 03] PRITZ T., "Five-parameter fractional derivative model for polymeric damping materials", *Journal of Sound and Vibration*, vol. 265, pp. 935–952, 2003.

[PRÜ 93] PRÜSS J., *Evolutionary Integral Equations and Applications*, Birkhäuser Verlag, Basel, 1993.

[PSH 05] PSHU A.V., *Partial Differential Equations of Fractional Order*, Nauka, Moscow, 2005. [In Russian]

[QAS 10] QASEM M., AL-MDALLAL Q. M., SYAM M.I., *et al.*, "A collocation-shooting method for solving fractional boundary value problems", *Communications in Nonlinear Science and Numerical Simulation*, vol. 15, pp. 3814–3822, 2010.

[QI 11] QI H., JIANG X., "Solutions of the space-time fractional Cattaneo diffusion equation", *Physica A*, vol. 390, pp. 1876–1883, 2011.

[QIN 08] QIN H.Z., LU Y.M., "A note on an open problem about the first Painlevé equation", *Acta Mathematica Sinica* (English series), vol. 24, pp. 203–210, 2008.

[RAB 48] RABOTNOV Y.N., "Equilibrium of an elastic medium with after effect" *Prikladnaja Matematika i Mehanika*, vol. 12, pp. 81–91, 1948. [In Russian]

[RAB 80] RABOTNOV Y.N., *Elements of Hereditary Solid Mechanics*, Mir Publishers, Moscow, 1980.

[RAB 07] RABEI E.M., NAWAFLEH K.I., HIJJAWI R.S., *et al.*, "The Hamilton formalism with fractional derivatives", *Journal of Mathematical Analysis and Applications*, vol. 327, pp. 891–897, 2007.

[RAJ 11] RAJA M.A.Z., KHAN J.A., QURESHI I.M., "Solution of fractional order system of Bagley-Torvik equation using evolutionary computational intelligence", *Mathematical Problems in Engineering*, vol. 2011, Article ID 675075 pp. 1–18, 2011.

[RAM 07] RAMIREZ L.E.S., COIMBRA C.F.M., "A variable order constitutive relation for viscoelasticity", *Annalen der Physik* (Leipzig), vol. 16, pp. 543–552, 2007.

[RAM 10] RAMIREZ L.E.S., COIMBRA C.F.M., "On the selection and meaning of variable order operators for dynamic modeling", *International Journal of Differential Equations*, vol. 2010, pp. 1–16, 2010.

[RAP 10] RAPAIC M.R., JELICIC Z.D., "Optimal control of a class of fractional heat diffusion systems", *Nonlinear Dynamics*, vol. 62, pp. 39–51, 2010.

[RAY 05] RAY S.S., BERA R.K., "Analytical solution of the Bagley-Torvik equation by Adomian decomposition method", *Applied Mathematics and Computation*, vol. 168, pp. 398–410, 2005.

[RAY 12] RAY S.S., "On Haar wavelet operational matrix of general order and its application for the numerical solution of fractional Bagley-Torvik equation", *Applied Mathematics and Computation*, vol. 218, pp. 5239–5248, 2012.

[REK 04] REKHVIASHVILI S.S., "The Lagrange formalism with fractional derivatives in problems of mechanics", *Technical Physics Letters*, vol. 30, pp. 55–57, 2004.

[RES 04] RESCIGNO A., *Foundations of Pharmacokinetics*, Kluwer, New York, 2004.

[RIE 96] RIEWE F., "Nonconservative Lagrangian and Hamiltonian mechanics", *Physical Review E*, vol. 53, pp. 1890–1899, 1996.

[RIE 97] RIEWE F., "Mechanics with fractional derivatives", *Physical Review E*, vol. 55, pp. 3582–3592, 1997.

[ROB 85] ROBERTS D., "The general Lie group and similarity solutions for one-dimensional Vlasov-Maxwell equations", *Journal of Plasma Physics*, vol. 33, pp. 219–236, 1985.

[ROS 93] ROSS B., SAMKO B.S., "Integration and differentiation to a variable fractional order", *Integral Transforms and Special Functions*, vol. 1, pp. 277–300, 1993.

[ROS 01b] ROSSIKHIN Y.A., SHITIKOVA M.V., "Analysis of dynamic behavior of viscoelastic rods whose rheological models contain fractional derivatives of two different orders", *Zeitschrift für Angewandte Mathematik und Mechanik*, vol. 81, pp. 363–376, 2001.

[ROS 01b] ROSSIKHIN Y.A., SHITIKOVA M.V., "A new method for solving dynamic problems of fractional derivative viscoelasticity", *International Journal of Engineering Science*, vol. 39, pp. 149–176, 2001.

[ROS 10a] ROSSIKHIN Y.A., "Reflections on two parallel ways in the progress of fractional calculus in mechanics of solids", *Applied Mechanics Reviews*, vol. 63, 010701 pp. 1–12, 2010.

[ROS 10b] ROSSIKHIN Y.A., SHITIKOVA M.V., "Application of fractional calculus for dynamic problems of solid mechanics: novel trends and recent results", *Applied Mechanics Reviews*, vol. 63, 010801 pp. 1–52, 2010.

[RUG 94] RUGGERI T., MURACCHINI A., SECCIA L., "Continuum approach to phonon gas and shape changes of second sound via shock waves theory", *Il Nuovo Cimento*, vol. 16, pp. 15–44, 1994.

[RUN 96] RUNST T., SICKEL W., *Sobolev Spaces of Fractional Order, Nemytskij Operators, and Nonlinear Partial Differential Equations*, Walter de Gruyter & Co, Berlin, 1996.

[SAB 07] SABATIER J., AGRAWAL O.P., TENREIRO MACHADO J.A. (eds), *Advances in Fractional Calculus: Theoretical Developments and Applications in Physics and Engineering*, Springer, Dordrecht, the Netherlands, 2007.

[SAM 87] SAMKO S.G., KILBAS A.A., MARICHEV O.I., *Integrals and Derivatives of Fractional Order*, Nauka i tehnika, Minsk, 1987. [In Russian]

[SAM 93] SAMKO S.G., KILBAS A.A., MARICHEV O.I., *Fractional Integrals and Derivatives*, Gordon and Breach, Amsterdam, 1993.

[SAM 95] SAMKO S.G., "Fractional integration and differentiation of variable order", *Analysis Mathematica*, vol. 21, pp. 213–236, 1995.

[SAP 13] SAPORA A., CORNETTI P., CARPINTERI A., "Wave propagation in nonlocal elastic continua modelled by a fractional calculus approach", *Communications in Nonlinear Science and Numerical Simulation*, vol. 18, pp. 63–74, 2013.

[SCH 51] SCHWARTZ L., *Théorie des distributions*, vol. I–II, Hermann, Paris, 1950– 1951.

[SCH 61] SCHWARTZ L., *Méthodes mathématiques pour les sciences physiques*, Hermann, Paris, 1961.

[SCH 00] SCHIESSEL H., FRIEDRICH C., BLUMEN A., "Applications to problems in polymer physics and rheology", in HILFER R. (ed.), *Applications of Fractional Calculus in Physics*, World Scientific, Singapore, 2000.

[SER 00] SEREDYŃSKA M., HANYGA A., "Nonlinear Hamiltonian equations with fractional damping", *Journal of Mathematical Physics*, vol. 41, pp. 2135–2156, 2000.

[SEY 02] SEYRANIAN A.P., "Bifurcation of eigenvalues and stability problems in mechanics", in SEYRANIAN A.P., ELISHAKOFF I. (eds), *Modern Problems of Structural Stability*, vol. 436 of *CISM International Centre for Mechanical Sciences*, Springer, Wien/New York, 2002.

[SIL 97] SILHAVY M., *The Mechanics and Thermodynamics of Continuous Media*, Springer, Berlin, 1997.

[SMI 72] SMITH T.E., HERRMANN G., "Stability of a beam on an elastic foundation subjected to a follower force", *Journal of Applied Mechanics – Transactions ASME*, vol. 39, pp. 628–629, 1972.

[SMI 02] SMIRNOV G.V., *Introduction to the Theory of Differential Inclusions*, vol. 41 of *Graduate Studies in Mathematics*, American Mathematical Society, Providence, RI, 2002.

[SOK 04] SOKOLOV I.M., CHECKIN A.V., KLAFTER J., "Distributed-order fractional kinetics", *Acta Physica Polonica A*, vol. 35, pp. 1323–1341, 2004.

[SOO 05] SOON C.M., COIMBRA C.F.M., KOBAYASHI M.H., "The variable viscoelasticity oscillator", *Annalen der Physik (Leipzig)*, vol. 14, pp. 378–389, 2005.

[STA 01] STANKOVIĆ B., ATANACKOVIĆ T.M., "On a model of a viscoelastic rod", *Fractional Calculus and Applied Analysis*, vol. 4, pp. 501–522, 2001.

[STA 02] STANKOVIĆ B., ATANACKOVIĆ T.M., "Dynamics of a rod made of generalized Kelvin-Voigt visco-elastic material", *Journal of Mathematical Analysis and Applications*, vol. 268, pp. 550–563, 2002.

[STA 04a] STANKOVIĆ B., ATANACKOVIĆ T.M., "On a viscoelastic rod with constitutive equation containing fractional derivative of two different orders", *Mathematics and Mechanics of Solids*, vol. 9, pp. 629–656, 2004.

[STA 04b] STANKOVIĆ B., ATANACKOVIĆ T.M., "On an inequality arising in fractional oscillator theory", *Fractional Calculus and Applied Analysis*, vol. 7, pp. 11–19, 2004.

[STA 06] STANKOVIĆ B., "Equations with left and right fractional derivatives", *Publications de l'Institut Mathématique de l'Académie Serbe des Sciences et des Arts*, vol. 80, pp. 259–272, 2006.

[STA 07] STANKOVIĆ B.. "A method to solve an equation with left and right fractional derivatives". *Bulletin de l'Académie Serbe des Sciences et des Arts, Sciences Mathématiques*, vol. 32, pp. 59–67, 2007.

[STA 09] STANKOVIĆ B., ATANACKOVIĆ T.M., PILIPOVIC S., "Weak solutions to differential equations with left and right fractional derivatives defined on R., *Bulletin de l'Académie Serbe des Sciences et des Arts, Sciences Mathématiques*, vol. 139, pp. 75–88, 2009.

[STR 11] STRAUGHAN B., *Heat Waves*, Springer, New York, 2011.

[SUN 09] SUN H.G., CHEN W., CHEN Y.Q., "Variable-order fractional differential operators in anomalous diffusion modeling", *Physica A*, vol. 388, pp. 4586–4592, 2009.

[SWE 78] SWENSON R.J., "Generalized heat conduction equation", *American Journal of Physics*, vol. 46, pp. 76–77, 1978.

[TAR 06] TARASOV V.E., ZASLAVSKY G.M., "Dynamics of low-level fractionality", *Physica A*, vol. 368, pp. 399–415, 2006.

[TAR 11] TARASOV V.E., *Fractional Dynamics: Applications of Fractional Calculus to Dynamics of Particles, Fields and Media*, Nonlinear Physical Science, Springer Verlag and Higher Education Press, Berlin/Beijing, 2011.

[TIK 77] TIKHONOV A.N., SAMARSKII A.A., *Equations of Mathematical Physics*, Nauka, Moscow, 1977. [In Russian]

[TOF 08] TOFIGHI A., GOLESTANI A., "A perturbative study of fractional relaxation phenomena", *Physica A*, vol. 368, pp. 1807–1817, 2008.

[TOR 84] TORVIK P.J., BAGLEY R.L., "On the appearance of the fractional derivative in the behavior of real materials", *Journal of Applied Mechanics – Transactions of ASME*, vol. 51, pp. 294–298, 1984.

[TRE 75] TREVES F., *Basic Linear Partial Differential Equations*, Academic press, New York, 1975.

[TRO 12] TROYER K.L., ESTEP D.J., PUTTLITZ C.M., "Viscoelastic effects during loading play an integral role in soft tissue mechanics", *Acta Biomaterialia*, vol. 8, pp. 234–243, 2012.

[TRU 99] TRUJILLO J.J., RIVERO M., BONILLA B., "On a Riemann-Liouville generalized Taylor's formula", *Journal of Mathematical Analysis and Applications*, vol. 231, pp. 255–265, 1999.

[TUR 01] TURNER J.D., "On the simulation of discontinuous functions", *Journal of Applied Mechanics – Transactions of ASME*, vol. 68, pp. 751–757, 2001.

[UCH 08] UCHAIKIN V.V., *Fractional Calculus*, Artishok, Ulyanovsk, 2008. [In Russian]

[UCH 13] UCHAIKIN V.V., *Fractional Derivatives for Physicists and Engineers, Background and Theory*, vol. I, Higher Education Press and Springer Verlag, Beijing/Berlin, 2013.

[UMA 09] UMAROV S., STEINBERG S., "Variable order differential equations with piecewise constant order-function and diffusion with changing modes", *Zeitschrift für Analysis und iher Anwendungen*, vol. 28, pp. 431–450, 2009.

[VIN 87] VINOGRADOV A.M., "Buckling of viscoelastic beam columns", *AIAA Journal*, vol. 25, pp. 479–483, 1987.

[VLA 73] VLADIMIROV V.S., *Generalized Functions in Mathematical Physics*, Mir Publishers, Moscow, 1973.

[VLA 84] VLADIMIROV V.S., *Equations of Mathematical Physics*, Mir Publishers, Moscow, 1984.

[VLA 88] VLADIMIROV V.S., DROSHINOV Y.N., ZAVIALOV B.I., *Tauberian Theorems for Generalized Functions*, Kluwer Academic Publishers, Maine, 1988.

[VOL 77] VOLOSHIN I., GROMOV V.G., "Stability criterion for a bar on an elastic base acted on by a follower force", *Izvestiia Akademii nauk SSSR. Mekhanika tverdogo tela.*, vol. 12, pp. 169–171, 1977.

[VUJ 89] VUJANOVIC B.D., JONES S.E., *Variational Methods in Nonconservative Phenomena, Mathematics in Science and Engineering*, vol.182, Academic Press, London, 1989.

[VUJ 04] VUJANOVIC B.D., ATANACKOVIĆ T. M., *An Introduction to Modern Variational Techniques in Mechanics and Engineering*, Birkhäuser, Boston, MA, 2004.

[WAN 10] WANGA Z.H., WANG X., "General solution of the Bagley-Torvik equation with fractional order derivative", *Communications in Nonlinear Science and Numerical Simulation*, vol. 15, pp. 1279–1285, 2010.

[WEL 99] WELCH S.W.J., RORRER R.A.L., DUREN R.G. JR., "Application of time-based fractional calculus methods to viscoelastic creep and stress relaxation of materials", *Mechanics of Time dependent Materials*, vol. 3, pp. 279–303, 1999.

[WES 03] WEST B.J., BOLOGNA M., GRIGOLINI P., *Physics of Fractional Operators*, Springer, New York, 2003.

[WHI 74] WHITHAM G.B., *Linear and Nonlinear Waves*, John Wiley, New York, 1974.

[WID 75] WIDDER D.V., *The Heat Equation*, Academic Press, New York, 1975.

[YU 06] YU R., ZHANG H., "New function of Mittag-Leffler type and its application in the fractional diffusion-wave equation", *Chaos, Solitons & Fractals*, vol. 30, pp. 946–955, 2006.

[YUA 98] YUAN L., AGRAWAL O.P., "A numerical scheme for dynamic systems containing fractional derivatives", *Proceedings of DETC'98, ASME Design Engineering Technical Conferences*, 13–16 September, Atlanta, GA, 1998.

[YUA 02]. YUAN L., AGRAWAL O.P.,“ A numerical scheme for dynamic systems containing fractional derivatives”, *Journal of Vibration and Acoustics*, vol. 124, pp. 321–324, 2002.

[ZHU 09] ZHUANG P., LIU F., ANH V., *et al.*, “Numerical methods for the variable-order fractional advection-diffusion equation with a nonlinear source term”, *SIAM Journal on Numerical Analysis*, vol. 47, pp. 1760–1781, 2009.

[ZIE 52] ZIEGLER H., “Die Stabiliatskriterien der Elastomechanik”, *Ingenieur-Archiv*, vol. 20, pp. 49–56, 1952.

[ZOL 09] ZOLFAGHARI M., GHADERI R., SHEIKHOL-ESLAMI A., *et al.*, “Application of the enhanced homotopy perturbation method to solve the fractional-order Bagley-Torvik differential equation”, *Physica Scripta*, vol. T136, pp. 014032–014039, 2009.

# Index

**P**



**R**



**S**



**T**



**V**

**W**

Printed and bound by CPI Group (UK) Ltd, Croydon, CR0 4YY

13/09/2026

14961081-0005